BIOHAZARDOUS WASTE

BIOHAZARDOUS WASTE

Risk Assessment, Policy, and Management

Wayne L. Turnberg
Washington Department of Health
University of Washington

A Wiley-Interscience Publication
JOHN WILEY & SONS, INC.
New York • Chichester • Brisbane • Toronto • Singapore

Library of Congress Cataloging in Publication Data:

Turnberg, Wayne L. (Wayne Lawrence)
    Biohazardous Waste: Risk Assessment, Policy and
  Management / Wayne L. Turnberg.
      p.   cm.
  "A Wiley-Interscience publication."
  Includes index.
  ISBN 0-471-59421-0
  1. Medical wastes. 2. Hazardous wastes. 3. Infectious wastes.
4. Medical wastes—United States. 5. Hazardous wastes—United
States. 6. Infectious wastes—United States. I. Title.
RA567.7.T87 1996
363.72'88—dc20   95-36613

Printed in the United States of America

10 9 8 7 6 5 4 3 2 1

To the memory of my father, Lawrence Turnberg, and brother-in-law, Richard Shillinger, for their inspiration; to my mother, Gloria, for her guidance; and to my wife, Patricia, and daughter, Sophia, for their patience.

CONTENTS

PART II   POLICY

PREFACE

The need to manage biohazardous waste safely has been recognized for decades. Yet the advent of the acquired immunodeficiency syndrome (AIDS) epidemic in conjunction with the 1988 washups of medical waste along beaches of the United States has elevated this issue to one of greater concern in our society. Since the beach washups, this field has emerged as a new, sometimes costly, and often confusing issue for generators, regulators, and researchers as they strive to identify and understand risks posed by this waste stream and the best management strategies to minimize these risks.

I have written this book to address issues and concerns relating to risk assessment, policy, and management of the biohazardous waste stream. The book is intended for professionals working in the medical, public health, and waste management fields; for regulatory officials and planners in federal, state, and local government; and for the community of scientists and engineers who are monitoring and researching the problems of biohazardous waste. It addresses issues of concern to the regulator, waste generator, waste collector, and the public who share an interest in safe and legal management of this waste stream.

Issues relating to biohazardous waste are examined in three parts. Part I (Risk Assessment) describes human infection risks associated with the biohazardous waste stream and the health implications of pathogenic microorganisms in the environment. Part II (Policy) reviews federal and state regulations, guidelines, and industry standards that may apply when managing this waste stream. Part III (Management) examines recognized biohazardous waste management practices and options for its treatment. The review of treatment options highlights two historically based technologies, incineration and steam sterilization, as well as the many newly developing alternative treatment technologies that are available today. In-depth reviews of individual alternative treatment systems are presented to provide an understanding of the wide range of options that are currently being marketed and their associated costs. A process to evaluate the effectiveness and safety of these new technologies, which was finalized by the State and Territorial Association on Alternate Treatment Technologies in 1994, has been included to assist both prospective purchasers and government regulators responsible for approving these systems.

This field continues to experience rapid changes both in the laws and regulations that must be followed and the technologies being developed for its management. The information presented in this book is based on published research in the scientific literature, federal and state policies and reports, product information from equipment manufacturers, and my personal research and experiences working in this field. Although the book focuses on policies and technologies of the United States, this information can readily be applied to international management issues. The biohazardous waste field continues to evolve, and with this evolution comes constant change. Resources and contacts have been included that will assist the biohazardous waste professional to identify and understand new technologies, policies, and management strategies as they develop. It is my goal for this book to present information based on what we know today, yet provide the tools for professionals to maintain up-to-date knowledge in this ever-changing field.

I gratefully acknowledge the members of the State and Territorial Association on Alternate Treatment Technologies, Mr. Roger Greene of the Rhode Island Department of Environmental Management, Ms. Lauri Lowen of the Washington Department of Ecology, Dr. Kathryn Kelly of Delta Toxicology Inc., and Ms. Diann Miele of the Rhode Island Department of Health for their contributions to this work. I also wish to thank Major Laura Carlan Battle of the Air Force Environmental Law and Litigation Division, Mr. Chris James of the Connecticut Department of Environmental Protection, Ms. Shelley Kneip of the Buck and Gordon Law Firm, Ms. Jill Trohimovich of the Seattle-King County Department of Public Health, and Dr. Kathryn Wagner, environmental consultant, for their critical reviews and suggestions. I wish to express my appreciation to Mr. Jack Hatlen of the University of Washington and Dr. Jaime Kooser of the Washington Department of Ecology for their interest and support as this book was being developed. Lastly, I thank the staff of John Wiley & Sons, Inc. for their guidance and expertise throughout this process.

WAYNE L. TURNBERG

Seattle, Washington

BIOHAZARDOUS WASTE

1

INTRODUCTION

On November 1, 1988, the Medical Waste Tracking Act (MWTA) was signed into law. This action, in part, required the administrator of the Environmental Protection Agency (EPA) to publish an interim final rule for a two-year demonstration medical waste management and tracking program. The MWTA was enacted in an attempt to end the washups of medical waste along beaches of the Atlantic coast and Great Lakes that had occurred during the summer of 1988. The act established a cradle-to-grave medical waste tracking program that was ultimately implemented in four states and one territory of the United States. Of note, the beach washups from which this issue arose consisted primarily of garbage and other debris. Only a very small part of the waste consisted of syringes, medical vials, or other wastes of medical origin. Nevertheless, the perception of medical waste littering the beaches raised public fears about the spread of the human immunodeficiency virus (HIV) through contact with such waste. The presence of garbage on the beaches and the fear that much of that waste was medical in origin resulted in beach closures and a fearful public often unwilling to use the beaches when open. As a result, public confidence was shaken and local shoreline economies suffered.

In the heat of public concern, the U.S. Congress enacted quick passage of House Bill 3515, the Medical Waste Tracking Act of 1988 (MWTA), codified at 42 U.S.C. 6992 and amending the Resource Conservation and Recovery Act of 1976 by adding a new Subtitle J. In its enactment, Congress decried the appearance of such waste on the beaches to be "repugnant, intolerable and unacceptable," which was being attributed by Congress to medical mismanagement and "midnight dumping." The sponsor of the bill, Representative Tom Luken of Ohio, declared: "Medical waste pollution is lethal and grotesque. It is sickening and frightening the public, fouling our coastlines, and crippling local tourist economies. We have a critically ill medical waste disposal system—and the hypodermic needles and vials of infected

1

blood washing ashore from Massachusetts to Florida and New York to Ohio are the leading symptoms of this illness." Of note, subsequent investigation found the beach washups to be largely attributed to a poorly functioning solid and wastewater management system rather than to illegal dumping.

Although the medical waste tracking program that was implemented by the EPA had only limited participation, and although the program expired in June 1991 without being reauthorized by Congress, the course of biohazardous waste management was forever changed in the United States. Prior to the MWTA of 1988, federal agencies with the authority to establish biohazardous waste regulations, such as the EPA and the Centers for Disease Control, often directed their efforts toward developing guidelines or "best management practices," rather than regulations for managing this waste stream, based on the viewpoint that regulations were not warranted or necessary. However, the public's growing fear of the AIDS epidemic in conjunction with the washups of medical waste on the nation's beaches led to a national shift in biohazardous waste policy in response to perceived risks associated with this waste stream.

In this book I address the biohazardous (e.g., infectious) component of medical waste. Today, the healthcare industry faces a significant challenge in understanding risks associated with biohazardous waste and the maze of federal, state, and local laws, regulations, and guidelines that apply to this waste stream. It is the purpose of this book to serve as a starting point and foundation for understanding the risks, policies, and management strategies in this continually changing field. It is the responsibility of the reader to use the information and contacts provided in this book to maintain a current and updated understanding of risks, policy, and management requirements associated with the biohazardous waste stream.

This book has been written for use by generators of biohazardous waste, government agencies responsible for developing and enforcing biohazardous waste management regulatory programs, the waste management industry, the community of scientists and engineers who monitor and research the problems of medical waste, professionals working in the medical, public health, and toxicology fields, and concerned members of the public. It can also serve as an educational tool in undergraduate and graduate curricula.

For biohazardous waste generators, the book provides the necessary tools to develop and implement a successful biohazardous waste management program. Today, biohazardous waste generators are no longer viewed as comprising only the larger hospitals and laboratories, but include smaller generators, such as dental offices, clinics, research facilities, surgery centers, nursing homes, veterinary offices, and funeral homes. Waste management activities conducted during the rapidly growing field of home healthcare and self-care (e.g., diabetes care) must also be considered. It has become increasingly important for providers of health-related services to be recognized in the community as responsible partners concerned with the safety and well-being of neighbors and of the environment as a whole. This book will assist in that effort through its description of what is known about risk; analysis of key requirements, guidelines, and standards; identification of commonly accepted waste management practices; and review of many present and emerging waste treatment options.

For government regulatory agencies the book provides factual information about risk and management practices associated with the biohazardous waste stream to assist regulators who are developing or revising biohazardous waste regulations. Government regulatory agencies are often required to justify the need for and benefit of new and revised regulations, particularly those that are difficult and costly to implement, such as many of today's local, state, and federal biohazardous waste regulations. This book provides the science behind this field and identifies the risks as they are known today, to assist those regulatory agencies charged with establishing or revising biohazardous waste regulations in factual decision making.

New methods to manage biohazardous waste safely are continually being developed, particularly in the field of waste treatment and destruction. This book provides a current overview of the state of the waste treatment technology and the direction that it is taking by reviewing many of the systems on the market today. This is important information for engineers, research scientists, toxicologists, public health officials, and entrepreneurs working in the field of biohazardous waste management.

CONTENT

The book is divided into three parts. Part I addresses risks associated with the biohazardous waste stream, Part II is concerned with policy (regulations, guidelines and industry standards), and Part III covers biohazardous waste management. These parts are addressed in Chapters 2 to 12. Chapter 2 covers what is known about the risks of infectious disease transmission to human populations resulting from improperly managed biohazardous waste, by exploring the elements required for infectious disease transmission to occur and the role of communicable diseases in human populations. Chapter 2 deals with the findings of the human health risk report published by the Agency for Toxic Substances and Disease Registry in response to the Medical Waste Tracking Act of 1988. Chapter 2 also addresses the role of solid waste and wastewater in the transmission of infectious agents from biohazardous waste. This is accomplished through a review of the scientific literature and key government studies.

The biohazardous waste policy environment is presented in Chapters 3 to 5, with key regulations, guidelines, accreditation standards, and federal initiatives described. Agency contacts for federal and state government officials are presented. The reader is encouraged to use these contacts as a means for remaining current in the field and for ensuring that all requirements are being observed.

Chapters 6 and 7 identify biohazardous waste management strategies to consider when designing and implementing the process of rule writing or facility program development and implementation for medical facilities and home healthcare activities. Management suggestions presented in these chapters are given only in general terms. A distinction between facility waste management and home healthcare waste management programs is that home care activities are largely unregulated (although this is changing) and the people involved are often untrained laypersons such as family members or friends. Readers must be aware that biohazardous waste management requires compliance with all applicable local, state, and federal regulations, current copies of which must be maintained on file by generators of this waste stream. Nevertheless, the general principles of effective biohazardous waste management identified in these chapters can be considered within the context of applicable regulations as biohazardous waste management plans are being established.

Chapters 8 to 11 address the treatment aspect of biohazardous waste management. Biohazardous waste incineration is described in Chapter 8 from the perspective of what is known about risk to human and environmental health. The general principles of steam sterilization are described in Chapter 9. Technical aspects associated with the various alternative (nonincineration) biohazardous waste treatment technologies available today are discussed in Chapter 10. Several treatment systems commercially available in 1995 are reviewed to provide readers with an understanding of the principles behind system operation and such critical factors as system cost. Although the most temporal of all the discussions, due to the highly volatile nature of the alternative treatment technology market, it will be of value to readers in terms of overall understanding of the many treatment options that are available. A system for evaluating the effectiveness of alternative treatment technologies is described in Chapter 11. This chapter is reprinted in its entirety from a manual entitled "Technical Assistance Manual: State Regulatory Oversight of Medical Waste Treatment Technologies," developed by government representatives from several U.S. states. The committee of state representatives, which termed itself the State and Territorial Association on Alternate Treatment Technologies (STAATT), developed the manual specifically to address in a uniform manner how alternative treatment technologies should be evaluated by state agencies responsible for approving or recommending use of a particular technology. The manual is recognized as a key guidance document for effective evaluation of emerging alternative medical waste treatment systems. It is a critical document for state and local regulatory agencies responsible for approving treatment technologies, for those designing and testing new systems, and for those interested in purchasing a system.

the Management of waste antineoplastic drugs is described in Chapter 12. The predominant characteristic of these agents is one of toxicity rather than infectiousness. However, the discussion is presented because these agents frequently are associated with biohazardous waste, and there is a considerable degree of uncertainty about their proper management.

BIOHAZARDOUS WASTE

Biohazardous waste is most simply defined as waste that is capable of producing infectious disease. It represents that subset of the overall medical waste stream with infectious characteristics and is often referred to in both regulations and the scientific literature by a number of different terms, such as *infectious waste, red bag waste, biomedical waste,* and *regulated medical waste.* Depending on how it is defined, it may include such wastes as (1) cultures and stocks of infectious agents and associated biologicals; (2) human blood and blood products; (3) pathological wastes; (4) contaminated sharps; (5) contaminated animal carcasses, body parts, and bedding; (6) isolation waste; (7) wastes from surgery and autopsy; (8) contaminated laboratory wastes; (9) dialysis unit wastes; or (10) contaminated equipment.

Because there is no nationally mandated definition of biohazardous waste, significant definitional changes in the waste stream occur routinely as one crosses borders between state and local jurisdictions. However termed or defined, proper and safe management of biohazardous waste must be an integral part of any healthcare strategy to protect the safety and health of healthcare providers and support staff, patients and their families, waste industry workers, and the general public.

Biohazardous waste is generated primarily in the course of healthcare or research by both medical institutions and home healthcare activities and to a lesser extent by illegal drug users. The primary medical institutions generating biohazardous waste in the United States include hospitals, laboratories, physicians, dentists, veterinarians, long-term healthcare facilities, clinics, blood establishments, and funeral homes. The estimated number of generators in each of these categories and volumes of waste generated are presented in Table 1.1.

Following is a fictitious account that raises several questions concerning the safe and proper management of biohazardous waste and includes incidents that have actually occurred during the past few years. Suppose that a hospital administrator is faced with a regulatory and public relations dilemma stemming from an incident involving a group of children found by the police playing with used syringes which they discovered laying on the ground in the facility's dumpster area. State regulators were quick to arrive on the scene. Inspections revealed not only the expected problems in the waste storage area, but other safety and health violations within the facility as well. Citations were issued. The parents of the children, terrified that their children might contract AIDS, are bringing legal action against the hospital. The news media quickly published and broadcast the story, headlining the name of the hospital. Despite years of excellent public relations in the community and recognized diligence to excellence in healthcare administration, the hospital's image as a good neighbor and sound provider of professional healthcare is now on shaky ground.

This fictitious account is extreme, of course, but raises several fundamental questions. What is biohazardous waste, and what are accepted and legal management practices for this waste stream? How is it regulated, who are the regulators, and how should they be approached? What is known about the health risks involved with this waste stream? Could these children really contract AIDS? How could the facility's biohazardous waste be rendered safe for disposal? What can a healthcare facility do to enhance its image in the community as an environmentally responsible neighbor? And perhaps most important, what can a healthcare facility do to ensure that a scenario such as the one described above never occurs?

This book is written to address these and other questions about the biohazardous

Table 1.1 Estimated U.S. Medical Facilities and Annual Waste Generated

Generator Category	Number of Facilities	Annual Infectious Waste Generated (tons)	Annual Total Waste Generated (tons)
Hospitals	7,000	360,000	2,400,000
Laboratories			
Medical	4,900	17,600	117,500
Research	2,300	8,300	55,500
Total	7,200	25,900	173,000
Clinics (outpatient)	41,300	26,300	175,000
Physicians' offices	180,000	35,200	235,000
Dentists' offices	98,000	8,700	58,000
Veterinarians	38,000	4,600	31,000
Long-term care facilities			
Nursing homes	18,800	29,700	198,000
Residential care	23,900	1,400	9,000
Total	42,700	31,100	207,000
Free-standing blood banks	900	4,900	33,000
Funeral homes	21,000	900	6,000
Health units in industry	221,700	1,400	9,000
Fire and rescue	7,200	1,600	11,000
Corrections	4,300	3,300	22,000
Police	13,100	<100	<1,000
Total	682,400	504,000	3,361,100

Source: Reference 1, page 3.

waste stream by describing human health risks, the regulatory environment, approaches to waste stream management, and waste treatment options to render biohazardous waste safe for disposal. Legal and sound biohazardous waste management practices are essential from the standpoint of human health protection, liability, and the need to maintain a positive image in the community.

REFERENCE

1. U.S. Environmental Protection Agency. Medical Waste Incinerators—Background Information for Proposed Guidelines: Industry Profile Report for New and Existing Facilities. EPA-453/R-94-042a. July 1994.

PART I

RISK ASSESSMENT

2

HEALTH HAZARD ASSESSMENT

During the past few years, a significant effort has been directed toward proper and safe management of biohazardous waste streams by government agencies and the regulated industry, but without a clear understanding of the disease transmission risks associated with improper disposal practices. This chapter, parts of which are based on work published previously by the author (1–5), addresses such risks based on the existing state of knowledge in this field. Topics explored include the elements necessary for infectious disease to occur, communicable diseases of most concern to the public, and studies found in the scientific literature that shed light on the potential for human disease to result from improperly managed biohazardous waste.

ELEMENTS OF HUMAN INFECTION

Microorganisms are ubiquitous in nature and play a major role in the world ecosystem. They include bacteria, protozoans, fungi (yeasts and molds), and viruses (which require other living cells to reproduce). Most found in nature are harmless to humans. Nevertheless, microbial agents capable of causing infectious disease in humans abound in the environment.

During the past few years, public concern has risen regarding the presence of microorganisms in medical waste and the potential for these microorganisms to cause infectious disease in the community. Fueled by the public's fear of the human immunodeficiency virus (HIV), and the perception, although never identified, that HIV transmission can occur from exposure to medical waste, lawmakers have promulgated statutes and regulations throughout the country at federal, state, and local levels of government in an effort to ease public fears, often without a true understanding of risks associated with infectious disease transmission from waste stream sources. In this section we address the presence of microorganisms in the environment, the elements necessary for infection and disease to occur in humans, and the modes of microbial transmission to humans.

Existing with Microorganisms

Many microorganisms grow and reside harmlessly on the internal and external surfaces of the human body as the normal body flora. The microorganisms comprising the body's normal flora are found in the eyes, nose, throat, mouth, skin, intestine, urinary tract, and genital systems. It has been estimated that the human body may harbor up to 1×10^{14} bacterial cells as part of its normal flora (6). These microorganisms coexist with the host in a constantly changing balance of forces as a result of human host activities and other external influences (7). For example, a change in diet may alter the existing microbial populations found in the intestine.

The term *symbiosis* is used to describe this coexistence between host and normal flora. Symbiotic relationships can be divided into *commensalism*, in which the microbe benefits but the host neither benefits nor is harmed; *mutualism*, in which the host and the microbe both benefit; and *parasitism*, in which the microbe benefits at the expense of the host (7). Many microorganisms of the normal body flora may fall into different symbiotic categories, depending on a number of factors. For example, the enteric bacterium *Escherichia coli* exists in a state of mutualism with its human host when in the large intestine, synthesizing vitamin K and some B vitamins that may benefit a host in need of such nutrients. However, if the same microorganism is introduced to other body sites, such as the urinary tract, it can infect the host at that site (in this example, result in a urinary tract infection), which may lead to a state of disease or pathological abnormalities.

A *pathogenic microorganism* is an agent capable of causing disease. The term *infection* is used to describe the colonization of the body by pathogenic microorganisms. *Infectious disease* describes the state of abnormal body functioning resulting from infection, although infection does not always result in a state of disease. An *opportunistic pathogen* is a microorganisms capable of causing infectious disease in a host with an impaired defense mechanism, such as a break in the skin or a suppressed immune system. A *nonpathogen* is a microorganism that never or only very rarely causes disease in a host. However, terms of pathogenicity cannot be used in any absolute sense, due to the many variables associated with host and microorganism interactions. Except in the rare cases of obligate pathogens (e.g., Tularemia, caused by *Francisella tularensis*, or syphilis, caused by *Treponema pallidum*), there is no clear division between microorganisms that are capable of causing infectious disease and those that are not (8).

The term *virulence* is used to describe the relative power and degree of pathogenicity possessed by a microorganism to produce disease. For some microbes, the responsible attribute is known (e.g., the presence of a cell capsule or toxins); in others, it is not. Relative virulence among certain microorganisms has been compared by determining the minimum infectious dose required to cause infection in laboratory animals or human volunteers. The human body has the ability to restrict or fight the growth of microorganisms by a number of mechanical and physiological defense mechanisms that help to maintain a favorable balance between host and microorganisms. However, when the microbial dose or degree of virulence surpasses these defense mechanisms, an imbalance between the host and the microorganism can occur, which can result in infection and a state of disease.

Elements of Infection

Infection requires the occurrence of a specific series of events. There must be a reservoir of an infecting organism, a susceptible host, and a mode of transmission to the host. There must be an infective dose of the pathogen to cause infection in the susceptible host and a portal of entry for the pathogen to enter the host. Each of these elements is described below.

Reservoir. The *reservoir* of disease-causing microorganisms can be either a living or an inanimate object that provides the microbe with conditions that allow the organism to live and multiply and provides an opportunity for transmission (6). Reservoirs of human disease include:

Human Reservoir. The human body represents the primary source or reservoir of microorganisms that are infectious to humans. The most obvious source are people exhibiting overt signs and symptoms of infectious disease. However, the appearance of good health offers little guarantee that infectious microorganisms are not present. Many people are carriers of human infectious agents without showing signs or symptoms of disease. Others carry and spread the disease-causing agents during the period before symptoms appear, or during their convalescent stage, after symptoms have disappeared. Infectious diseases are often spread from healthy-appearing carriers to their unsuspecting hosts.

Animal Reservoir. Both wild and domestic animals can harbor microorganisms that can be infectious to humans. About 150 animal diseases that can infect humans, called *zoonoses*, are known (6).

Nonliving Reservoirs. A major nonliving reservoir that has contributed extensively to human infectious disease is water. Human society has made considerable efforts to ensure that its drinking and recreational water sources are free of infectious agents. The soil represents another major nonliving reservoir for pathogenic agents.

Susceptible Host. Human resistance to disease-causing agents is extremely variable among different individuals and within a given individual depending on a broad range of factors. Some individuals may be immune or be able to fight-off the infectious agent; others exposed to the same agent may become carriers of the organisms without showing signs of illness; while others may become infected and develop signs and symptoms of disease.

The *susceptibility*, or lack of resistance, of a host to an infectious agent involves many complex factors. A host's resistance mechanisms may be categorized into two types: *nonspecific resistance* (directed at all pathogens) and *specific resistance* (directed against specific pathogens). Nonspecific resistance mechanisms include the barrier action of the skin and mucous membranes against the entrance of microorganisms; phagocytosis, or the ingestion and elimination of microorganisms by special host cells; local inflammation to destroy the agent if possible or limit its effects and repair the damaged area; fever, a systemic response to infection; and the production of certain antimicrobial agents, such as interferon. Specific resistance mechanisms come from the immune system when the host is invaded by foreign microbial agents. The human body has the ability to recognize foreign substances as not belonging to itself, which in healthy individuals activates an immune response to ward off the invader.

Acquired immunity is developed in the course of a person's life. When the human body is challenged by a foreign invader (antigen), the immune system responds by producing specialized lymphocytes and proteins (antibodies) to inactivate the invader. This immunity can be acquired by natural exposure to an infectious agent, or artificially, by vaccination with a specially prepared antigen. Acquired immunity can last for the life of the host (e.g., measles, chickenpox) or for a few years. For other infectious agents, there may be no immunity to reinfection (e.g., gonorrhea).

Mode of Transmission. For infection to occur in a susceptible host, an infectious agent must have a mechanism of transmission from the source of the infectious agent to the host. The four main routes of transmission are by contact, by vehicle, by air, and by vector.

Contact Transmission. Contact transmission involves the transfer of an infectious agent by (1) direct transmission from reservoir to host (e.g., person to person), (2) indirect transmission from the source to the host by means of a nonliving object (e.g., tissues or contaminated syringes), and (3) droplets that travel less than 1 meter through the air from the reservoir to the host (e.g., from a sneeze).

Vehicleborne Transmission. Vehicleborne transmission occurs when infectious agents are transmitted by inanimate reservoirs of infectious agents (food, water, drugs, blood) (9).

Airborne Transmission. Airborne transmission involves the spread of infectious agents on droplet nuclei of dust particles. Droplet nuclei are the residues of evaporated droplets. Infectious agents carried by droplet nuclei or dust can remain suspended in the air for long periods of time before being inhaled or landing on a susceptible host.

Vectorborne Transmission. Vectorborne transmission occurs from animals, primarily arthropods, that spread infectious agents from one host to another. Transmission can occur by mechanical means (e.g., on the feet of an insect) or by biological transmission (e.g., mosquitoes biting an infected host and transferring infected blood to another host).

Infective Dose. The *infective dose*, or number of organisms required to cause human infection, is widely variable, depending on the virulence of the organism, the portal of entry, and the susceptibility of the host. An infective dose can require only a few cells (e.g., Q-fever, caused by inhalation of *Coxsiella burnetti*) or millions of cells (e.g., cholera, caused by ingestion of *Vibrio cholera*) (10). An infective dose of *Francisella tularensis* to cause tularemia when inhaled has been reported to be as few as 10 cells, but when ingested may require about 100 million cells (10). There is limited information on the infective dose for many pathogens. Some information has been gathered from experiments with human subjects and from animal studies, although it is reported that animals often do not respond to infective agents in the same way that humans do.

Portal of Entry. Microorganisms may gain entry to the body through the respiratory tract, by inhalation; the gastrointestinal tract, by ingestion; the skin and mucous membranes, through breaks (some organisms also appear to be able to penetrate the intact skin); the genitourinary system; and the blood, by direct inoculation (e.g., insects that penetrate the skin or hypodermic needle-stick injuries). Microorganisms usually exit the body by the same routes (8).

COMMUNICABLE DISEASES IN HUMANS

Infectious agents are prevalent in the human population. The extent of infection and disease in a community is controlled largely through primary prevention mechanisms of *general health promotion* (e.g., conditions that favor healthy living, such as good nutrition, clothing, shelter, and heat) and *specific protective measures* (e.g., immunizations and environmental sanitation). Secondary protective measures include early detection of infection and prompt treatment of disease. Early diagnosis and treatment may also have the effect of breaking the disease transmission chain to other healthy individuals.

Well-known communicable diseases in our society include influenza, chickenpox, and rubella as well as sexually transmitted diseases such as gonorrhea, syphilis, hepatitis, or the acquired immunodeficiency syndrome (AIDS). The infectious agents of communicable disease are found throughout our population, whether a person is in the community or is a patient in a hospital. Infections that occur within hospitals are either preexisting (community acquired) or hospital acquired (*nosocomial*). Most nosocomial infections are caused by opportunistic pathogens in hosts with compromised immune systems. Such pathogens, which are normally nonpathogenic to healthy individuals, become pathogenic due to the altered physiological state of the host. Nosocomial infections have been observed to affect about 5 percent of hospitalized individuals (11).

Isolation Precautions

Various systems have been developed to protect healthcare providers, patients, and visitors from infectious agents in healthcare facilities. In 1983, the Centers for Disease Control (CDC) published its recommendation for isolation in the hospital setting in the manual "Guideline for Isolation Precautions in Hospitals". (9). The isolation precautions guidelines represent a compilation of prudent practices recommended by CDC personnel and a panel of outside experts to prevent the spread of microorganisms among patients, personnel, and visitors in the hospital setting. Interruption of the chain of infection is directed at transmission rather than at agent and host factors, which are more difficult to control. A weakness in the system is that it is diagnosis driven and does not address the asymptomatic carrier.

In the guidelines, two isolation systems are described. System A (*category*-specific isolation precautions) is based on isolation categories developed through grouping diseases that call for similar isolation techniques. In system A, isolation procedures have been developed for seven categories. These include (1) strict isolation, (2) contact isolation, (3) respiratory isolation, (4) tuberculosis (AFB) isolation, (5) enteric precautions, (6) drainage/secretion precautions, and (7) blood/body fluid precautions. System B (*disease*-specific isolation precautions) considers each specific disease individually so that only precautions to interrupt disease transmission are recommended (e.g., gloves, gowns, and masks).

Techniques for isolation precautions are presented in the guidelines. Examples include proper handwashing; use of a private room; use of masks, gowns, and/or gloves; bagging of articles; disposal of patient-care equipment and waste; management of linen, dishes, drinking water, dressings, tissues, urine and feces, and laboratory specimens; and routine and terminal cleaning.

Universal Precautions

The CDC 1983 guideline for isolation precautions presented "blood and body fluids precautions" as a category-specific isolation recommendation, which was developed to prevent infections that are transmitted by direct or indirect contact with infective blood or body fluids (9). With the advent of the AIDS epidemic, a disease in which the infective status of individuals is often unknown, and the concern with hepatitis B infection (HBV) among medical personnel, the CDC recommended that blood and body fluid isolation precautions be applied to *all* patients, whether or not infection or disease had been diagnosed. This extension of precautions to all patients is known as *universal blood and body fluid precautions* or *universal precautions* (12).

The system of universal precautions was established by the CDC to protect healthcare workers from occupationally acquired infection from HIV, HBV and other bloodborne pathogens (13). By this system, blood and certain body fluids of all patients are considered to be potentially infectious for these pathogenic agents. Methods to prevent exposure to these body substances would be applied to all patients, whether or not they have been diagnosed with infectious bloodborne disease.

Universal precautions apply to blood and to other body fluids containing visible blood, semen and vaginal secretions, tissues, and the following body fluids: cerebrospinal fluid, synovial fluid, pleural fluid, peritoneal fluid, pericardial fluid, and amniotic fluid. Universal precautions do not apply to feces, nasal secretions, sputum, saliva, sweat, tears, urine, and vomitus, unless visibly containing blood (13).

Body Substance Isolation

As described earlier, the primary weakness with the 1983 CDC isolation precautions guidelines was that they are *diagnosis driven*. In 1986, Lynch and Jackson[1] reported an *interaction-driven* system called *body substance isolation* (14). As described by the authors:

> Body substances such as feces, airway secretions and wound drainage always contain potentially infectious organisms, and blood, urine and other moist body substances sometimes do as well. In fact, many infectious and communicable diseases are characterized by a large proportion of undiagnosed cases, and all communicable diseases are infectious before the diagnosis is established. Thus, a system that is initiated only after a diagnosis is made is less likely to prevent transmission than a system for certain procedures or interaction where precautions are implemented for all patients, regardless of their diagnosis.

This approach extends beyond that recommended by the CDC in its universal precautions by applying to isolation precaution techniques for *all* body substances rather than just to blood and a specific but limited list of body substances.

Bloodborne Pathogens

Various isolation precaution techniques have been developed to protect the healthcare worker, patient, and visitor from infectious disease transmission in the medical setting. In the community, concern for proper and safe disposal of medical waste arose from the public's fear of bloodborne infectious diseases, particularly of HIV and HBV transmission through waste stream sources, although such transmission has never been identified. HIV and HBV and the diseases that they cause are discussed in the following sections.

Human Immunodeficiency Virus. In June 1981, several cases of *Pneumocystis carinii* pneumonia and a rare cancer of the skin, Karposi's sarcoma, were observed among young homosexual men. This condition, normally seen only among immunosuppressed individuals, would later become known as *acquired immunodeficiency syndrome* (AIDS). During 1983 and 1984, a human virus associated with AIDS was identified which would later be referred to as the *human immunodeficiency virus* (HIV) (15). Human immunodeficiency virus is a retrovirus responsible for

[1]From Lynch P and Jackson MM. Isolation practices: How much is too much or not enough. ASEPSIS The Infection Control Forum, 8(4):2–5, copyright 1986 by Ad/Com Incorporated Publishing. Reprinted by permission.

acquired immunodeficiency syndrome. Two viral types have been identified, type 1 (HIV-1) and type 2 (HIV-2), which are serologically and geographically relatively distinct but have similar epidemiologic and pathologic characteristics (16).

Lacking the ability to reproduce independently, viruses must replicate within other cells. In the human host, the CD4+ T-lymphocyte is the primary target for HIV infection. CD4+ T-lymphocyte cells are responsible for coordinating a number of immune system functions. Loss of CD4+ T-lymphocyte cells due to HIV infection leads to impairment of the immune system, leaving the infected host susceptible to opportunistic infections and other clinical disorders. Such disorders have been observed ranging from asymptomatic infection to life-threatening conditions characterized by severe immunodeficiency, serious opportunistic infections, and cancers (17). Risk and severity of opportunistic diseases increases with the depletion of CD4+ T-lymphocytes.

Initial symptoms of the disease generally appear within a month following exposure (19). Symptoms include fever, fatigue, lymphadenopathy, myalgia, diarrhea, and rash, similar to a mononucleosis-like syndrome. Following infection by HIV, the patient undergoes seroconversion as the body produces antibodies to fight the invading organism. Seroconversion generally occurs within 6 to 12 weeks following initial exposure but may not occur for up to 6 to 14 months (18). Most individuals remain asymptomatic for months to years following infection, although their blood and certain body fluids can transmit HIV to others during this time. Many individuals survive for only about two to four years following AIDS diagnosis, although this is variable. Prophylaxis and treatment to control opportunistic infections has been observed to prolong life.

Transmission of the human immunodeficiency virus (HIV) has been shown to occur from sexual contact, parenteral exposure to blood (needle-sticks, mostly intravenous drug abusers), exposures to blood (e.g., mucous membrane or nonintact skin contact), and from an infected mother to a fetus or infant (19). Transplants of HIV-infected organs such as bone (20,21) and transfusions of infected blood (13) have also been implicated. HIV has been recovered from human blood, semen, vaginal secretions, cerebrospinal fluid, amniotic fluid, breast milk, saliva, tears, and urine, although only blood, semen, vaginal secretions, and possibly breast milk have been linked epidemiologically to transmission (12,22). HIV is not transmitted through casual contact, such as shaking hands; sharing food, eating utensils, plates, or glasses; hugging; or kissing. There is no evidence to suggest HIV transmission by insects.

HIV transmission has been observed in the occupational setting from percutaneous inoculation or contact with an open wound, nonintact skin, or mucous membranes to blood, blood-contaminated body fluids, or concentrated virus (23). Blood is recognized as the single most important source of HIV infection in the workplace. As of late 1989, it was estimated by the U.S. Public Health Service that 1,000,000 people in the United States were infected with HIV (regardless of clinical manifestations), with a range of 800,000 to 1,200,000 (24). As of December 31, 1994, 441,528 cases of AIDS had been diagnosed in the United States.

Several studies have been conducted to examine the survival of HIV in the environment. Resnick et al. examined the HIV survival in human plasma when subjected to various environmental conditions that may be encountered in natural, clinical, or laboratory settings (25). HIV concentrations used for the experiments were 7 to 10 logs of viral activity greater than what would be found in human blood. Nevertheless, infectious cell-free virus was recovered for up to three days from dried material held at room temperature, and for longer than 15 days in aqueous solution at room temperature, although the rate of inactivation was rapid. Studies conducted by the CDC found that drying HIV causes a rapid (within several hours) 1 to 2 log (90 to 99 percent) decrease in concentration. Tissue culture fluid studies observed cell-free HIV detection for up to 15 days at room temperature and up to 11 days at 37°C (98.6°F) (12).

In a study involving the decay of HIV in plasma, whole blood, and peripheral blood mononuclear cells (PBMCs), Cao et al. observed the time decay in infectivity to be slow, with three of five whole-blood and plasma samples showing almost no loss over 96 hours, and two PBMC samples remaining infective over a four-day period (26). Moudgil and Daar found that infectious viral decay in plasma is slow at room temperature and even less so at 4°C, with all samples showing less than 80 percent decay over 48 hours when held at 4°C (27). Complete decay of infectious HIV-1 in plasma can require more than seven days. These findings raise the possibility of shipping fresh samples for quantitative analysis. As reported by the CDC, HIV is much less able than the hepatitis B virus to survive and remain infective outside the human body (28).

Bankowski et al. examined the recovery of HIV from the plasma and mononuclear cell (MNC) fractions of postmortem blood samples from HIV-positive patients (29). HIV was recovered from 21 of 41 cadavers; no HIV was recovered from cadavers after 21.25 hours following death. Although the study was not designed to examine HIV survival in postmortem tissue, it was reported that the viability of HIV appeared to decrease over time during the postmortem period.

There are published reports of up to 25 healthcare workers occupationally infected with HIV (30). HIV-positive needle-stick injuries have been estimated to result in infection at a rate of three to five infections per 1000 persons injured (30). HIV transmission from exposure to contaminated blood is reported to be low (31). Based on the existing body of evidence, nosocomial/occupational HIV transmission among healthcare workers has been concluded to be extremely small, much smaller than the transmission of HBV in this setting (32).

Hepatitis B Virus. *Hepatitis B*, formerly called *serum hepatitis*, is described as a major cause of acute and chronic hepatitis, cirrhosis, and primary hepatocellular carcinoma in the United States and the world (33). It is also a major occupational infection hazard among healthcare workers and those in other occupations who routinely come into contact with human blood and other body fluids. The CDC's Hepatitis Branch has estimated that approximately 12,000 of these infections occur annually among healthcare workers due to occupational exposure to blood (30). It is estimated that these will result in 2500 to 3000 cases of clinical acute hepatitis, 500 to 600 hospitalizations, and over 200 deaths (30).

After introduction into a susceptible host, the hepatitis B virus (HBV) begins replication in liver cells. Infection can result in two outcomes: self-limited acute HBV and chronic HBV. The most common response for healthy persons is the self-limited acute hepatitis response, in which the body attempts to eliminate the virus from the body. This results in destruction of liver cells containing the virus but provides lifetime immunity against the virus. About one-third of those infected with the virus experience no disease symptoms, with such cases detectable only by liver dysfunction tests. Approximately one-third develop mild flulike symptoms, and one-third develop a much more severe disease resulting in jaundice (from liver dysfunction) and other symptoms, including extreme fatigue, vomiting, anorexia, nausea, abdominal pain, rash, or fever (30). For these cases, convalescence can require several weeks to months. Overall fatality rates are generally reported below 2 percent.

About 6 to 10 percent of those infected with the virus are unable to clear the virus from their liver cells and become chronic HBV carriers. These individuals are at high risk of developing a chronic persistent state of HBV infection, or chronic active disease, which often results in cirrhosis of the liver and death. HBV can be spread through breaks in the skin (e.g., cuts or needle-stick injuries), mucous membranes (e.g., eyes or mouth), sexually, or from mother to infant at birth. One milliliter of blood from an HBV-positive person may contain up to 100 million infectious doses of virus (30).

HBV infection does not occur uniformly in the U.S. population. Ethnic and racial differences are evident. Certain high-risk activities have also been identified: homosexual activities, intravenous drug use, heterosexual sexual contact with an HBV carrier or multiple partners, and occupational exposure to human blood (34). Those in the general population who have ever been infected with HBV approach about 3 to 4 percent for whites, about 13 to 14 percent for blacks, and about 50 percent for foreign-born Asians (34). Chronic carriers of the virus represent about 0.2 percent for whites, 0.7 percent for blacks, and up to 13 percent for foreign-born Asians.

Several studies have defined the risk of HBV infection to healthcare workers (35–41). Risk of infection is closely associated with occupational contact with blood and, to a lesser extent, with needle-stick injuries. Transmission of HBV from healthcare workers to patients has been documented, but only among workers who have been identified as exhibiting extremely high concentrations of HBV virus in their blood (at least 100,000,000 infectious virus particles per milliliter of serum) (28). Transmission from exposure to contaminated environmental surfaces has been documented as well, due to the ability of the virus to survive in infective doses on dried surfaces for up to one week or longer at room temperature (42). The risk of acquiring HBV infection due to a needle-stick injury from an HBV carrier ranges between 6 and 30 percent within healthcare settings (28).

An HBV vaccine is available that can induce protection from HBV infection for over 90 percent of healthy adults (43). Current vaccines are given in three doses over a six-month period and last for at least seven years for most recipients. Preexposure vaccination is recommended by the Occupational Safety and Health Administration

(OSHA) for people working in occupations in which contact with human blood and blood products is routine. Hepatitis B immune globulin is recommended for postexposure protection (e.g., following a needle-stick injury) if given within seven days of exposure. HBV vaccine can also be prescribed for postexposure protection (43).

Other Bloodborne Pathogens. Other bloodborne diseases include hepatitis (e.g., viral hepatitis C), malaria, syphilis, babesiosis, brucellosis, leptospirosis, arboviral infections, relapsing fever, Creutzfeldt–Jacob disease, human T-lymphotropic virus type I, and viral hemorrhagic fever (30).

HEALTH RISK: ATSDR REPORT

The most comprehensive examination of the role that biohazardous waste may play in human infection and disease was published by the Agency for Toxic Substances and Disease Registry (ATSDR) in November 1990 (44). The study was mandated under Section 11009 of the Medical Waste Tracking Act of 1988 (MWTA), which was enacted into law on November 1, 1988, and codified at 42 U.S.C. 6992 et seq. Section 11009 of the MWTA required the ATSDR administrator to prepare a report on the health effects of medical waste that includes the following information:

1. A description of the potential for infection or injury from the segregation, handling, storage, treatment, or disposal of medical wastes
2. An estimate of the number of people injured or infected annually by sharps and the nature and seriousness of those injuries or infections
3. An estimate of the number of people infected annually by other means related to waste segregation, handling, storage, treatment, or disposal and the nature and seriousness of those infections
4. For diseases possibly spread by medical waste, including acquired immune deficiency syndrome and hepatitis B, an estimate of what percentage of the total number of cases nationally may be traceable to medical wastes

Information Development

In developing the report, information was collected by the ATSDR from a wide range of sources, as depicted in Figure 2.1. An announcement was also published in the *Federal Register* requesting any information on infection and injury associated with medical waste management. In addition, a federal advisory panel was established to assure that the report was based on the best contemporary science. This panel was comprised of representatives from the Centers for Disease Control, the National Institutes of Health, the Environmental Protection Agency, the Food and Drug Administration, the Indian Health Service, the Alcohol, Drug Abuse, and Mental Health Administration, the Health Resources and Services Administration, and the Health Care Financing Administration. In addition, the Department of

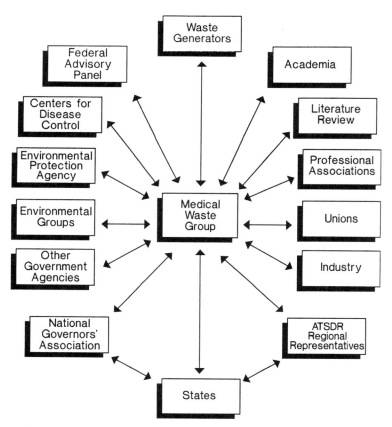

Figure 2.1   Data-gathering scheme for the ATSDR Medical Waste Tracking Act Report. (From reference 47, page 1.5.)

Defense, the Department of Veterans Affairs, and the Occupational Safety and Health Administration were requested to provide relevant information.

Data Analysis

Populations under consideration in the study for estimating medical waste injuries, infection, and disease included healthcare providers (those involved with direct patient care, such as nurses or laboratory technicians, and nonpatient-care workers such as janitors and housekeeping staff); waste-handling workers (e.g., trash collectors and handlers, landfill operators, and incinerator operators); and the general public. The strategy for estimating injury, infection, and disease related to medical waste exposure involved estimating the probabilities of occurrence of each step in the chain of events necessary for injury, infection, and disease to occur. These events require that an injury occur (e.g., physical injury or mucous membrane contamination allowing entry of infectious agents), that the mechanism of infectious disease be completed, and that the infection result in a state of disease. These steps are outlined in Figure 2.2. Formulas used for determining estimates for the number of medical waste–related injuries, infections, or disease are presented as follows:

Estimated number of medical waste–related injuries
= (number of persons in each occupational subgroup) × (medical waste–related injury rate)

Estimated number of medical waste–related infections
= (number of medical waste injuries) × (prevalence of contaminated material) × (seroconversion rate)

Estimated number of medical waste–related diseases
= (number of medical waste–related infections) × (clinical disease rate among seroconverted persons)

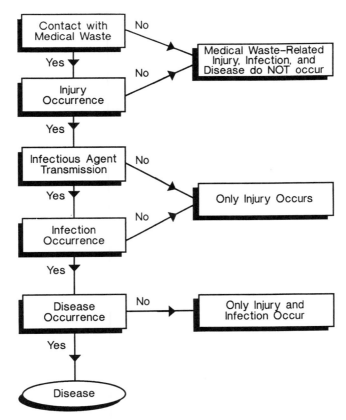

Figure 2.2   Chain of events required for medical waste–related injury, infection, and disease. (From reference 47, page 1.6.)

ATSDR notes:

1. Each calculation must be made separately for each occupational subgroup and for each route of contact.
2. Prevalence of contaminated material was based on the seroprevalence of infected persons in the healthcare setting.
3. Prevalence of infected persons was based on seroprevalence surveys conducted in hospitalized populations.
4. Seroconversion rates were obtained from case studies of exposed workers.
5. Clinical disease rates among seroconverted persons were obtained from case studies.

Prior to the ATSDR effort, no estimates of the number of persons suffering injury and infection related to contact with medical waste had been made. Injury data estimates were based on information derived from the medical literature or special surveys. When information was not available, estimates were made from similar situations. Due to the lack of information and methods to collect or calculate data, certain biases and inaccuracies were recognized. Data collected from multiple sources may not reflect identical degrees of involvement with medical waste. In addition, individual studies always contain biases, which will probably remain when the studies are combined. Therefore, the ATSDR chose to base estimates on extremes: worst-case situations that reflect the highest estimates from the information available, and best-case situations that reflect the lowest estimates from the information available.

Defining the Medical Waste Stream

Estimation of injuries, infections, and disease resulting from exposure to the medical waste stream requires defining the elements of the waste stream under review. Section 11002 of the Medical Waste Tracking Act of 1988 defined 10 categories of medical waste. The ATSDR was given no flexibility under the MWTA to change the types of waste listed by Congress and therefore assessed the health impacts of all 10 categories.

- *Category 1: Cultures and Stocks.* Cultures and stocks of infectious agents and associated biologicals, including cultures from medical and pathological laboratories, cultures and stocks of infectious agents from research and industrial laboratories, wastes from the production of biologicals, discarded live and attenuated vaccines, and culture dishes and devices used to transfer, inoculate, and mix cultures.
- *Category 2: Pathological Wastes.* Pathological wastes, including tissues, organs, and body parts that are removed during surgery or autopsy.
- *Category 3: Waste Human Blood and Blood Components.* Waste human blood and products of blood, including serum, plasma, and other blood components.

- *Category 4: Sharps.* Sharps that have been used in patient care or in medical, research, or industrial laboratories, including hypodermic needles, syringes, Pasteur pipettes, broken glass, and scalpel blades.
- *Category 5: Animal Waste.* Contaminated animal carcasses, body parts, and bedding of animals that were exposed to infectious agents during research, production of biologicals, or testing of pharmaceuticals.
- *Category 6: Surgery or Autopsy Waste.* Wastes from surgery or autopsy that were in contact with infectious agents, including soiled dressing, sponges, drapes, lavage tubes, drainage sets, underpads, and surgical gloves.
- *Category 7: Laboratory Wastes.* Laboratory wastes from medical, pathological, pharmaceutical, or other research, commercial, or industrial laboratories that were in contact with infectious agents, including slides and coverslips, disposable gloves, laboratory coats, and aprons.
- *Category 8: Dialysis Wastes.* Dialysis wastes that were in contact with the blood of patients undergoing hemodialysis, including contaminated disposable equipment and supplies such as tubing, filters, disposable sheets, towels, gloves, aprons, and laboratory coats.
- *Category 9: Discarded Medical Equipment.* Discarded medical equipment and parts that were in contact with infectious agents.
- *Category 10: Isolation Wastes.* Biological waste and discarded materials contaminated with blood, excretions, exudates, or secretions from human beings or animals that are isolated to protect others from communicable diseases.

Sources of Medical Waste

Based on the MWTA definition of medical waste, sources, types, and estimated quantities were identified in the report based on estimates derived from various sources. The sources of waste were identified to be (1) hospitals, (2) physicians' offices, (3) dentists' offices, (4) biomedical research facilities, (5) clinical laboratories, (6) manufacturing facilities, (7) veterinary offices and clinics, (8) funeral homes, (9) in-home medical care, (10) other healthcare and residential care facilities, (11) illicit intravenous drug use, and (12) other sources (e.g., cruise ships and naval vessels).

Nonsharp Medical Waste

The ATSDR addressed questions posed by Congress in the MWTA by categorizing medical waste as being either "nonsharp" (representing all medical waste categories except category 4 waste) or "sharp" (representing only category 4 waste) in nature. When conducting its analysis, the ATSDR distinguished between those injuries and infections related to medical waste from those occurring in the course of patient care that are unrelated to medical waste. Additionally, injuries and infections related to medical waste before it is discarded are identified in the report as not related to medical waste.

The ATSDR described conditions necessary for infection to occur and noted that nonsharp medical waste generally does not create a portal of entry upon contact. Therefore, for transmission to occur, a portal of entry would already have to exist at the time of contact. Mechanisms for entry into the body by infectious agents associated with nonsharp waste include transmission via contact with breaks in the skin (e.g., cuts and abrasions) or contact with mucous membranes (e.g., contact with eyes or inner lining of the nose). This could occur if infectious agents are suspended in blood.

In its review, the ATSDR noted several cases which suggested that contact with blood other than via percutaneous routes (e.g., needle-stick or other injury-producing entry) may produce HIV infection during patient care. However, upon examination of eight longitudinal studies that evaluated healthcare workers who had been exposed to blood or blood-containing body fluids through skin or mucous membrane contact from an HIV-positive patient, no HIV infections were observed. This is based on evaluations of 538 healthcare workers who had experienced up to 921 mucous membrane contacts with HIV-positive blood or blood-contaminated body fluids. Findings are summarized in Table 2.1.

Based on this information, the ATSDR estimated the probability for transmission following mucous membrane exposure (e.g., splash to the eyes or other mucous membrane) to blood or blood-contaminated body fluids from an HIV-positive patient to be less than 0.1 percent, with an upper bound for the 95 percent confidence interval for the rate of transmission of approximately 0.33 percent. Based on literature reports, the ATSDR noted that the probability of transmission following a percutaneous contact (e.g., needle-stick or other skin-breaking injury) with HIV-contaminated blood is approximately 0.36 percent, and that it seems reasonable that the probability following a mucous membrane contact would be less. The ATSDR also noted that no HIV or HBV infections associated with medical waste were reported in the scientific literature and that no injury data for medical waste management had ever been reported in the literature except as related to sharps waste. The ATSDR concluded: "Based on the principles of infectious disease transmission, the potential for infection as the result of contact with non-sharp medical waste is likely to be significantly less than that for injuries related to medical waste sharps. The primary reason for the lower potential is, in contrast to sharps, a portal of entry must exist prior to contact with non-sharp medical waste."

Sharps Medical Waste

In contrast to nonsharp medical waste, sharps (i.e., category 7 wastes) have the ability to create a portal of entry through the skin. As required by the MWTA, the ATSDR estimated the number of medical waste injuries that occur for various occupational subgroups based on available data sources. Limiting factors included "(1) the general unavailability of overall injury data in specific occupations at the national level; (2) the use of workers' compensation claims in selected states to calculate overall injury rates, and (3) the fact that needle-stick injuries generally are not reported."

Table 2.1 Longitudinal Studies of the Potential for HIV Transmission in the Healthcare Workplace

Author	Number of Percutaneous Exposures	Personnel Experiencing Percutaneous Exposures	Number Seroconverting (% per exposure)	Number of Mucous Membrane Exposures	Personnel Experiencing Membrane Exposures	Number Seroconverting (% per exposure)
Marcus	582	582	3 (0.52)	38	38	0 (0)
Gerberding	224	180	1 (0.45)	401	168	0 (0)
Henderson	126	108	0 (0)	337	234	0 (0)
Elmslie	115	115	0 (0)	24	24	0 (0)
Kuhls	52	45	0 (0)	81	34	0 (0)
McEvoy	76	76	0 (0)	24	24	0 (0)
Ramsey	31	31	1 (3.23)	13	13	0 (0)
Wormser	48	48	0 (0)	3	3	0 (0)
Hernandez	58	58	0 (0)	—	—	—
Pizzocolo	77	77	0 (0)	—	—	—
Total	1389	1320	5 (0.36)	921	538	0 (0)

From Henderson DK. HIV-1 in the health-care setting. In: Mendell GL, Douglas RG, and Bennett JE, Eds. *Principles and Practice of Infectious Diseases*, 3rd ed. New York, New York. Churchill Livingstone, Inc. 1990:2225. Reprinted by permission.

Transmissions of infectious agents have been reported in the literature by contaminated sharps, although almost all transmissions occurred in the course of patient care or laboratory procedures and were not associated with medical waste. ATSDR identified only one medical waste–related occurrence, which involved a hospital housekeeper who developed staphylococcal bacteremia and endocarditis following a needle-stick injury, reported by Jacobson et al. (45). Except for this case, the ATSDR could find no information in the scientific literature that involved transmission from exposure to a contaminated waste needle.

Calculating Theoretical Transmission Rates

To calculate a theoretical infectious disease transmission rate, the ATSDR estimated the probability that any medical waste sharp would be contaminated with an infective dose (e.g., a sufficient number of infectious agents to cause infection in a susceptible host), that infection will occur following injury, and that the infection will result in disease. The formula used by the ATSDR to estimate the number of medical waste–related infections is

(number of medical waste injuries) × (prevalence of contaminated material)
× (seroconversion rate)

The formula to estimate the number of medical waste–related diseases is

(number of medical waste–related infections) × (clinical disease rate
among seroconverted persons)

In its study, the ATSDR developed infection and clinical disease estimates for HIV and HBV only. Information needed to calculate estimated transmission rates for other infectious diseases is not available.

Estimating Injury Numbers. To determine estimates of the number of medical waste injuries, the ATSDR relied on information that reported injuries resulting from improper disposal of medical sharps in the waste stream for several occupational groups. Four sources of information were used: (1) a medical waste–related injury survey conducted by 17 health departments for the ATSDR report, (2) data provided by the solid waste industry, (3) data provided by the Department of Defense, and (4) the scientific literature. Recognizing the limitations of these data, and adjusting data as necessary, the injury rates were used by the ATSDR to calculate the number of medical waste injuries for each occupational category. Table 2.2 presents estimates of the annual range of medical waste–related injuries from sharps for nonhospital and hospital employees.

Estimating Contaminated Sharps Prevalence. Information on the prevalence of contaminated sharps is not available. To provide an estimate, the ATSDR relied on

Table 2.2 Estimated Annual Range of Medical Waste–Related Injuries from Sharps for Non-hospital and Hospital Employees

Subgroup	Non-hospital Employees		Hospital Employees	
	Number of Employees	Injury Range	Number of Employees	Injury Range
Physicians	642,100	500–1,700	—	—
Physicians/dentists/interns	—	—	131,000	100–400
Registered nurses	1,542,300	17,800–32,500	841,400	9,800–17,900
Licensed practical nurses	723,000	10,200–15,400	201,200	2,800—4,300
Emergency Medical Personnel[a]	400,000	12,000	—	—
Laboratory workers	—	—	250,200	800–7,500
Janitorial/laundry workers	—	—	281,500	11,700–45,300
Hospital engineers[a]	—	—	198,100	12,200
Veterinarians	63,300	50–200	—	—
Animal technicians	10,000	400–1,600	—	—
Dentists	126,000	100–300	—	—
Dental assistants	181,000	2,600–3,900	—	—
Refuse workers	200,000	500–7,300	—	—

Source: Reference 44, table 5.6.

[a]Only one injury rate was reported in the scientific literature.

a surrogate of the rate based on the prevalence of individuals with an infectious disease entering a healthcare facility. This estimate assumes that the use of sharps is equal among all patients (e.g., one infected person generates one contaminated sharp). The ATSDR used available information on the prevalence of individuals with HIV and HBV in the healthcare setting.

ATSDR based the prevalence of HIV-infected persons in the healthcare setting on sentinel hospital studies conducted by the Centers for Disease Control. Recognizing the limitations of these studies, the ATSDR used a range to characterize the prevalence of HIV-infected persons, these figures being used to estimate the probability that a random sharp inside a hospital would be HIV contaminated. The seroprevalence values used in the sentinel hospital studies to make this determination are the 25th and 75th percentile values (0.3 to 2.0 percent) and the crude median (0.8 percent).

For outpatient estimates (e.g., to estimate HIV-contaminated sharps in non-hospital medical waste), seroprevalence in outpatients were derived from an ongoing laboratory-based survey of primary care outpatients also conducted by the CDC. Considering the limitations of the data, the ATSDR chose a probability of 0.8 percent for the prevalence of HIV-contaminated sharps outside a hospital. The seroprevalence rate of persons entering a hospital is approximately 1 percent, which represents a median value. The abundance of information related to HBV seroprevalence allows the use of a median value rather than a range as is used for HIV based on sentinel hospital information.

Estimating Seroconversion Rates. Seroconversion rates resulting from sharps injuries were determined from studies reported in the scientific literature on cases related to patient care. A common characteristic of these injuries is that they occurred with a needle containing freshly drawn blood or other body fluids contaminated with a sufficient amount of viable HIV to cause infection. Because injuries involving medical waste would presumably occur at a much later time (e.g., several hours to days later), use of this information is expected to overrepresent the potential for infections to occur from medical waste because of the fragile nature of HIV when removed from an infected person.

The seroconversion rate used by the ATSDR for HIV infection following a sharps injury in the course of patient care is 0.36 percent. This is based on a study reported by ATSDR in which all cases related to HIV seroconversion following injury from a sharp or significant blood exposure are summarized (44). The seroconversion rate for HBV infection following injury from a sharp known to be contaminated with HBV is 20 percent, which represents a median value.

Estimating Clinical Disease Rates. Of those infected with HIV, all are expected to develop clinical AIDS, for a clinical disease rate of 100 percent. Of those infected with HBV, 50 percent will develop clinical disease. Probability parameters presented in previous paragraphs for prevalence of sharp contamination, seroconversion rate, and clinical disease rate for HIV and HBV are grouped together in Table 2.3. By applying the figures presented in Tables 2.2 and 2.3 into the applicable formulas developed for the study, the ATSDR provided estimates of the annual number of HIV and HBV infections and disease in non-hospital employees and hospital employees resulting from medical waste–related injuries from sharps. These estimates, which represent upper-limit estimates, are presented in Tables 2.4

Table 2.3   Probability Parameters to Calculate a Theoretical Infectious Disease Transmission Rate as a Result of Injuries from Sharps

	Prevalence of Sharp Contamination	Seroconversion Rate	Clinical Disease Rate
Hepatitis B virus	0.01[a]	0.2[a]	0.50[a]
HIV (inside hospitals)	0.003–002[b]	0.0036[c]	1.0[b]
HIV (outside hospitals)	0.008[b]	0.0036[c]	1.0[b]

Source: Reference 44, table 5.1.

[a]From Maynard JE. Viral hepatitis as an occupational hazatd in the health care profession. In: Vyans GH, Cohen SN, and Schmid E, Eds. *Viral Hepatitis: A Contemporary Assessment of Etiology, Epidemiology, Pathogenesis and Prevention*. Philadelphia: The Franklin Press, 1978:321–331.

[b]Personal communication from staff, Center for Infectious Diseases, CDC, November 28, 1989.

[c]From Henderson DK. HIV in the health-care setting. In: Mendell GL, Douglas RG, and Bennett JE, Eds. *Principles and Practice of Infectious Diseases*, 3rd ed. New York: Churchill Livingstone, 1990:2221–2236.

Table 2.4   Theoretical Estimate of the Annual Number of HBV Infections and Hepatitis B Disease in Non-hospital and Hospital Employees as a Result of Medical Waste–Related Injuries from Sharps[a]

	Non-hospital Employees		Hospital Employees	
	HBV Infections	Hepatitis B Cases	HBV Infections	Hepatitis B Cases
Physicians	1–3	<1–2	<1	<1
Physicians/dentists/interns	—	—	<1	<1
Registered nurses	36–65	18–33	20–36	10–18
Licensed practical nurses	20–31	10–15	6–9	3–4
Emergency medical personnel[b]	24	12	—	—
Laboratory workers	—	—	2–15	1–8
Janitorial/laundry workers	—	—	23–91	12–45
Hospital engineers[b]	—	—	24	12
Dentists	<1	<1	—	—
Dental assistants	5–8	3–4	—	—
Refuse workers	1–15	<1–7	—	—

Source: Reference 44, table 5.7.

[a]*Example calculation:* janitorial/laundry workers:

$$11{,}700 \times 0.01 \times 0.2 = 23.4 \text{ HVB infections annually}$$
$$23.4 \times 0.50 = 11.7 \text{ hepatitis B cases annually—round}$$
to the nearest whole number; therefore, 12

[b]Only one rate was available.

and 2.5. The report notes that the actual number of infections and disease is probably lower than the theoretical estimates. Additionally, estimates could not be made for those occupational subgroups in which sufficient injury information was not available.

Based on the theoretical calculations, between 162 and 321 HBV medical waste–related infections from sharps injuries could occur annually, which would account for 0.05 to 0.1 percent of the total number of HBV infections that occur annually in the United States. Of these, an estimated 81 to 160 would proceed to hepatitis B disease, which also accounts for approximately 0.05 to 0.1 percent of HBV clinical disease cases that occur annually in the United States. For HIV infection, there is a theoretical possibility that fewer than one to four cases of AIDS per year could be attributed to injury from medical waste sharps which, based on 1989 statistics, would account for <0.003 to 0.01 percent of all the 1989 AIDS cases in the United States. Contact with nonsharp medical waste may also contribute to infections, although no information was available on which to base theoretical calculations. This contribution, however, is expected to be much less than that associated with medical sharps waste when considering the principles of infectious disease transmission.

Table 2.5　Estimate of the Annual Number of HIV Infections in Hospital Employees as a Result of Medical Waste–Related Injuries from Sharps[a]

	Number of HIV Infections Based on Healthcare Worker Seroconversion Rate of 0.36% and Patient Seroprevalence Rate of:		
Subgroup	0.3%[b]	0.8%[c]	2.0%[d]
Physicians/dentists/interns	<1	<1	<1
Registered nurses	<1	<1	<1 to 1
Licensed practical nurses	<1	<1	<1
Laboratory workers	<1	<1 to 1	<1 to 3
Hospital engineers	<1	<1	<1

Source: Reference 44, table 5.8.

[a]Assuming an HIV clinical disease rate of 100%, the estimated annual number of AIDS cases as a result of medical waste–related injuries from sharps would be equivalent to the estimated number of HIV infections.

　Example calculation: janitorial/laundry workers:

$$11,700 \times 0.02 \times 0.0036 = 0.84 \text{ HIV infections annually}$$

[b]The lower (25th percentile) seroprevalence rate determined by the 32 sentinel hospitals surveyed by the Centers for Disease Control.

[c]The crude median seroprevalence rate determined by the 32 sentinel hospitals surveyed by the Centers for Disease Control.

[d]The upper (75th percentile) seroprevalence rate determined by the 32 sentinel hospitals surveyed by the Centers for Disease Control.

Conclusions and Recommendations

Sixteen conclusions and 11 recommendations were presented in the final report to Congress. The report's conclusions, which are presented below in their entirety, are as follows:

　1. The general public's health is not likely to be adversely affected by medical waste generated in the traditional healthcare setting.

　2. Outside the healthcare setting, the potential for hepatitis B virus (HBV) or human immunodeficiency virus (HIV) infection in the general public following medical waste–related injuries is not likely to be a health concern. However, needle-stick injuries may cause local or systemic secondary infections, similar to injuries from nails.

　3. The increase of in-home healthcare provides opportunities for the general public to contact medical waste. In addition, other sources of non-regulated medical waste may also present opportunities for medical waste contact.

　4. Based on estimates of the number of medical waste–related HIV and HBV infections and disease cases (Section V of this Report), occupational health con-

cerns exist for selected occupations involved with medical waste. Those populations include janitorial and laundry workers, nurses, emergency medical personnel, and refuse workers.

5. When in effect, the recently proposed regulations by the Department of Labor's Occupational Safety and Health Administration (OSHA), "Occupational Exposure to Bloodborne Pathogens; Proposed Rule and Notice of Hearing" (*Federal Register*, 54:23042–139, 1989), should decrease workplace medical waste–related injuries and infections nationwide. This decrease should be achieved through increased awareness, regulatory control, and immunization.

6. From a public health standpoint, medical waste should include the following categories of waste material as defined in Section IX.E of this Report: Cultures and Stocks, Pathological Wastes, Blood and Blood Products, Sharps, Animal Waste, Selected Isolation Waste, and Unused Discarded Sharps.

7. The amount of medical waste generated by in-home healthcare and hospice care is under-appreciated—and is expected to increase, because treating individuals in those settings is becoming more and more common. As a result, refuse workers may experience an increase in needle-stick injuries caused by medical waste discarded with residential waste, resulting in an increased opportunity for infection and disease in this occupational subgroup.

8. Illicit intravenous drug users (IVDUs)—who have high rates of HIV and HBV infection—are a significant source of discarded sharps. (It is thought there are approximately 1.1 to 1.3 million illegal IVDUs nationwide.) The general public could come in contact with these discarded sharps and thus have an increased opportunity for injury and infection. A lack of data prevents estimating the potential HIV and HBV infection rates from IVDU-related waste.

9. Scientific studies indicate that, outside a living host, numbers of the human immunodeficiency virus (HIV) rapidly decline, and the virus does not remain viable after a few days. (HIV is the virus that causes acquired immunodeficiency syndrome, or AIDS.) Thus, persons coming in contact with medical waste outside the healthcare setting have a very low potential for HIV infection. The hepatitis B virus (HBV), however, does remain viable for an extended time outside a host. Consequently, the potential for HBV infection following contact with medical waste is likely to be higher than that associated with HIV.

10. The number of persons infected with the human immunodeficiency virus is anticipated to increase in the future. Based on the data analyzed and the methods of calculating the estimates for medical waste–related injuries, infections, and disease developed in this Report, a maximum of <1 to 4 cases of AIDS per year (<0.0003 to 0.01 percent of all the 1989 AIDS cases in the United States) are estimated to occur in healthcare workers as a result of contact with medical waste sharps. However, the increase in the number of persons infected with HIV is expected to increase the potential for medical waste–related HIV transmission in the health care setting.

11. Based on the data analyzed and the methods of calculating the estimates for medical waste–related injuries, infections, and disease developed in the Report, a

maximum of approximately 162 to 321 HBV infections and 81 to 160 hepatitis B disease cases related to medical waste sharps could occur annually. The 162 to 321 HBV infections and 81 to 160 hepatitis B disease cases estimated to occur as a result of contact with medical waste would account, respectively, for 0.01 to 0.1 percent of the total number of HBV infections and 0.05 to 0.1 percent of hepatitis B clinical disease cases occurring annually in the United States.

12. Communicable disease spread within medical facilities are usually the result of community-acquired (pre-existing) or nosocomial (hospital-acquired) infections. Although, theoretically, communicable diseases may be transmitted by medical waste, the probability of such transmission is generally considered to be remote. Appropriate preventive health measures and personal hygiene practices have controlled and should continue to successfully control the incidence of medical waste–related disease transmission within medical facilities.

13. Medical waste can be effectively treated by chemical, physical, or biological means, such as chemical decontamination, autoclaving, incineration, irradiation, and sanitary sewage treatment. Research indicates medical waste does not contain any greater quantity or different types of microbiological agents than does residential waste, and viruses present in solid waste tend to adsorb to organic matter and deactivate. Additionally, properly operated sanitary landfills provide microbiological environments hostile to most pathogenic agents. Therefore, untreated medical waste can be disposed of in sanitary landfills, provided that procedures to prevent worker contact with this waste during handling and disposal operations are strictly employed. It is worth noting, however, that 158 million tons of municipal solid waste are created yearly nationwide. Medical waste is a part, albeit a small one at 0.3 percent, of the overall problem of solid waste management. Clearly, the most effective way to deal with this issue is to strive to reduce the amount of waste created, on a small scale in homes or on a large scale in industrial operations. Simultaneously, the impetus to recycle, reuse, and reclaim products is paramount to adequately manage solid waste, including medical waste, now and in the future.

14. Based on the principles of infectious disease transmission, the potential for infection resulting from contact with non-sharp medical waste is likely to be significantly less than that related to contact with medical waste sharps. The primary reason for the reduced potential is, in contrast to sharps, that a portal of entry must exist prior to contact with non-sharps for infection or disease to occur.

15. Medical waste adversely affects the environment. Generally, this waste stream contributes to the overall environmental problem of solid waste disposal in the United States. Specifically, beach wash-ups and products of incomplete combustion are among the adverse environmental effects of inadequate medical waste management. Most environmental concerns related to beach wash-ups are associated with medical waste primarily generated by non-regulated sources.

16. Several datá needs and limitations were identified during the Report's data analysis phase:

a. Limited data are available on communicable diseases potentially attributable to medical waste. Theoretical estimates for the number of HBV and HIV infections that could possibly occur were developed using a wide variety of injury data. Most of these data were obtained from scientific literature. However, literature studies may vary due to differing study designs, case-finding methods, sources of records or case information, completeness of case information, and institutional policies on injury reporting and medical waste handling.

b. Because the probability of infection is based on case studies of persons coming in contact with freshly drawn blood or other body fluids, this Report's calculations are upper theoretical estimates of the potential for infections related to HIV and HBV. Health care providers are more likely to contact freshly drawn blood or body fluids during patient care than they are during medical waste management.

c. Insufficient information was available to determine the number of infections or infectious diseases other than AIDS and hepatitis B related to medial waste.

d. Limited information is available on the number of medical waste injuries or infections sustained by refuse workers and other occupational subgroups outside the healthcare setting. The estimates derived in this Report are based on data provided to ATSDR by the New York City Department of Sanitation, Browning-Ferris Industries (BFI), 17 state health departments, and the Department of Defense. These data are self-reported and, therefore, may be biased estimates of the actual number of medical waste–related injuries.

e. Information on the number of medical waste–related injuries incurred by nurse's aides, porters, therapists, morticians, and wastewater workers was not readily available. Preliminary information indicates that medical waste–related injuries do not occur frequently in these occupational subgroups.

f. For some occupational subgroups (janitorial and laundry workers, laboratory workers, and building engineers), the total number of persons who might contact medical waste was not available. Because the number of people in these subgroups employed in hospitals was known, only the number of medical waste–related injuries and infections for hospital-based janitorial and laundry workers, laboratory workers, and hospital engineers could be calculated.

g. During the data analysis, it became clear that several participating facilities in the 17-state health department medical waste injury survey reported injuries that were not related to medical waste. This source of bias is the result of the survey's self-reporting nature and could not be corrected entirely by subsequent error-checking because, in many cases, records were not detailed enough to establish medical waste-relatedness. Attributable intervening factors included the extensiveness of the survey and resource limitations within the states to perform error-checking.

h. Limited information is available on the amount of medical waste generated in the United States and on the number and type of generators. This information would have been useful to better determine populations potentially involved with medical waste. For example, the amount and types of medical waste generated by in-home healthcare are not thoroughly documented. This information is needed to adequately determine the public health implications of medical waste produced by small-quantity generators and, specifically, the degree to which that waste affects refuse workers.

i. Insufficient epidemiologic, health effects, stack, and ash data are available to determine if there are public health implications associated with medical waste incineration. What is known is that the old retort (pre-1970s) pathological incinerators currently being used to incinerate medical waste were not designed to burn plastics. Accordingly, they have the potential to produce incompletely combusted chlorinated products, such as dioxins and furans. Exposure to these substances may result in adverse health effects. The potential adverse health effects resulting from these exposures depend on several factors, including the type of chlorinated product; route and duration of exposure; and exposure absorbed, and effective dose.

The 11 recommendations presented in the final report are as follows:

1. To improve the understanding of what constitutes medical waste, every effort should be made to adopt a regulatory definition based on sound public health principles that is consistent throughout the United States. The definition presented in this Report is proposed as a basis for the national definition.

2. Work practices of the occupational subgroups frequently contacting medical waste (e.g., janitorial and laundry workers, nurses, emergency medical personnel, and refuse workers) should be evaluated by relevant government and private-sector organizations to determine appropriate protective measures and to develop effective personal protective equipment.

3. Occupational health and surveillance data should be collected for occupational subgroups whose frequency of injury, infection, and disease related to medical waste are of public health concern and are not well described—for example, janitorial and refuse workers, waste site workers, and home healthcare providers.

4. Within five years following the promulgation of the OSHA regulations, an evaluation of those regulations' effectiveness in preventing workplace medical waste–related injuries and infections is recommended.

5. A determination of the number of janitorial and laundry workers, laboratory workers, and building engineers employed at medical waste–generating facilities nationwide is recommended. Currently, employment figures are available only for the hospital setting. In addition, the frequency of medical waste–related injuries to nurse's aides, porters, therapists, morticians, and wastewater workers should be further evaluated.

6. Guidelines for in-home healthcare medical waste management should be developed by relevant government and private-sector organizations. As much as possible, these guidelines should also address the management of other sources of non-regulated medical waste. These guidelines may assist in alleviating the negative environmental impact of this waste stream.

7. Intervention strategies relating to the health effects of medical waste should include continual education and training in proper medical waste management.

Specifically, this training should be provided to persons in occupational subgroups who frequently contact medical waste (e.g., janitorial and laundry workers, nurses, emergency medical personnel, and refuse workers). In addition, in-home healthcare providers should be trained in the proper management of medical waste that will become part of residential waste.

8. Relevant information on the potential for infection and disease related to medical waste should be made available to the public, including in-home healthcare patients.

9. Efforts should be undertaken to determine the amount and disposition of medical waste generated from in-home health care and other sources of non-regulated medical waste and the implications of these waste streams.

10. To fill data gaps identified in this report, research should be undertaken in the following areas:

a. To evaluate the probability of infection resulting from contact with dried blood or blood products. Specifically, the research should evaluate the probability of infection following contact with body fluids previously exposed to the environment. This research should also evaluate the difference between sharp and non-sharp contacts.

b. To evaluate the likelihood of medical waste–related disease caused by infectious agents other than HIV and HBV.

c. To determine the chemical constituents of incinerator stack emissions and ash, their concentrations under typical and worst-case conditions, and their mutagenicity. Such studies would assist in developing a scientifically sound and structured medical waste disposal system.

11. The development of new technologies for medical waste management should be encouraged. New treatment technologies should effectively disinfect medical waste with minimal negative impact on the environment. These new treatment technologies should be developed for situations where incineration may not be the preferred treatment method. In addition, technology development in the areas of medical waste reduction, recycling, reuse, and reclamation should be undertaken.

INFECTIOUS AGENTS IN SOLID WASTE

In this section we examine the potential role of the solid waste stream in transmitting human infectious disease. In the review we focus on the waste stream from the point of waste collection to final landfill disposal and examine the presence of infectious agents in the waste stream, including potential disease transmission through direct contact, aerosols, ground and surface water, and biological vectors.

Presence of Infectious Agents

Tortora et al. define the criteria for a reservoir of infection as "a living organism or an inanimate object that provides a pathogen with adequate conditions for survival and multiplication and an opportunity for transmission" (6). In the broadest sense, the solid waste stream should not be viewed as a reservoir of infection, although exceptions to this can be found. Nevertheless, the discussion that follows is based on the potential for the solid waste stream to act as an intermediary fomite (nonliving object) in transmitting infectious agents from reservoirs of infectious agents to human hosts.

The presence of human infectious microbial agents in the solid waste stream is well established in the literature (46–53). Peterson reports that three 30-gram samples of solid waste (in duplicate) are required to yield a pathogen-positive sample (53). Lynch and Jackson report that "body substances such as feces, airway secretions and wound drainage always contain potentially infectious organisms, and blood, urine and other moist body substances sometimes do as well" (14). General wastes, whether from medical institutions or homes, may contain human pathogens entered into the waste stream from human excreta from disposable diapers, animal excrement, blood, exudates, or secretions from dressings, bandages, or feminine hygiene products, facial tissue, condoms, bandages, home-used syringes, or other inanimate objects that have come into contact with human body substances. Gaby reports that the microbiological flora of solid waste has been judged to be equal to that of sewage and the human upper respiratory tract (51).

Few studies have been conducted to compare the microbial loads of waste from residential sources with those of hospital sources. Of the studies that have been conducted, taking into account sampling and assay limitations, the microbial loads of general household waste equal and exceed microbial loads observed in hospital wastes (54–56). It is evident from a review of the literature that any waste source, whether of medical or residential origin, can contain microorganisms that could cause infection in a susceptible host under specific circumstances.

Contact with Infectious Agents

In 1987, Turnberg examined the waste stream to determine common routes of exposure to humans (57). The waste stream was observed from waste collection areas through solid waste transfer stations to final landfill disposal to identify exposure patterns. Public exposure to the waste stream was concluded to be low if solid waste storage and disposal regulations are observed. Occupational exposure by waste worker groups, particularly by direct contact, was observed as a routine part of employment.

In 1982, the National Institute for Occupational Safety and Health reported on the potential for infectious disease transmission to residential waste collectors from direct contact with the waste stream (58). The report stressed personal hygiene as a health matter of considerable importance to this group and recommended that employees bathe daily and wash before eating during the day and before going home. The report recommended that proper techniques for cleaning and covering wounds be addressed by employers and that clean gloves and coveralls be provided daily by employers.

In a 1987 publication, the Health Division of the Oregon Department of Human Resources further recommended that safety protective equipment be employed by waste workers to add further protection against potential infectious disease transmission (59). Protective measures include use of safety glasses, hard hats with chin straps, coveralls, waterproof gauntlet gloves, boots with sufficient thickness and strength to protect the wearer from injury from sharp objects, and NIOSH-approved dust masks when working indoors (e.g., at a transfer station) or whenever necessary. The report also urged waste worker employees to report all injuries and illness to the person responsible for employee health. Proper employment of hygiene practices and safety equipment by waste industry workers will significantly reduce direct contact exposure to infectious agents in the waste stream. Orientation and ongoing educational programs by employers are advised.

Ground and Surface Water Contamination

The presence of pathogenic microorganisms and microbes of sanitary significance in solid waste has been established in the literature. Concern has been raised regarding potential human health implications regarding the release of these agents from sanitary landfills to the environment. In this section we examine the literature regarding the survival and release of these agents to ground and surface water. The health significance of pathogenic microorganisms in landfills has been described as related to:

- The concentration and nature of the pathogen
- The pathogen's ability to survive and retain its infectious properties in the landfill environment
- The pathogen's ability to migrate through the landfill into the surrounding environment and be a potential human hazard (60)

Concentration and Nature. The presence of infectious agents and microorganisms of sanitary significance in the waste stream has been established. However, studies that specifically characterize the concentration and nature of pathogens in the waste stream and landfills are limited. Several studies have been published that have quantified indicator organisms of fecal pollution (total coliform, fecal coliform, and fecal streptococcus) from leachate derived from field or simulated sources (50,51,60–66). The substantial presence in landfills of fecal waste from human and animal sources has been suggested based on indicator bacterial recoveries (50,63). By this, the presence of pathogenic microorganisms associated with fecal waste is indicated.

Enteric pathogenic bacteria have been identified from leachate sources, but not enumerated (51,66). Enteric virus particles, including poliovirus, have also been isolated from landfill leachates, although one study that examined leachate samples collected from 22 landfill sites isolated enteric virus particles (poliovirus types 1 and 3) from only one site which was described by the author as having deficient sanitary landfill practices (67). Further study to characterize pathogenic microorganism populations in solid waste, landfills, and landfill leachates are needed.

Pathogen Survival in the Landfill Environment. Several studies have been published in the literature that examine the survival of microorganisms in landfills and landfill leachates. In 1972, Peterson published pioneering research examining poliovirus survival in the landfill environment (48). The author seeded an area of new landfill material at various depths and observed that the virus was inactivated at all depths within 2 to 4 days. This was attributed to the sharp temperature increases of up to 140°F that is typically associated with the initial aerobic decomposition of waste in a landfill that did not receive daily cover. This phenomenon has been observed by other researchers (51,61,68,69). Other potential antagonisms to the survival of microorganisms in a landfill were identified by Peterson as chemical contaminants such as pesticides, drugs, or heavy metals, although none could be correlated with the inactivating process (48).

Englebrecht et al. observed a significant decrease in bacteria and an absence of virus from leachate derived from a laboratory lysimeter simulating a landfill environment. These results indicated that the harsh conditions of landfill leachate have an inactivating effect upon certain microorganisms (60). However, the authors could not correlate the inactivating constituents. In a field examination of leachates from 22 landfills, Sobsey could isolate virus from only one site, which was described as poorly operated (67). Based on these findings, the author suggested that leachates from properly run sanitary landfills pose little threat to the public with regard to infection by enteric virus.

In 1981, Donnelly et al. published a report on the recovery of fecal indicator and pathogenic microbes from landfill leachates and laboratory lysimeters simulating landfill conditions (54). The authors demonstrated that leachate contained gram-negative rods, all potentially present in human feces, and several potentially pathogenic agents. Fecal indicator bacteria were isolated from older landfill sites, indicating their potential survival for long periods of time.

Pathogen Migration to Groundwater.[2] Although researchers indicate that the chemical and physical characteristics of the landfill environment have an inactivating effect on viruses and bacteria, it is still possible that microbial pathogens can survive these conditions. However, movement of these microorganisms through soils of the landfill would depend on many factors, including soil texture and composition, soil moisture, salt concentrations, pH, climate (rainfall and temperature), nutrient availability, and antagonisms (52). Absorption of viruses onto fill material is likely and may also explain the low recovery of viruses from landfill leachate studies.

Sobsey conducted an examination of viral particles in leachate obtained from a seeded laboratory lysimeter simulating laboratory conditions. No viral particles could be recovered, although more than 80 percent of the test leachate had been

[2]Adapted from Turnberg WL. Infectious waste disposal—An examination of current practices and risks posed. Reprinted from the *Journal of Environmental Health*, 53(6):21–25, copyright 1991 by the National Environmental Health Association, Denver, Colorado. All rights reserved. Reprinted by permission.

tested. The author suggested that the viruses were either inactivated or adsorbed onto refuse components (68). Based on a review of literature sources, Ware reports that the adsorption of virus particles onto fill material is probable and may partly explain low recovery rates from leachate (52).

Novello observed that 10 centimeters of gravelly, silty sand could remove from 80 to 98 percent of poliovirus in leachate. The author suggested that if viral particles were to survive and contaminate a leachate, the soil underlying a landfill could filter these particles out (69). This premise is supported by Rogers (70).

Federal Solid Waste Regulations. Landfilling of our society's solid waste remains the most widely used and accepted method for waste disposal. In 1991, the U.S. Environmental Protection Agency promulgated minimum landfill standards that must be adopted and implemented by each state (71). The standards address how municipal solid waste landfills are sited, constructed, operated, closed, and cared for during a 30-year postclosure period to ensure that the environment is protected.

The standards were written primarily to protect ground and surface water from leachate pollutants. In doing so, EPA established requirements for landfill liners, leachate collection systems, compaction and covering of waste, and environmental monitoring. The operation of sanitary landfills has become far more sophisticated in terms of protecting the health of the public and environment in comparison to the open dump sites of only a few years ago.

In conclusion, the risk associated with ground and surface water contamination by infectious microorganisms in a properly operated landfill appears to be low. This premise is supported in the literature, although further examination of this issue is needed.

Biological Vectors

Concern has been expressed regarding the potential role of biological vectors in the spread of infectious disease from landfill sites (49). Potential vectors include rodents, insects, birds and wild or domestic animals. The role of biological vectors in disease transmission and human morbidity is well established (46). Reports in the literature of biological vector disease transmission from solid waste stream sources are virtually nonexistent, based on this review.

The control of vectors at landfill sites is addressed by current federal solid waste standards (71). On-site containerized storage, collection, and transportation standards for solid waste disposal have addressed the prevention of vector harborage, proliferation, and access. Landfill sites must be fenced to prevent unauthorized entrance by the public or entry by animals. Controls must also be in place to prevent the harborage and presence of vectors such as rats, insects, birds, and burrowing animals. Risks associated with vectorborne disease transmission are addressed by federal solid waste standards. The absence of morbidity statistics or documentation in the literature regarding vector disease transmission associated with waste disposal indicates the risk to be low.

Infectious Aerosols

It is well established that aerosols containing pathogenic microorganisms can cause infectious disease, particularly in laboratory settings (72–75). The ability of an aerosol to cause infection by inhalation is related to the susceptibility of the host and the infective dose and virulence of the pathogen. With regard to indoor microbial aerosols, Spendlove and Fannin report that despite the considerable volume of data available, little is known about the true significance to health except in terms of overt epidemic disease (76).

The role of infectious aerosols originating from the solid waste stream in human morbidity is less well understood and supporting studies are limited. Two studies have been identified that focus on the subject. Ducel et al. conducted an examination of aerosols associated with waste collection and their effects on workers (77). The authors concluded that reports of the incidence and continuation of chronic bronchitis could be associated with exposure to airborne infectious agents.

Fiscus et al. examined airborne levels of bacteria and viruses at a refuse processing plant, a municipal incinerator, a waste transfer station, a landfill, and a wastewater treatment plant (78). No viruses were isolated during the study, which may actually reflect more on the limitations of the assay recovery system than on the absence of airborne virus particles. The highest levels of airborne bacteria colonies were observed at the refuse-derived fuel facility, both within the facility and at the property line. The health significance of any of the levels observed could not be determined.

The study also involved a comprehensive literature review to determine existing information regarding bacteria and virus emissions from waste-handling facilities. The authors identified studies observing airborne bacterial colonies ranging from 200 per cubic meter in a laboratory, up to 700,000 per cubic meter in a sewage treatment plant and between 2000 and 4000 per cubic meter in offices, factories, and streets. Again the authors stated that the health significance of any of these levels could not be judged based on the existing literature.

Disease Transmission[3]

The Centers for Disease Control reports that "there is no epidemiologic evidence to suggest that most hospital waste is any more infective than residential waste. Moreover, there is no epidemiologic evidence that hospital waste disposal practices have caused disease in the community" (79). Rutala states further that with the exception of sharps, there is only one instance of infectious waste associated with in-hospital transmission of infection (55). That case involved disease transmission from a chute-hydro pulping waste system in a hospital (80).

Medical waste disposal has emerged as a concern for waste industry workers.

[3]Adapted from Turnberg WL and Frost F. Survey of occupational exposure of waste industry workers to infectious waste in Washington state. *American Journal of Public Health*, 80(10):1262–1264, copyright 1990 by the American Journal of Public Health, Washington, D.C. Reprinted by permission.

The AIDS epidemic has elevated awareness by and fear within this group regarding potential disease transmission through waste stream sources, particularly hospital waste, although such disease transmission has not been demonstrated epidemiologically (12). Few studies have been conducted to examine waste industry worker exposure to potentially infectious agents in the waste stream. During January and February 1968, Gellin and Mitchell examined for skin disorders 97 waste workers employed by the city of Cincinnati (81). Forty-one cases of bacterial, viral, or fungal dermatitis were observed in this group, but all were classified as nonoccupational in origin. The authors reported that no systemic infectious diseases had been diagnosed in the Cincinnati Division of Waste Collection at the time of their study and that only one claim had been filed for occupational skin disease, and that was later judged to be nonoccupational in origin.

Cimino examined New York City Sanitation Department health records of waste workers employed between 1968 and 1969 (82). Needle-stick injuries were reported due to the presence of uncontained needles in waste collected from hospitals, doctors', and dentists' offices, and discarded needles from drug addicts. All workers reporting needle-stick injuries were given gamma globulin prophylaxis, and no cases of hepatitis were reported. Cimino's subsequent 1987 publication of occupational hazards from New York City sanitation workers examined death records between 1975 and 1984 for those employed as solid waste collectors in January 1973 (83). Of those 10,565 persons, 511 died during the period. The author did not report any deaths or illness due to infectious disease.

A 1979 report by Clark et al. examined the incidence of viral infection among 43 waste collection workers (84). Sera antibody levels for 18 viruses were examined from blood samples collected during the spring and fall. The authors found no evidence for an increased occupational risk from bloodborne viral infections.

In a 1990 report by Turnberg and Frost, 940 waste industry workers were surveyed to evaluate occupational exposure to potentially infectious materials in the municipal waste stream (3). Responses were received from 438 (47 percent) of the 940 workers surveyed. Waste worker employee job safety training rates were ascertained as well as occurrences of occupationally incurred cuts and scratches. The prevalence of exposure to blood-contaminated waste and injury from hypodermic needles was also estimated. Sixty-nine percent of respondents reported having received job safety training, but only 26 percent were trained specifically to deal with safety hazards associated with medical waste. Seventy-four percent of respondents reported having received cuts and scratches on the job, and 32 percent of respondents reported direct contact with waste blood on their clothing or shoes. Thirteen percent of respondents reported blood exposure on their skin, and 5 percent reported blood exposure on their face or eyes. Occupational needle-stick injuries were reported by 21 percent of respondents overall, with 10 percent of 240 responding waste collectors reporting having sustained a needle-stick injury in the year preceding the survey. Needle-sticks were reported from both residential and commercial waste collectors as well as by landfill/transfer station operators. Although injuries were reported, none was linked to infectious disease transmission.

Summary

In summary, the published literature provides no information linking the solid waste stream with infectious disease transmission, although studies on this subject are limited. To date, no comprehensive examination of disease transmission among waste industry workers has been published. The lack of information regarding infectious disease outbreaks among waste worker groups may be significant in itself. If occupationally acquired infectious disease is occurring among this group, it appears to be occurring in ways that are difficult to readily identify.

INFECTIOUS AGENTS IN WASTEWATER

During the past few years, concerns have been raised regarding the disposal into the sanitary sewage system of blood and other human body substances potentially containing pathogenic microorganisms. The Centers for Disease Control recognizes the disposal of blood and body fluids into a municipal sewage system as a feasible disposal option (79). The Environmental Protection Agency also recognizes this option but cautions that the receiving system should have secondary sewage treatment capabilities (85). In this section we explore this issue by examining studies conducted to identify infection risks associated with the operation of sewage treatment plants.

It is well known among health officials that improperly managed wastewater is a direct cause of human disease. Wastewater can contain human pathogenic microorganisms such as bacteria, viruses, protozoan cysts, helminth ova, and fungi (86). Wastewater can also contain a wide range of toxic chemical contaminants (87). Wastewater treatment systems have been developed to protect the public's health by preventing the contamination of ground and surface water. Municipal wastewater systems with either primary or secondary sewage treatment capabilities treat wastewater from residential, commercial, and industrial sources. Smaller on-site septic systems have been designed to treat wastewater loads generated beyond the service boundaries of the larger municipal systems.

Municipal wastewater treatment plant workers face potential occupational exposure to pathogenic microorganisms and chemicals through contact with aerosols generated during wastewater treatment or contact with the wastewater stream, either directly or indirectly. The health significance of these exposures has been the subject of research during the past few years and to date remains unclear.

Enteric Pathogens

In 1975, Hickey and Reist (88) conducted a review of the literature to summarize the health significance of airborne microorganisms generated in the course of sewage treatment. Although the literature was observed to contain studies implying health risks resulting from exposure to aerosols to be both potentially significant and minimal, no study had actually demonstrated a host–wastewater–air–recipient transmission route. However, the authors cautioned that the evidence also failed to negate the presence of potential health risks and recommended that further study was needed.

Lundholm and Rylander (89) conducted a study of employees at six sewage treatment plants in Sweden and concluded that this group observed a significantly higher incidence of medical symptoms than that of the control group. These symptoms included skin disorders, diarrhea, and other gastrointestinal abnormalities. The authors indicated that the symptoms may be due to airborne gram-negative bacterial enterotoxins rather than infection from airborne microorganisms.

In 1984, Clark et al. conducted an examination of U.S. wastewater workers to investigate occupationally acquired parasitic infection (90). The study results did not indicate an increased risk of parasitic infection among this group. As reported by the authors, these findings are not supported by two studies conducted in Europe in which the researchers observed increased incidences of *Entamoeba histolytica* and *Giardia lamblia* infection among sewage workers (91,92). From these studies it was concluded that amebiasis and giardiasis should be considered occupational hazards of sewage-exposed workers. Clark et al. observed that these conflicting study findings may be due to differences in degree of wastewater exposure and/or the prevalence of parasites in the general population.

Several serological studies have been conducted to investigate occupationally acquired enteric viral infections among wastewater workers. Iftimovici et al. observed a statistically significant difference between wastewater workers in Romania and a control group regarding the prevalence of serologic markers to adenovirus and parainfluenza virus, suggesting the possibility of viral contamination (93).

A similar serological examination of wastewater workers and controls in the United States conducted by Clark et al. failed to identify excess seroconversion to rotavirus or Norwalk virus (94). In this study inexperienced wastewater workers were observed with higher Norwalk virus antibody levels than those of their experienced counterparts. Higher antibody titers to Norwalk virus were also observed in wastewater workers with medium to high aerosol exposures than in those with low exposures. A serological examination of sewer workers in Copenhagen reported sewer workers with increased hepatitis A infection compared to the control population (95). The authors concluded that occupational exposure to sewage provides a limited risk of enteric infections.

Although in the literature a majority of wastewater worker symptoms have been suggested to result from occupational exposure to microbiological agents, Scarlett-Kranz examined the possible role of chemicals as a cause of these symptoms (87). Chemical exposures may originate from contact with aerosolizing wastewater or from dusts or vapors of wastewater treatment chemicals. In this study the presence of mutagenic agents in urine was used as one indicator of chemical exposure. The researchers observed a significantly higher frequency of symptoms among the study group than in the control population. Symptoms of itching and burning eyes were reported to be associated significantly with mutagen exposure. Diarrhea, dizziness, and headache among sewage plant workers were also reported to be associated significantly with sewage plant employment, although symptom causes could not be determined.

In 1986, McCunney published a literature review regarding adverse health effects associated with wastewater treatment plant employment (96). The author summarized that although available information fails to indicate serious health effects, a cautious approach is advised until more information becomes available. Methods to prevent enteric disease transmission include frequent handwashing and the use of protective clothing that is not worn home: boots, coveralls, gloves (where practical), and plastic face shields (when appropriate) (97).

Bulk Blood Disposal[4]

In 1987, several maintenance workers at the University of California in Los Angeles were exposed to sewage containing bulk blood when the university's plumbing system broke unexpectedly. Although this incident represented a very rare occurrence in plumbing history, it again focused attention on the practice of disposing of bulk blood into the sanitary sewage system and the potential for this practice to spread bloodborne pathogens to humans in this occupational setting. In this discussion we address bulk blood disposal by reviewing general sewage disposal practices, bulk blood disposal policies and guidelines, health studies conducted on sewage treatment plant workers, and studies examining the presence, survival, and infectivity of HIV in water and wastewater.

Sewage Disposal Practices

Publicly Operated Treatment Works. Discharges of waste to the sanitary sewer are regulated under the Clean Water Act. Both general and specific pretreatment requirements have been developed for indirect dischargers to publicly owned treatment works (POTWs) if the discharge would adversely affect operation of the POTW. No categorical pretreatment standard has been developed for the release of bulk blood to a POTW by the federal government, although such practices may be restricted through state or local regulations.

Under the Clean Water Act, all POTWs in the United States are required to conduct secondary treatment of sewage. By this system, organic materials are microbiologically digested through a biological process followed by removal of solids through settling. Depending on the POTW, various degrees of disinfection of the wastewater are achieved, with the level of wastewater disinfection dependent on many factors, such as disinfectant concentration and contact time, pH, temperature, turbidity, inorganic compounds and ions, and dissolved organic matter. The goal of wastewater disinfection is to destroy microbial pathogens to reduce the risk of

---

[4]Adapted from Turnberg WL. Bulk blood disposal. Should HIV be considered? *Medical Waste Analyst*, 2(7):12–15, copyright 1994 by Technomic Publishing Company, Lancaster, Pennsylvania. Reprinted by permission.

waterborne or related foodborne illness. In general, the order of microbiological sensitivity to disinfection is vegetative bacteria > viruses > bacterial spores, acid-fast bacteria, and protozoan cysts.

On-Site Sewage Disposal. Domestic and commercial sewage disposal outside the boundaries of POTW service areas often rely on on-site sewage disposal systems, which are systems that ultimately dispose of sewage effluent to the ground. On-site sewage disposal is not subject to the federal Clean Water Act, but rather, must adhere to state health codes developed by each state's respective board of health or equivalent regulatory entity.

Generally, on-site disposal systems accumulate and digest waste solids in a septic tank, draining the liquid effluent to a drain field, which discharges to the ground. Removal of microbial particles is accomplished in part by filtration through the soils underlying the drain field and through the action of environmental conditions. Risks to human health result when the drain field fails and effluent is discharged to surface soils, surface water, or groundwater.

Blood Disposal Practices and Policies

Bulk human blood is generated in hospitals, particularly in the operating room and emergency room areas, and to a lesser extent from patient floors. Other sources of bulk human blood disposal include funeral homes during embalming procedures, or clinics, dental and medical offices, and ambulatory care facilities during patient care. Its proper and safe disposal must be assured to protect healthcare workers and the public from potential infection risks.

Bulk blood was included within the definition of *regulated medical waste* by the Environmental Protection Agency during its two-year medical waste tracking program (40 CFR Part 259.30). From a federal perspective, the EPA considers the waste that enters a sanitary sewage system leading to a POTW to be domestic sewage. Domestic sewage has been excluded from the definition of solid waste under Section 1004(27) of the Resource Conservation and Recovery Act. Therefore, during its two-year medical waste tracking program, the EPA did not regulate bulk blood as medical waste once it had been disposed to a sanitary sewage system (98).

Various guidance for bulk blood disposal has been published. In its guidelines for infectious waste management published in 1986, the EPA advised that bulk blood only be disposed to a POTW with secondary sewage treatment capabilities (85). The Centers for Disease Control identified bulk blood disposal to a sanitary sewage system as a disposal alternative but did not specify a final sewage treatment process (79). In a study on infectious waste management conducted by the Washington State Department of Ecology, the environmental health offices of the state's 32 local health districts responsible for permitting on-site sewage treatment systems were surveyed to determine risks to public health due to medical waste handling practices (99). None of the health districts reported local restrictions on the practice of bulk blood disposal, and none reported system failures or adverse public health incidents due to bulk blood disposal practices.

Sewage Worker Health Studies

Many studies have been reported in the literature investigating the occupational infection risk of sewage treatment plant operators. From these studies an indirect occupational infection relationship has occasionally been reported, although overall, the available information has not identified serious health effects among sewage workers. Methods to prevent enteric disease transmission have been reported to include frequent handwashing and use of protective clothing that is not worn home: boots, coveralls, gloves (where practical), and plastic face shields (when appropriate) (97).

Although most sewage worker studies have focused on infection from enteric pathogens as described in the preceding section, two studies have been identified that addressed bloodborne infection among sewage treatment plant operators. The incidence of infection by the bloodborne pathogen hepatitis B was investigated among wastewater workers by Iftimovici et al. (93) and Skinhoj et al. (95). Neither study demonstrated an increased risk of HBV infection associated with wastewater plant employment. Although the hepatitis B surface antigen (HBsAg) virus has been identified from human feces (100), it has been reported as apparently nonviable when excreted (101).

HIV transmission via wastewater was addressed in a 1988 EPA memorandum written by Stephen P. Allbee, acting director of the Municipal Construction Division (102). In the memorandum, two facts were emphasized: (1) the Centers for Disease Control does not consider the waterborne route of HIV transmission a possibility, and (2) no cases of waterborne or water vapor transmission of HIV have been reported. Based on CDC information, the EPA "does not consider workers at municipal wastewater treatment facilities to be at risk of being infected with HIV through the water-borne route."

HIV Survival in Water and Wastewater

Early reports from the Centers for Disease Control indicated a rapid decay of HIV when in the environment (12). Four studies have since been conducted that indicate this decay not to be as rapid as initially suspected, although the health significance of this finding appears unchanged from earlier reports. The four studies, which address the presence, survival, or potential infectivity of HIV in water and wastewater, are reported below.

Study 1. In 1989, researchers in the United Kingdom published a study measuring the decay of HIV in sterilized samples of water, sewage, and seawater held at 16°C over a 30-day period (103). The decay of poliovirus 2 was also tested for comparison purposes. HIV concentrations were calculated to decay by 90 percent in 2.9 days in sewage, 1.8 days in tap water, and 1.6 days in seawater. By comparison, a

90 percent decline in poliovirus was calculated at 23 to 30 days in sewage. Poliovirus was described by the authors as an enterovirus widely used as an indicator of the effectiveness of waste treatment processes, and it was noted that the poor survival of HIV in the environment makes it an extremely unlikely health threat to disinfected water supplies.

Study 2. In 1991, David Preston and colleagues published research examining the presence of viral and proviral nucleic acids sequences with homology to HIV-1 in raw wastewater obtained from wastewater treatment plants of four Florida cities (104). In this study the authors noted the potential for HIV to be shed from the gastrointestinal tract based on other published studies. From their study, HIV–RNA and proviral–DNA sequences were detected in wastewater concentrates from only one of the Florida cities examined. This finding was correlated by the authors to the city, which they described as having an abnormally high incidence of AIDS patients, although no further information was given to support this statement. According to the authors, the presence of either viable or inactive HIV would explain the presence of HIV–RNA in raw wastewater concentrates, although the authors noted HIV survival outside the human host to be unlikely for extended periods, due to its fragile outer lipid membrane. The authors questioned the public health significance of the findings because the assay was not designed to detect infectious HIV-1. In addition, the study findings were characterized by other researchers as controversial.

Study 3. In 1992, Leonard W. Casson and his colleagues at the University of Pittsburgh School of Engineering published a report on the survival of HIV in wastewater and sterile water (105). Samples were inoculated with approximately 10+6 reverse transcriptase units of HIV and held at 25°C for up to 72 hours. Samples were extracted at 6, 12, 24, 48, and 72 hours to determine HIV survival, based on infectivity. The infectivity was measured in peripheral blood lymphocyte cultures by comparing the virus infectivity titer at various times with the initial titer, as described by the researchers.

In sterile water, HIV was observed to remain stable for 6 hours, followed by a 1-log reduction between 6 and 12 hours, 2 log units at 24 hours, and up to 3 log units after 72 hours. In primary effluent, HIV was found to be relatively stable during the first 6 hours. A 1-log infectivity reduction was observed at 12 hours, 2 logs after 48 hours, and more than 3 logs at 72 hours. When HIV stability was examined in nonchlorinated secondary effluent, no loss in infectivity was observed after 6 hours, a greater than 1-log reduction after 24 hours, and a reduction of between 2 and 3 logs at 48 and 72 hours.

From these results it was concluded that HIV was fairly stable in wastewater up to 12 hours, followed by a 2- to 3-log reduction after 48 hours. When comparing HIV stability in primary effluent and nonchlorinated secondary effluent at 25°C, no significant differences were observed. The researchers also compared HIV survival data with another study examining poliovirus and concluded that HIV survivability in wastewater is significantly less than that of poliovirus under similar conditions.

Study 4. In 1993, Barbara Moore of the University of Texas Department of Microbiology conducted a study to examine (1) the stability of cell-free HIV, and (2) the stability of peripheral blood lymphocytes actively replicating HIV in room-temperature tap water (106). Moore also attempted to recover HIV from tap water following introduction of HIV-contaminated blood and HIV-contaminated peripheral blood lymphocytes. Moore observed that the infectivity of cell-free HIV in dechlorinated tap water was reduced by 90 percent (1 log) during the first 30 to 60 minutes, and by 99.9 percent (3 logs) after 8 hours. A parallel study was conducted by the author using poliovirus, which was found to be stable in tap water, with no loss of infectivity over 24 hours.

Moore observed the fate of HIV-infected lymphocytes in tap water to be similar to that of uninfected lymphocytes, with cell viability lost rapidly, within 15 minutes. However, the reduction of HIV infectivity associated with infected lymphocytes was similar to that of free virus inactivation, with a 90 percent (1-log) reduction in 1 hour and a 99 percent (2-log) reduction in 8 hours. No infectious virus could be recovered from tap water after 24 hours.

In an experiment to isolate HIV from tap water following introduction of up to 2 percent HIV-infected whole blood, no infectious HIV was detected as free virus in any sample tested. On the other hand, HIV was recovered from peripheral blood lymphocytes introduced to tap water at 1 minute (1 percent vol/vol) and 5 minutes (2 percent vol/vol).

Of note, the rate of HIV inactivation observed by Moore was substantially more rapid than the inactivation rates previously published by Slade et al. (103) and Casson et al. (105). Moore identified several differences in the studies. One was the variation in test water temperature (25°C for Moore as opposed to 16°C for Slade). Another was the presence of suspended solids in the water (105). Both factors would be expected to account for greater HIV stability.

Moore found that HIV is not particularly stable in tap water compared to human enteric viruses such as poliovirus. Furthermore, the presence of a residual chlorine disinfectant would be expected to compromise viral survival further. Nevertheless, the author recommended that prudence dictates that any contaminated blood products be adequately disinfected before disposal to a sewage treatment system.

Moore concluded that the concentration of HIV in blood is substantially lower than that of enteric viruses, and that the amount of infectious HIV would be reduced further over a period of 8 to 12 hours. In light of epidemiological evidence supporting transmission only through sexual contact or by direct contact with infected blood, Moore concluded that HIV poses a minimal risk to workers at publicly operated sewage treatment works.

Conclusion

In conclusion, data from four recently published studies generally support the original statement made by the EPA in 1988 that POTW operators are not at risk from HIV transmission via the waterborne route. Therefore, the practice of disposing blood to a sanitary sewage system appears to be an effective solution to bulk blood disposal provided that (1) the blood is handled and poured such that it does not create an occupational exposure to the handler, (2) the receiving POTW does not object to the practice, (3) the blood does not contain regulated contaminants, and (4) no other regulatory restrictions apply.

Bulk blood disposal to an on-site sanitary sewage system also appears to be a feasible alternative. Based on a survey of Washington State's 32 local health departments, no system failures or human exposure risks were reported due to bulk blood disposal to on-site sewage disposal systems in Washington. Decisions regarding the disposal of bulk blood to an approved on-site sewage treatment system should be in conjunction with the responsible health authority.

Although one researcher advised prudence to dictate that contaminated blood be disinfected prior to sewage disposal, the need for such additional handling practices appears unsupported. The activities associated with disinfecting or sterilizing bulk blood either on-site or off-site from the point of generation may have the negative effect of increasing human exposure during transport to a treatment area or during the treatment process itself. Ultimately, bulk blood disposal should be conducted to minimize human exposure events.

REFERENCES

1. Turnberg WL. Human infection risks associated with infectious disease agents in the waste stream. In: Washington State Infectious Waste Project: Report to the Legislature, Attachment 1. 89-62. Washington Department of Ecology, Olympia, Washington. December 1989.

2. Turnberg WL. Infectious waste: An examination of practices and risks posed. *J Environ Health*, 53(6):21–25. 1991.

3. Turnberg WL and Frost F. Survey of occupational exposure of waste industry workers to infectious waste in Washington State. *Am J Public Health*, 80(10):1262–1264. 1990.

4. Turnberg WL. Bulk blood disposal—Should HIV be considered? *Med Waste Anal*, 2(7):12–15. April 1994.

5. Turnberg WL and Lowen L. Home syringe disposal: Practice and policy in Washington State. *Diabetes Educ*, 20(6):489–492. November/December 1994.

6. Tortora GJ, Berdell, RF and Case CL. *Microbiology: An Introduction*, 3rd ed. The Benjamin-Cummings Publishing Co., Redwood City, California. 1989.

7. Nester EW, Roberts CE, McCarthy BJ, and Pearsall NN. *Microbiology: Molecules, Microbes and Man*. Holt, Rinehart and Winston, Inc., Orlando, Florida. 1973.

8. Volk WA, Benjamin DC, Kadner RJ, and Parsons JT. *Essentials of Medical Microbiology*, 3rd ed. J.B. Lippincott Company, Philadelphia. 1986.

9. Garner JS and Simmons BP. CDC Guideline for Isolation Precautions in Hospitals. HHS Publication (CDC) 83-8314. U.S. Department of Health and Human Services, Public Health Service, Centers for Disease Control, Atlanta, Georgia. July 1983.

10. Barkley WE, Wedum AG, and McKinney RW. The hazard of infectious agents in microbiological laboratories. In: Block S, Ed. *Disinfection, Sterilization, and Preservation*, 3rd ed., pp. 566–576. Lea & Febiger, Philadelphia. 1983.

11. Mandell G., Dougals RG, and Bennett JE. *Principles and Practice of Infectious Diseases*, 3rd ed. Churchill Livingstone, Inc., New York. 1990.

12. Centers for Disease Control. Recommendations for prevention of HIV transmission in healthcare settings. *MMWR*, 36(2S):1S–18S. August 21, 1987.

13. Centers for Disease Control. Update: Universal precautions for prevention of transmission of human immunodeficiency virus, hepatitis B virus, and other bloodborne pathogens in healthcare settings. *MMWR*, 37(24):377. June 24, 1988.

14. Lynch P and Jackson MM. Isolation practices: How much is too much or not enough. *ASEPSIS: Infect Control Forum*, 8(4):2–5. Fourth Quarter 1986.

15. Centers for Disease Control. Provisional public health service inter-agency recommendations for screening donated blood and plasma for antibody to the virus causing acquired immunodeficiency syndrome. *MMWR*, 34(1):1–5. January 11, 1985.

16. American Public Health Association. *Control of Communicable Diseases in Man*, 15th ed. Abram S. Berenson, Washington, DC. 1990.

17. Centers for Disease Control. 1993 revised classification system for HIV infection and expanded surveillance case definition for AIDS among adolescents and adults. *MMWR*, 41(RR-17):1–19. December 18, 1993.

18. Centers for Disease Control. Classification system for human T-lymphotropic virus type III/lymphadenopathy-associated virus infections. *MMWR*, 35(20):334–339. 1986.

19. Centers for Disease Control. Public health service guidelines for counseling and antibody testing to prevent HIV infection and AIDS. *MMWR*, 36(31):509–515. August 14, 1987.

20. Centers for Disease Control. Semen banking, organ and tissue transplantation, and HIV antibody testing. *MMWR*, 37(4):57–63. February 5, 1988.

21. Centers for Disease Control. Transmission of HIV through bone transplantation: Case report and public health recommendations. *MMWR*, 37(39):597. October 7, 1988.

22. Lifson A. Do alternate modes of transmission of human immunodeficiency virus exist? *JAMA*, 259(9):1353–1356. March 4, 1988.

23. Centers for Disease Control. Guidelines for prevention of transmission of human immunodeficiency virus and hepatitis B virus to health-care and public-safety workers. *MMWR*, 38(S-6):111–155. June 23, 1989.

24. Centers for Disease Control. Projections of the number of persons diagnosed with AIDS and the number of immunosuppressed HIV-infected persons—United States, 1992–1994. *MMWR*, 41(RR-18):1–29. December 25, 1992.

25. Resnick L, Veren K, Salakuddin SZ, Tondreau S, and Markham PD. Stability and inactivation of HTLV-III/LAV under clinical and laboratory environments. *JAMA*, 255(14):1887–1891. April 11, 1986.

26. Cao Y, Ngai H, Gu G, and Ho D. Decay of HIV-1 infectivity in whole blood, plasma and peripheral blood mononuclear cells. *AIDS*, 7(4):596–597. 1993.

27. Moudgil T and Daar E. Infectious decay of human immunodeficiency virus type 1 in plasma. *J Infect Dis*, 167:210–212. 1993

28. Centers for Disease Control. Recommendations for preventing transmission of infection with human T-lymphotropic virus type III/lymphadenopathy-associated virus in the workplace. *MMWR* 34(45):682–695. November 15, 1985.

29. Bankowski M, Landay A, Staes B, Shuburg R, Kritzler M, Hajakian V, and Kessler H.

Postmortem recovery of human immunodeficiency virus type 1 from plasma and mononuclear cells. *Arch Pathol Lab Med*, 116:1124–1127. 1992.

30. U.S. Department of Labor, Occupational Safety and Health Administration. Occupational Exposure to Bloodborne Pathogens; Proposed Rule and Notice of Hearing. *Federal Register*, 29 CFR Part 1910. May 30, 1989.

31. Marcus R. Surveillance of healthcare workers exposed to blood from patients infected with the human immunodeficiency virus. *N Engl J Med*, 319(17):1118–1123. October 27, 1988.

32. Henderson DJ. HIV infection: Risks to healthcare workers and infection control. *Nurs Clin North Am*, 23(4):767–777. December 1988.

33. Centers for Disease Control. Update on hepatitis B prevention. *MMWR*, 36(23):354–360. June 19, 1987.

34. Centers for Disease Control. Changing patterns of groups at high risk for hepatitis B in the United States. *MMWR*, 37(28):429. July 22, 1988.

35. Bock KB, Tong MJ, and Bernstein S. The risk of accidental exposure to hepatitis B virus via blood contamination in medical students. *J Infect Dis*, 144:604. 1981.

36. Dienstag JL and Ryan D. Occupational exposure to hepatitis B virus in hospital personnel: Infection or immunization? *Am J Epidemiol*, 115(1):26–39. 1982.

37. Jovanovich JF, Saravolatz LD, and Arking LM. The risk of hepatitis B among select employee groups in an urban hospital. *JAMA*, 250(14):1893–1894. October 14, 1983.

38. Parry MF, Brown AE, Cobbs LG, Gorke DJ, and New HC. The epidemiology of hepatitis B infection in housestaff. *Infection*, 6(5):204–206. 1978.

39. Tong MJ, Howard AM, Schatz GC, Kane MA, Roskamp DA, Co RL, and Boone C. A hepatitis B vaccination program in a community teaching hospital. *Infect Control*, 8(3):102–107. 1987.

40. Pantelick E, Steere A, Lewis H, and Miller D. Hepatitis B infection in hospital personnel during an eight-year period. *Am J Med*, 70:924–927. April 1981.

41. Sienko DG, Anda RF, McGee HB, Weber JA, Remington PL, Hall WN, and Gunn RA. Hepatitis B vaccination programs for hospital workers: Results of a statewide survey. *Am J Infect Control*, 16(5):193–197. 1988.

42. Bond WW, Favero MS, Peterson NJ, Gravelle CR, Ebert JW, and Maynard JE. Survival of hepatitis B virus after drying and storage for one week [Letter]. *Lancet*, 1:550–551. 1981.

43. Centers for Disease Control. Recommendations for protection against viral hepatitis. *MMWR*, 34(22):313–335. June 7, 1985.

44. U.S. Department of Health and Human Services, Agency for Toxic Substances and Disease Registry, Public Health Service. The Public Health Implication of Medical Waste: A Report to Congress. 1990.

45. Jacobson T, Burke JP, and Conti MT. Injuries of hospital employees from needles and sharp objects. *Infect Control*, 4:100–2. 1983.

46. Hanks TG. Solid Waste/Disease Relationships: A Literature Survey. U.S. Department of Health, Education, and Welfare, Public Health Service, Cincinnati, Ohio. 1967.

47. Cook H, Cromwell D, and Wilson H. Microorganisms in household refuse and seepage water from sanitary landfills. *Proc W Va Acad Sci*, 39:107–114. 1967.

48. Peterson M. The occurrence and survival of viruses in municipal solid waste. Ph.D. thesis, University of Michigan, Ann Arbor, Michigan. 1972.

49. Peterson M. Soiled disposable diapers: A potential source of viruses. *Am J Public Health*, 64(9):912–914. September 1974.

50. Cooper R, Klein S, Leong C, Potter J, and Golueke C. Effect of Disposable Diapers on the Composition of Leachate from a Landfill. SERL Report 74-3. Sanitary Engineering Research Laboratory, College of Engineering and School of Public Health, University of California, Berkeley, California, p. 93. 1974.

51. Gaby W. Evaluation of Health Hazards Associated with Solid Waste/Sewage Sludge Mixtures. EPA-670/2-75-023. U.S. Environmental Protection Agency, National Environmental Research Center, Cincinnati, Ohio. 1975.

52. Ware S. A Survey of Pathogen Survival During Municipal Solid Waste and Manure Treatment Processes. EPA-600/8-80-034. U.S. Environmental Protection Agency, Cincinnati, Ohio. 100 pp. August 1980.

53. Peterson ML. Pathogens Associated with Solid Waste Processing. U.S. Environmental Protection Agency, SW-49r. 1971.

54. Donnelly J, Scarpino P, and Brunner D. Recovery of fecal indicator and pathogenic microbes from landfill leachate. In: Land Disposal: Municipal Solid Waste. EPA-600/9-81-002a. U.S. Environmental Protection Agency, Cincinnati, Ohio, pp. 37–54. 1981.

55. Rutala WA. Cost effective application of the Centers for Disease Control Guideline for Handwashing and Hospital Environmental Control. *Am J Infect Control*, 13(5):218–224. 1985.

56. Kalnowski B, Wiegant H, and Ruden H. The microbial contamination of hospital waste. *Zentralbl Bakt Hyg*, I (Abt Orig B). 1983.

57. Turnberg WL. An Examination and Risk Evaluation of Infectious Waste in King County, Washington. Seattle–King County Department of Public Health, Seattle, Washington. March 18, 1988.

58. National Institute for Occupational Safety and Health. Residential Waste Collection: Hazard Recognition and Prevention. U.S. Department of Health and Human Services, Centers for Disease Control, Atlanta, Georgia. March 1982.

59. Oregon Health Division. Recommendations for Preventing Disease Transmission While Handling Solid Waste in Oregon. Oregon Department of Human Resources, Health Division, Portland, Oregon. September 1988.

60. Englebrecht R, Weber M, Amirhor P, Foster D, and La Rossa D. Biological properties of sanitary landfill leachate. In: Malina JF and Sagick BP, Eds., *Virus Survival in Water and Waste Water Systems*. Water Resources Symposium 7. Center for Research in Water Resources, University of Texas, Austin, Texas, pp. 201–217. 1974.

61. Englebrecht R. Survival of Viruses and Bacteria in a Simulated Sanitary Landfill. NTIS/PB-234 589. National Technical Information Service, Springfield, Virginia. December 14, 1973.

62. Blannon J and Peterson M. Survival of fecal coliforms and fecal streptococci in a sanitary landfill. News of Environmental Research in Cincinnati. U.S. Environmental Protection Agency, Cincinnati, Ohio. April 12, 1974.

63. Cooper R, Potter J, and Leong C. In: Malina JF and Sagick BP, Eds., *Virus Survival in Water and Waste Water Systems*. Water Resources Symposium 7. Center for Research in Water Resources, University of Texas, Austin, Texas, pp. 218–232. 1974.

64. Englebrecht R and Amirhor P. Inactivation of Enteric Bacteria and Viruses in Sanitary Landfill Leachate. NTIS/PB-252 973/AS. National Technical Information Service, Springfield, Virginia. 1975.

65. Cameron R and McDonald E. Coliforms and municipal landfill leachate. *JWPCF*, 49:2504–2506. 1977.

66. Donnelley F and Scarpino P. Isolation, Characterization and Identification of Microorganisms from Laboratory and Full-Scale Landfills. EPA-600/2-84-119, PB84-212 737, 492 pp. July 1984.

67. Sobsey M. Field survey of enteric viruses in solid waste leachates. *Am J Public Health*, 68(9):858–863. 1978.

68. Sobsey M, Wallis C, and Melnich J. Studies on the survival and fate of enteroviruses in an experimental model of a municipal solid waste landfill leachate. *Appl Microbiol*, 30(12):565–574. 1975.

69. Novello A. Polio survival in landfill leachates and migration through soil columns. Master's thesis, University of Cincinnati, Cincinnati, Ohio. 1974.

70. Statement of Harvey Rogers, Chief of the Environmental Protection Branch, Division of Safety, National Institutes of Health, U.S. Department of Health and Human Services before the Subcommittee on Transportation, Tourism, and Hazardous Materials, Committee on Energy and Commerce, U.S. House of Representatives, October 21, 1987.

71. U.S. Environmental Protection Agency. Municipal solid waste landfill standards. *Federal Register*. October 9, 1991.

72. Dimmick RL, Vogl WF, and Chatigny MA. Potential for accidental microbial aerosol transmission in the biological laboratory. In: Hellman A, Oxman MN, and Pollack R, Eds., *Biohazards in Biological Research*, pp. 246–266. Cold Spring Harbor Laboratories, Cold Spring Harbor, New York. 1973.

73. Pike RM. Laboratory-associated infections: Summary and analysis of 3921 cases. *Health Lab Sci*, 13(2):105–114. 1976.

74. Pike RM. Laboratory-associated infections: Incidence, fatalities, causes, and prevention. *Annu Rev Microbiol*, 33:41–66. 1979.

75. Richardson JH and Barkley WE, Eds. Biosafety in Microbiological and Biomedical Laboratories. HHS Publication (CDC) 84-8395. U.S. Department of Health and Human Services, Public Health Service, Washington DC. 1984.

76. Spendlove JC and Fannin KF. Source, significance and control of indoor microbial aerosols: Human health aspects. Public Health Rep, 98(3):229–244. May/June 1983.

77. Ducel G, Pitteloud J, Rufener-Press M, Bky M, and Rey P. Importance de l'exposition bactérienne chez les employés do la voirie chargés de la favee des ordures. *Soz Praventivmed*, 21:136–138. 1976.

78. Fiscus D, Gorman P, Schrag M, and Shannon L. Assessment of Bacteria and Virus Emissions at a Refuse Derived Fuel Plant and Other Waste Handling Facilities. EPA-600/2-78-152. August 1978.

79. Garner JS and Favero MS. Guideline for Handwashing and Hospital Environmental Control, 1985. HHS Publication 99-1117. Public Health Service, Centers for Disease Control, Atlanta, Georgia. 1985.

80. Grieble HG, Bird TJ, Nidea HM, and Miller CA. Chute-hydropulping waste disposal system: A reservoir of enteric bacilli and *Pseudomonas* in a modern hospital. *J Infect Dis*, 130(6):602–607. December 1974.

81. Gellin GA and Mitchell ZR. Occupational dermatoses of solid waste workers. *Arch Environ Health*, 20:510–515. 1970.

82. Cimino JA. Health and safety in the solid waste industry. *Am J Public Health*, 65(1):38–46. January 1975.

83. Cimino JA and Mamtani R. Occupational hazards for New York City sanitation workers. *J Environ Health*, 50(1):8–12. July/August 1987.

84. Clark C, Van Meer G, Bjornsen A, Linneman C, Schiff G, and Gartside P. Incidence of Viral Infections Among Waste Collection Workers. University of Cincinnati Medical Center, Institute of Environmental Health Kettering Laboratory, Cincinnati, Ohio. January 19, 1979.

85. U.S. Environmental Protection Agency. EPA Guide for Infectious Waste Management. EPA/530-SW-86-014. May 1986.

86. Fradkin L, Goyal SM, Bruins RJF, Gerba CP, Scarpino P, and Stara JF. Municipal wastewater sludge: The potential public health impacts of common pathogens. *J Environ Health*, 51(3):148–152. 1989.

87. Scarlett-Kranz JM, Babish JG, Strickland D, and Lisk DJ. Health among municipal sewage and water treatment workers. *Toxicol Ind Health*, 3(3):311–319. 1987.

88. Hickey J and Reist P. Health significance of airborne microorganisms from wastewater treatment processes, Part II: Health significance and alternatives for action. *JWPCF*, 47:2758–2768. 1975.

89. Lundholm M and Rylander R. Work related symptoms among sewage workers. *Br J Ind Med*, 40:325–329. 1983.

90. Clark C, Linnemann C, Clark J, and Gartside P. Enteric parasites in workers occupationally exposed to sewage. *J Occupat Med*, 26(4):273–275. April 1984.

91. Doby JM, Duval JM, and Beaucournu JC. Amoebiasis, an occupational disease of sewer workers? *Nouv Presse Med*, 9:532–533. 1980.

92. Knobloch J, Bialek R, and Hasemann J. Intestinal protozoal infestation in persons with occupational sewage contact. *Dtsch Med Wochenschr*, 108:57–60. 1983.

93. Iftimovici R, Iacobescu V, Copelovici Y, Dinga A, Iordan L, Niculescu R, Teleguta L, and Chelaru M. Prevalence of antiviral antibodies in workers handling wastewaste and sludge. *Rev Roum Med Virol*, 31(3):187–189. 1980.

94. Clark CS, Linnemann CC, Gartside PS, Phair JP, Blacklow N, and Zeiss CR. Serologic survey of rotavirus, Norwalk agent and *Prototheca wickerhamii* in wastewater workers. *Am J Public Health*, 75(1):83–85. 1985.

95. Skinhoj P, Hollinger FB, Hovind-Hougen K, and Lous P. Infectious liver diseases in three groups of Copenhagen workers: Correlation of hepatitis A infection to sewage exposure. *Arch Environ Health*, 36(3):139–143. 1981.

96. McCunney RJ. Health effects of work at waste water treatment plants: A review of the literature with guidelines for medical surveillance. *Am J Ind Med*, 9:272–279. 1986.

97. State of California Department of Health Services. Illness in sewage treatment workers and recommendations for prevention. California Morbidity, Weekly Report from the Infectious Disease Section, State Department of Health Services, No. 33. August 24, 1984.

98. U.S. Environmental Protection Agency. First Interim Report to Congress—Medical Waste Management in the United States, EPA/530-SW-90-051A. May 1990.

99. Turnberg WL. A characterization of infectious waste issues as experienced by local

environmental health jurisdictions in Washington State. In: Washington State Infectious Project: Report to the Legislature, Attachment 5. 89–62. Washington Department of Ecology, Olympia, Washington. December 1989.

100. Tiku ML, Beutner KR, Ramirez RI, Dienstag JL, Sultz MA, and Ogra PL. Distribution and characteristics of hepatitis B surface antigen in body fluids of institutionalized children and adults. *J Infect Dis*, 134:342–347. 1976.

101. Villarejos MV, Visona KA, Gutierrex A, and Rodriguez A. Role of saliva, urine and feces in the transmission of type B hepatitis. *N Engl J Med*, 291:1375–1378. 1974.

102. EPA memorandum from Stephen P. Allbee, Acting Director of the Municipal Construction Division. 1988.

103. Slade JS, Pike EB, Eglin RP, Colbourne JS, and Kurtz JB. The survival of human immunodeficiency virus in water, sewage and sea water. *Water Sci Technol*, 21(3):55–59. 1989.

104. Preston DR, Farrah SR, Bitton G, and Chaudhry GR. Detection of nucleic acids homologous to human immunodeficiency virus in wastewater. *J Virol Methods*, 33:383–390. 1991.

105. Casson LW, Sorber CA, Palmer RH, Enrico A, and Gupta P. HIV survivability in wastewater. *Water Environ Res*, 64(3):213–215. 1992.

106. Moore BE. Survival of human immunodeficiency virus (HIV), HIV-infected lymphocytes, and poliovirus in water. *Appl Environ Microbiol*, 59(5):1437–1443. 1993.

PART II

POLICY

3

BIOHAZARDOUS WASTE REGULATION

A principal consideration in managing biohazardous waste is an understanding of applicable federal, state, and local regulations. To date there is no national regulation that comprehensively addresses how biohazardous waste is to be managed from cradle to grave, nor does it appear that such a mandate is forthcoming from the U.S. Congress. Nevertheless, at least five federal agencies have developed or are in the process of developing requirements that address biohazardous waste that are specific to their respective jurisdictional responsibilities. These agencies are the Occupational Safety and Health Administration (OSHA), the Public Health Service (PHS), the Department of Transportation (DOT), the U.S. Postal Service (USPS), and the U.S. Environmental Protection Agency (EPA). The OSHA, PHS, DOT, and USPS programs are described in this chapter. The EPA's interim medical waste tracking program, which expired in 1991, is described in Chapter 5, and its current rule development effort addressing standards for medical waste incinerators is addressed in Chapter 8. Although these federal agencies focus only on how biohazardous waste affects their specific areas of concern, the overall effect of each standard serves to protect public health and safety in a comprehensive sense. As an example, rules promulgated by OSHA to protect workers from exposure to biohazardous waste serve as a national standard for how this waste stream is handled. Inter- and intrastate transportation standards established by the PHS and DOT, and mailing standards by the USPS, also provide protection to the public. When taken as a whole, the federal regulations begin to take a comprehensive approach to biohazardous waste management.

In addition to the federal activities, most states and many local governments have also developed biohazardous waste regulations during the past few years. However, without a federal mandate or clearly understood regulatory need, state and local regulations are typically inconsistent and widely variable in approach.

The federal standards reviewed in this chapter include:

- *Safe Workplace Standards.* The Occupational Safety and Health Administration's "Occupational Exposure to Bloodborne Pathogens" standard was established in 1991. This standard, which addresses biohazardous waste handling, packaging, and labeling, is published at 29 CFR Part 1910.1030.
- *Transportation Standards.* Regulations for biohazardous waste transportation are currently under the jurisdiction of two federal agencies: (1) the Public Health Service (PHS) of the U.S. Department of Health and Human Services, and (2) the U.S. Department of Transportation (DOT). The PHS standard, published at 42 CFR Part 72, is currently being revised to address biohazardous waste transportation in a more realistic manner, which, in effect, would rely on regulations currently being developed by the U.S. Department of Transportation. The proposed DOT standard, which will be published at 49 CFR Part 171–180, is intended to address the inter- and intrastate transportation of infectious substances and regulated medical waste.
- *Mail System Standards.* In 1992 the U.S. Postal Service (USPS) published its rule entitled "Mailability of Sharps and Other Medical Devices" under Sections 8.1 through 8.10 of the USPS Domestic Mail Manual (DMM). The rule is incorporated by reference at 39 Part CFR 111.1.

In this chapter we also discuss and summarize regulations developed by the U.S. states and territories addressing biohazardous waste. However, when reviewing these summaries, the reader must be aware that such summaries are quickly outdated as state and local governments revise and update their regulations. Therefore, it is essential that biohazardous waste generators maintain on file copies of all current applicable standards. A list of state and federal contacts is presented in this chapter to assist in that effort.

OCCUPATIONAL SAFETY AND HEALTH ADMINISTRATION

The Occupational Safety and Health Administration (OSHA) published its final rule on bloodborne pathogen safety standards in the workplace on December 6, 1991 (1). The rule, entitled "Occupational Exposure to Bloodborne Pathogens" was developed to protect healthcare and other workers exposed to blood on the job from bloodborne infection. Such protection is based on the concept of *universal precautions*, defined by the rule as " . . . an approach to infection control. According to the concept of universal precautions, all human blood and certain human body fluids are treated as if known to be infectious for HIV, HBV and other bloodborne pathogens."

OSHA determined that occupational bloodborne infection hazards can be reduced substantially by adhering to a combination of engineering and workplace practice controls, use of personal protective clothing and equipment, employee training, medical surveillance, and hepatitis B vaccination, signs and labeling systems, and other control measures. In writing the rule, OSHA directly addressed how infectious waste should be defined and managed to protect workers from exposure to bloodborne pathogens, which is the subject of this review.

In this section we describe how the Occupational Safety and Health Act of 1972 (OSH Act) addresses worker safety and health protection in the workplace and how this protection extends to workers exposed to blood while on the job. The interaction between state and federal regulatory agencies while enforcing provisions of the OSH Act and its implementing rules is also discussed. Following the OSH Act overview, the bloodborne pathogen standard is described.

Federal and State OSHA Contacts

Further information about the bloodborne pathogen standards may be obtained from applicable state and federal resources. For information regarding the federal OSHA program, contact:

U.S. Department of Labor
Occupational Safety and Health Administration
Office of Information and Public Affairs
FP/DOL Building, Room N3647
200 Constitution Avenue, NW
Washington, DC 20210
Tel: (202) 219-8148

Additional information may be obtained by contacting OSHA through one of its regional offices for administrative and policy issues, or one of its area offices for enforcement and state program oversight issues. Contacts for the OSHA regional offices and OSHA area offices are presented in Appendix A. In addition to federal OSHA offices, safety and health information may be obtained by contacting the applicable state government office in those states with OSHA-approved programs. State OSHA program contacts are also presented in Appendix A.

Occupational Safety and Health Act

The OSH Act (29 U.S.C. 651 et seq.) was enacted by Congress in 1970 (2). The intent as stated by the act is to "assure safe and healthful working conditions for working men and women by authorizing enforcement of the standards developed under the act, by assisting and encouraging the states in their efforts to assure safe and healthful conditions, by providing for research, information, education, and training in the field of occupational safety and health, and for other purposes." The OSH Act is implemented and enforced by OSHA of the U.S. Department of Labor and by states that have been authorized to conduct their own programs by OSHA.

Effect on Biohazardous Waste Management. OSHA's primary function under the OSH Act is to adopt, implement, and enforce standards that protect the safety and health of the nation's workforce as defined by the OSH Act. During the 1980s the risk of acquiring bloodborne infection to healthcare and other workers exposed to blood on the job became apparent. Responding to this regulatory need, OSHA issued its "Occupational Exposure to Bloodborne Pathogens" standard on December 6, 1991 (1). The standard, which is summarized later in this chapter, directly addresses biohazardous waste management and disposal issues.

Coverage Under the OSH Act. The OSH Act was written to cover all employers and employees as defined by the act in the 50 states, the District of Columbia, Puerto Rico, and all other territories under federal jurisdiction. Under the OSH Act, the definition of employers and employees is limited to those working in businesses affecting commerce. The term *employer* is defined to mean "a person engaged in a business affecting commerce who has employees, but does not include the United States or any state or political subdivision of a state." An *employee* is defined to mean "an employee of an employer who is employed in a business of his employer which affects commerce." State and local government employers and employees are not included within these definitions and therefore are not covered under the act, except in the case where the occupational safety and health program is implemented by a state that has received OSHA approval to conduct a program. Although federal employees are not covered directly under OSHA's enforcement authority, the OSH Act requires that the head of each federal agency be responsible for establishing and maintaining an effective and comprehensive occupational safety and health program that is consistent with the safety and health standards under the act. Enforcement of such standards is conducted either by OSHA or by a state with an OSHA-approved occupational safety and health program.

OSHA-Approved States and Territories. Twenty-three states and two territories are currently conducting their own occupational safety and health programs through plans that have received federal OSHA approval under Section 18(b) of the OSH Act. Twenty-three of these cover both private- and public-sector employees, and two cover only public-sector employees, deferring private-sector program responsibilities to federal OSHA. States and territories with OSHA-approved programs are listed in Appendix A (3).

State OSHA Program Approval Requirements. States and territories are encouraged under the OSH Act to assume the full administration and enforcement responsibilities for occupational safety and health laws. States and territories seeking such authority must submit an occupational safety and health plan (OSH plan) to OSHA for federal review and approval. The OSH plan must demonstrate that its standards and enforcement program is at least as stringent and effective in protecting occupational safety and health as the federal program, and that the necessary statutes and rules to implement the program have been adopted by the state or territory. State and territory OSH plans must also provide voluntary compliance activities to employers. Unlike the federal OSHA program, states and territories seeking to have their plans approved must extend coverage to all employees of state and local public agencies.

Range of Federal OSHA Jurisdiction. The jurisdiction of federal OSHA within states and territories varies depending upon whether a state or territory has received federal approval of its occupational safety and health plan. For states and territories without such approval, OSHA conducts all enforcement activities within that jurisdiction. Two states, Connecticut and New York, have received federal approval of their occupational safety and health plan, although the plans address only employers and employees of state and local governments. In these two states, private-sector employees fall under the jurisdiction and enforcement authority of OSHA. In states and territories with full OSHA approval of their plans, OSHA retains some degree of jurisdictional authority for various occupations (e.g., OSHA's jurisdiction applies to shipbuilders when the ship is on water). Questions regarding occupational safety and health jurisdiction should be made either to the appropriate state or territorial agency that is responsible for implementing its OSHA-approved plan, to an OSHA area office, which is responsible for conducting enforcement activities for occupations within its jurisdiction, or to an OSHA regional office, which coordinates the administrative activities within its region.

OSHA Consultation Services. Consultation services to employers who seek assistance in meeting occupational safety and health requirements are provided by each state/territory, whether or not the state/territory is conducting its own federally approved program. Funded by OSHA, the service is conducted at no cost to the employer, nor are penalties proposed or citations issued for violations observed during consultation visits. Further information concerning consultation services is found in the OSHA document "Consultation Services for the Employer" (4). State/territorial consultation service phone numbers are listed in Appendix A (3).

History of Regulation Development

In 1983, OSHA first addressed the hazards and risks associated with hepatitis B infection among healthcare workers in an internal instruction, OSHA Instruction CPL 2-2.36, entitled "Hepatitis B Risks in the Healthcare System" (5). The instruction called for OSHA regional administrators and area directors to mail voluntary guidelines entitled "The Risk of Hepatitis B Infection for Workers in the Healthcare Delivery System and Suggested Methods for Risk Reduction" to all major healthcare facilities in the country (6). The guidelines described the disease, identified high-risk workers, and recommended work practice techniques to prevent the spread of infection. In the cover letter that accompanied the guidelines, OSHA described its recognition of the significant risk of contracting hepatitis B infection among healthcare delivery occupations.

The 1983 guidelines distributed by OSHA addressed infectious waste management in general terms. OSHA recommended that (1) contaminated items be placed in impervious, clearly labeled bags, and doubled bagged if the inner bag becomes punctured or contaminated; (2) contaminated needles and syringes be placed in labeled, puncture resistant containers dedicated solely for disposal purposes; and (3) contaminated items, such as dressings and paper tissues, be disposed of in accordance with local regulations.

In the months to follow, the hazards associated with occupational exposure to blood infected with the hepatitis B virus (HBV) and human immunodeficiency virus (HIV) became much more evident. On September 19, 1986, OSHA was petitioned by the American Federation of State, County and Municipal Employees (AFSCME) to address occupational bloodborne pathogen hazards immediately. The petition in part requested that OSHA issue an emergency temporary standard to protect healthcare workers from these hazards and that they proceed immediately with rules that would require healthcare employers to offer the HBV vaccine free of charge to all employees at risk. On September 22, 1986, the Service Employees International Union (SEIU), the National Union of Hospital and Healthcare Employees (NUHHE), and RWDSU Local 1199—Drug, Hospital and Health Union petitioned OSHA to take further steps to protect healthcare workers from occupationally acquired bloodborne pathogen infection through rule development.

The petition for an emergency temporary standard requested by AFSCME was denied on October 22, 1987 by the assistant secretary because it was believed that it failed to meet the criteria for such an action under the OSH Act. The assistant secretary did determine that OSHA issue an Advanced Notice of Proposed Rulemaking (ANPR) to collect information and to initiate the formal rulemaking process. The assistant secretary also committed OSHA to address the concerns of the unions immediately by conducting enforcement under General Industry Standards (e.g., general standards relating to personal protective equipment, housekeeping, sanitation and waste disposal, accident-prevention signs and tags), and the General Duty Clause of the Act [Section 5(a)(1)], which states that "each employer shall furnish to each of his employees employment and a place of employment which are free from recognized hazards that are causing or are likely to cause death or serious physical harm to his employees." OSHA also determined that it would initiate an educational program in conjunction with the Department of Health and Human Services in an effort to protect workers from occupational bloodborne pathogen hazards.

A Joint Advisory Notice entitled "Protection Against Occupational Exposure to Hepatitis B Virus (HBV) and Human Immunodeficiency Virus (HIV)" was published by the Department of Labor (DOL) and the Department of Human and Health Services (HHS) on October 30, 1987 (7). The precautions published in the Joint Advisory Notice were based on recommendations by the Centers for Disease Control and included engineering controls, work practices, and protective equipment. The notice with a cover letter was mailed to about 500,000 employers and associated groups. In the cover letter, the secretaries of DOL and HHS warned healthcare industry employers of their legal responsibility to provide appropriate safeguards for healthcare workers who may be exposed to HIV and HBV.

The advanced notice of proposed rulemaking (ANPR) for bloodborne pathogen protection standards was published in the *Federal Register* on November 27, 1987 (8). In the ANPR, OSHA requested relevant information relating to occupational exposure and protection from bloodborne pathogens in the workplace, and comments regarding OSHA's development of a proposed bloodborne pathogen standard. Over 350 comments were received from interested parties, which were analyzed and used to develop a proposed rule. OSHA also announced that it was implementing a targeted inspection program under the OSH Act to examine work practices at healthcare facilities and that it would be enforcing existing general standards under the general duty clause to protect workers from HIV and HBV transmission risks.

Following the publication of the ANPR for bloodborne pathogen protection standards, the assistant secretary of OSHA distributed OSHA instruction CPL 2-2.44, "Enforcement Procedures for Occupational Exposure to Hepatitis B Virus (HBV) and Human Immunodeficiency Virus (HIV)," on January 19, 1988 (9). The purpose of the instruction was to provide "uniform inspection procedures and guidelines to be followed when conducting inspections and issuing citations under Section 5(a)(1) of the OSH Act (General Duty clause) and pertinent standards for healthcare workers potentially exposed to HBV and HIV."

That instruction was later canceled and replaced on August 15, 1988 by CPL 2-2.44A (10), and subsequently, on February 27, 1990, CPL 2-2.44B (11). The instructions were based on guidelines developed by the Centers for Disease Control, which were published as appendices to the OSHA compliance directives. These two CDC publications included (1) "Update: Universal Precautions for Prevention of Transmission of Human Immunodeficiency Virus, Hepatitis B Virus, and Other Bloodborne Pathogens in Health-Care Settings" published in the *Morbidity and Mortality Weekly Reports* (*MMWR*) on June 24, 1988 (12), and (2) "Recommendations for Prevention of HIV Transmission in Healthcare Settings" published in *MMWR* on August 21, 1987 (13).

OSHA published a Proposed Rule and Notice of Hearing (PRNH) in the *Federal Register* on May 30, 1989 (14). In the proposal, the administrator requested written comments on the proposal and Notices of Intention to Appear at one of their public hearings. OSHA also requested comments relating to health effects, risk assessment, significance of risk determination, technological and economic feasibility, and provisions that should be included as a final rule and included a list of 40 specific issues of concern to OSHA. Public hearings were held in Washington, Chicago, New York, Miami, and San Francisco, between September 1989 and January 1990. Over 400 persons participated in the hearings.

The final rule on bloodborne pathogen safety standards in the workplace was published in the *Federal Register* on December 6, 1991 (1). The provision for information collection requirements under the rule was approved by the Office of Management and Budget on February 7, 1992 for a period of three years (15). Minor corrections to the rule were published on July 1, 1992 (16). The rule is codified at 29 CFR Part 1910.1030 under the title "Bloodborne Pathogens." OSHA policies and clarifications to ensure uniform inspections and enforcement under the rule were published in an internally distributed instruction, OSHA Instruction CPL

2-2.44C, "Enforcement Procedures for the Occupational Exposure to Bloodborne Pathogens Standard," 29 CFR Part 1910.1030, on March 6, 1992 (17).

OSHA Bloodborne Pathogen Regulation

The OSHA rule entitled "Occupational Exposure to Bloodborne Pathogens" and published at 29 CFR Part 1910.1030, was written as a general performance standard with various requirements pertaining to biomedical waste handling. The goal of a general performance standard such as the bloodborne pathogen standard is to state a general objective to be achieved without stating precise requirements for accomplishing the objective. In the case of the bloodborne pathogen standard, the employer must display judgment when determining which employees are exposed to infectious materials, the extent of the exposure and the protection necessary to prevent the transmission of bloodborne pathogens. Judgment must also be used by employers when establishing waste management protocols within a facility, based on the bloodborne pathogen standards.

Regulated waste management represents only one facet of the bloodborne pathogen rule and must be conducted in context with all rule provisions. Under the rule, each employer with occupational exposure is required to establish a written Exposure Control Plan. The plan must document the exposure determination for employees occupationally exposed to blood or other potentially infectious materials. For personnel involved with waste management activities, this would require identifying the frequency of exposure and the tasks and procedures in which exposure may occur. The plan must also address methods of compliance, such as engineering controls and personal protective equipment, hepatitis B vaccination and postexposure follow-up, communication of hazards to employees, record keeping, and procedures for evaluating exposure incidents.

Coverage Under the Bloodborne Pathogen Regulation. Worker protection under the bloodborne pathogen regulation must be provided to those in all occupations exposed to blood or other potentially infectious materials as defined by the rule. Under the rule, *occupational exposure* means "reasonably anticipated skin, eye, mucous membrane, or parenteral contact with blood or other potentially infectious materials that may result from the performance of an employee's duties." For many occupations within or external to the medical setting, categorizing employees based on exposure is self-evident. For example, physicians or nurses working in a hospital emergency room would meet the test of reasonably anticipated blood exposure. However, such categorization is often not so evident for other occupations. For example, janitorial, maintenance personnel, or other housekeeping staff within a medical facility, particularly those who may be involved in areas of patient care or microbiology laboratory maintenance, may be at risk of exposure to infectious materials (e.g., during infectious waste removal) and should be covered under the regulation. Housekeeping staff in a nonmedical setting would probably not face exposure to infectious materials, although such a determination should be made by the employer only after considering the specific tasks that each performs and determining whether such tasks may "reasonably anticipate exposures."

For example, laundry attendants may routinely come into contact with sheets and linens soiled with infectious human body substances such as feces, urine, sweat, sputum or saliva, or vomitus. However, in most cases laundry attendants would not be covered by the bloodborne pathogen standard because such materials do not present a risk of transmitting bloodborne pathogens. Furthermore, these materials do not fall within the OSHA definition of "regulated infectious waste." Conversely, if it could be shown that the attendants are exposed to blood or other OSHA-defined potentially infectious materials, such as syringes, with reasonable anticipation, these laundry attendants would be covered by the rule. If, during an inspection, a federal or state OSHA representative determines that there is sufficient evidence that an employee's job results in a reasonably anticipated exposure despite the opinion of his or her employer, the employer will be held responsible, will probably receive a citation, and must provide protection to that employee as required by the bloodborne pathogen standard.

The bloodborne pathogen regulations apply to many job classifications within the healthcare industry. Examples of healthcare facilities affected by the rule include hospitals, clinics, dental offices, surgery centers, nursing homes, medical laboratories, physicians' offices, medical research facilities, funeral homes, plasma centers, and blood banks. The regulation may also apply to other, non-healthcare industry groups, such as police or fire department employees or medical waste transporters. For each, applicability decisions must be made on a case-by-case basis against the test of reasonably anticipated occupational exposure as defined by OSHA. Defining job classifications to which these regulations apply is often cited as a difficult task for employers and can only be done with full recognition by the employer of all functions of a specific job.

Wastes Regulated by the Rule. 29 CFR Part 1910.1030, paragraph (b), defines *regulated waste.* specifically to include only wastes that meet the following test:

1. Liquid or semiliquid blood or other potentially infectious materials
2. Contaminated items that would release blood or other potentially infectious materials in a liquid or semiliquid state if compressed
3. Items that are caked with dried blood or other potentially infectious materials and are capable of releasing these materials during handling
4. Contaminated sharps
5. Pathological and microbiological wastes containing blood or other potentially infectious materials

Several elements within the definition of regulated medical waste are defined by the rule and must be considered by employers when making decisions regarding the applicability of the rule to their institution.

- *Blood* is defined to mean "human blood, human blood components, and products made from human blood."

- *Contaminated* means "the presence or the reasonably anticipated presence of blood or other potentially infectious materials on an item or surface."
- *Other potentially infectious materials* means "the following human body fluids: semen, vaginal secretions, cerebrospinal fluid, synovial fluid, pleural fluid, pericardial fluid, peritoneal fluid, amniotic fluid, saliva in dental procedures, any body fluid that is visibly contaminated with blood, and all body fluids in situations where it is difficult or impossible to differentiate between body fluids."
- *Contaminated sharps* means "any contaminated object that can penetrate the skin including, but not limited to, needles, scalpels, broken glass, broken capillary tubes, and exposed ends of dental wires."

Other potentially infectious materials that are not associated with bloodborne diseases have not been included within the scope of the bloodborne pathogen rule's definition of regulated waste. Examples include vomitus, sputum, feces, and urine. However, if these materials contain visible blood, they would be subject to the standards of the bloodborne pathogen rule.

Waste Management Requirements. Containment, handling, and labeling requirements are specified for regulated medical waste management under 29 CFR Part 1910.1030(d)(4)(iii)(A) to (C). The standard does not specify specific disposal requirements for regulated medical waste. It does, however, defer to the applicable territorial, state, or local medical waste disposal requirements within those jurisdictions, should they exist. The regulated medical waste management requirements addressed in the bloodborne pathogen standard are described in the discussion to follow.

Contaminated Sharps. The waste management requirements for contaminated sharps are addressed in 29 CFR Part 1910.1030(d)(4)(iii)(A). This section of the rule establishes performance standards for sharps containment, discarding, container use, and container transport. The standard requires that contaminated sharps be discarded either immediately or as soon as feasible in containers that are closable, puncture resistant, leakproof on sides and bottom, and either labeled or color coded as specified in 29 CFR Part 1910.1030(g)(1)(i) of the standard, which is described below.

Within a facility, sharps containers must be easily accessible to personnel and located as close as is feasible to the immediate area where sharps are used or can be reasonably anticipated to be found. The containers must be maintained upright throughout use, and replaced routinely. They must never be allowed to be overfilled.

When sharps containers are to be transported from their original area of use to another area in a facility, the containers must be closed immediately before they are removed or replaced, to prevent either spillage or protrusion of contents during handling, storage, transport, or shipping. If it is possible that the sharps container may leak for any reason, it must be placed in a second container before being moved. The secondary container must be closable, constructed to contain all contents, and prevent leakage during handling, storage, transport, or shipping. The secondary container must also be labeled or color-coded according to paragraph (g)(1)(i) of the standard as described below. If the sharps container is reusable, the rule prohibits the container from being opened, emptied, or cleaned manually or in any other manner that would expose employees to potential needle-stick or other sharps-related injuries.

Nonsharp Regulated Waste. Nonsharp regulated waste, such as dressings, bandages, or gauze that have been saturated with blood, must be placed in containers that are closable and constructed to contain all contents and prevent leakage of fluids during handling, storage, transport, or shipping. Containers must be closed prior to removal to prevent spillage or protrusion of contents during handling, storage, transport, or shipping. Containers must also be labeled or color-coded in accordance with paragraph 29 CFR Part 1910.1030(g)(1)(i), as described below.

If the outside of the container of regulated waste becomes contaminated, it must be placed in a second outer container. The second container must meet the specifications of the first container as described in the preceding paragraph.

Labelling Requirements. The labeling requirements of 29 CFR Part 1910.1030(g)(1)(i) of the rule were established to communicate the presence of hazards to employees. The standard requires that regulated waste be identified either with a warning label or by containment in a red bag or a red container, which may be used in lieu of labels. Labels must be affixed to containers of regulated waste, refrigerators, and freezers containing blood or other potentially infectious material. They must also be affixed to other containers used to store, transport, or ship blood or other potentially infectious materials.

Labels must be fluorescent orange or orange-red or predominantly so in color, with lettering and symbols in a contrasting color and must include the biohazard legend (see Figure 3.1). The standard also requires that labels be affixed as close as feasible to the container holding the regulated waste by either string, wire, adhesive, or other method that prevents their loss or unintentional removal.

The standard identifies two exemptions from the labeling requirements. Exempted are (1) containers of blood, blood components, or blood products that are labeled as to their contents and have been released for transfusion or other clinical use, and (2) individual containers of blood or other potentially infectious materials that are placed in a labeled container during storage, transport, shipment, or disposal.

Labeling required for contaminated equipment must be done as described in this section of the rule. Labels must also state which portions of the equipment remain contaminated. Regulated waste that has been decontaminated either by steam sterilization or other state-approved alternative biomedical waste treatment technology is not required to be either labeled or color-coded.

Waste Disposal Requirements. The OSHA bloodborne pathogen rule does not prescribe disposal requirements. Rather, it defers disposal requirements to other authorities, requiring that all disposal be in accordance with applicable regulations

Figure 3.1   Biohazard label.

of the United States, states and territories, and political subdivisions of states and territories.

HIV and HBV Research Laboratories and Production Facility Requirements.
HIV and HBV research laboratories and production facilities are defined by the bloodborne pathogen rule to include those facilities that are engaged in the culture, production, concentration, experimentation, and manipulation of HIV and HBV. Not included are clinical or diagnostic laboratories that are engaged solely in the analysis of blood, tissues, or organs. All regulated waste management requirements addressed above for both sharp and nonsharp waste also apply to regulated waste generated during the course of HIV and HBV research and production. Nevertheless, additional requirements identified in 29 CFR Part 1910.1030(e) must be addressed for these specialized facilities.

The rule establishes a requirement for all regulated waste, including all waste from work areas and from animal rooms, to be either incinerated or decontaminated by an infectious waste treatment method known to effectively destroy bloodborne pathogens, such as steam sterilization. All HIV and HBV research laboratories must have an autoclave for decontamination of regulated waste available within or as near as possible to the work area.

The rule requires extreme caution when handling needles and syringes. A needle must not be bent, sheared, replaced in the sheath or guard, or removed from the syringe following use. Used needles and syringes must be placed promptly in a puncture-resistant container and autoclaved or decontaminated before disposal.

PUBLIC HEALTH SERVICE

The Public Health Service (PHS) transportation of etiologic agents regulation is published at 42 CFR Part 72. The rule that is in effect today was last amended on July 21, 1980. The rule applies to inter- and intrastate transport biological products, diagnostic specimens, and any material that is known to contain, or is reasonably believed to contain, an etiologic agent. This broad definition includes etiologic agents contained in biomedical waste. Therefore, based on a strict interpretation of the rule language, biomedical waste must meet the packaging, labeling, and other transport requirements of the standard (18).

Contact Information

Further information about this PHS program may be obtained from the following source:

U.S. Department of Health and Human Services
Public Health Service
Centers for Disease Control and Prevention
Biosafety Department
1600 Clifton Road, NE, Mail Stop F05
Atlanta, GA 30333
Tel: (404) 639-3883

Rule Development History

In 1971 the responsibility for regulating the interstate shipment of etiologic agents was delegated to the U.S. Public Health Service (19). Following delegation, the PHS published standards regulating the interstate shipment of etiologic agents at 42 CFR Part 72. The regulation that is in effect today was last revised on July 21, 1980 (20). The intent of the regulation is to prevent the exposure and possible infection of transportation personnel and others to infectious materials in interstate transit by:

- Requiring containment packaging for known or potentially infectious agents and materials
- Identifying known or potentially infectious materials with a hazard warning label
- Providing a system for reporting and responding to damaged or leaking packages of etiologic agents
- Specifying specific requirements for the shipments of highly infectious agents

The standard was last revised before the emergence of large-scale interstate biomedical waste transportation destined for off-site treatment and disposal. Because of this, the regulations were not designed to address realistically biomedical waste transportation as it is conducted today.

On March 2, 1990, the PHS proposed revisions to the regulation in an effort to address concerns by those having to handle improperly packaged or damaged packages containing etiologic agents, to clarify definitions, and to update the list of etiologic agents requiring notification of receipt (21). To date, the regulations remain in a draft stage, with a projected final rule publication during or after 1995 (22). The delayed issuance of a final rule by the PHS is in large part due to the need to address the interstate transport of biomedical waste in a more realistic manner and to ensure that the rule does not conflict with other federal biomedical waste transportation initiatives.

When the PHS rules were adopted and last revised, off-site medical waste transportation was not an issue and the rules were not written with that purpose in mind. Although the current rules are technically applicable to biohazardous waste transportation, in practice they are rarely observed by the biohazardous waste transportation industry. This discussion addresses 42 CFR Part 72 as it currently applies to the interstate transportation of etiologic agents contained in biomedical waste. Rule changes proposed in 1990 are also described.

Interstate Shipment Regulation

The PHS rule on interstate shipment of etiologic agents, published at 42 CFR Part 72, regulates the shipment of any material known or reasonably believed to contain etiologic agents, including biological products and diagnostic specimens, by prescribing packaging, labeling, and other shipping standards.

Materials Regulated by the Rule. Part 72.3 of the standard presents a list of human pathogenic organisms under the headings of (1) bacterial agents, (2) fungal agents, and (3) viral and rickettsial agents. The rule requires any person who transports or causes to be transported any material in interstate traffic that is known to contain, or is reasonably believed by that person to contain, one or more of the listed pathogens to meet the stringent packaging, labeling, and shipping requirements identified under paragraphs (a) through (f) of the section. The term *interstate transport* is interpreted to mean also *intrastate transport*. Because biomedical waste is a material reasonably believed to contain human pathogens, such material when transported is technically covered under the PHS rule as currently written (18).

Transportation Requirements. Part 72.3 presents two sets of requirements for transportation of this material based on volumes of less than 50 milliliters (mL), or volumes of 50 mL or more of material.

A. *Volume Not Exceeding 50 mL*

1. Material shall be placed in a securely closed, watertight primary container (e.g., test tube, vial, etc.).

2. The primary tube must be enclosed in a secondary container that is durable and watertight.

3. Several primary containers may be enclosed in a single secondary container, as long as the total volume of all primary containers does not exceed 50 mL.

4. The space at the top, bottom, and sides between the primary and secondary containers must contain sufficient nonparticulate absorbent material (e.g., paper toweling) to absorb the entire contents of the primary container(s) in case of breakage or leakage.

5. Each set of primary and secondary containers must then be enclosed in an outer shipping container constructed of corrugated fiberboard, cardboard, wood, or other material of equivalent strength.

B. *Volume Greater Than 50 mL.* Packages of material of 50 mL or more must meet the packaging requirements identified in the preceding paragraphs required for those materials of less than 50 mL, and also these additional requirements:

1. A shock-absorbent material, in volume at least equal to that of the absorbent material between the primary and secondary containers, must be placed at the top, bottom, and sides between the secondary container and the outer shipping container.

2. Single primary containers must not contain more than 1000 mL of material. However, two or more primary containers whose combined volumes do not exceed 1000 mL may be placed in a single secondary container.

3. The maximum amount of etiologic agent that may be enclosed within a single outer shipping container must not exceed 4000 mL.

Part 72.3 also establishes other requirements when transporting materials containing etiologic agents. These requirements are:

A. *Dry Ice Packaging Requirement.* If dry ice is used as a refrigerant, it must be placed on the outside of the secondary container. If placed between the secondary container and the outer shipping container, the shock-absorbent material must be placed so that the secondary container does not loosen in the package as the dry ice sublimates.

B. *Labeling Requirement.* The outer shipping container must display a specific label. The label must be printed on a white material, the symbol must be red, and the printing must be either red or white. The label must be rectangular, measuring 2 inches high by 4 inches long. The red symbol must measure $1\frac{1}{2}$ inches in diameter and must be centered in a white square measuring 2 inches on a side. The type size of the letters of the label must be as follows:

Etiologic agents: 10 pt. rev.

Biomedical material: 14 pt.

In case of damage or leakage: 10 pt. rev.

Notify Director CDC, Atlanta, Georgia: 8 pt. rev.

C. *Damaged Packages.* If any package bearing the etiologic agents/biomedical material label is discovered by the carrier to have leaked or been otherwise damaged, the carrier is required promptly to notify both the director of the Centers for Disease Control (CDC) at (404) 633-5313 and the sender.

D. *Registered Mail or Equivalent System.* Part 72.3(f) of the standard contains a list of highly virulent etiologic agents that must be transported only by registered mail or an equivalent system that ensures notification of receipt to the sender immediately following delivery. Part 72.4 requires the sender of materials containing these etiologic agents to notify the director of the CDC at (404) 633–5313 if the notice of delivery has not been received within 5 days following the package's anticipated date of delivery.

E. *Variances.* Part 72.5 of the standard allows the director of the CDC to approve variations to these standards if found to provide equivalent protection and if the findings are made a matter of official record.

Proposed Regulation Amendments

In 1971 the PHS issued its first regulations that pertain to both inter- and intrastate transport of any material known to contain or reasonably suspected of containing etiologic agents. These requirements are currently being revised to address bio-hazardous waste transportation in a more realistic manner, which, in effect, would rely on regulations currently being developed by the U.S. Department of Transportation (18).

In a notice of proposed rulemaking published in the *Federal Register* on March 2, 1990, the PHS identified proposed amendments to the Interstate Shipment of Etiologic Agents standard (21). The amendments as proposed would have the effect of eliminating applicability of the standard to biomedical waste transportation. Based on the proposed amendments, the scope of the revised standard would be more clearly defined, limiting applicability to (1) biological products that contain etiologic agents, (2) all clinical specimens, and (3) etiologic agent preparations. The revision would not include the broad applicability of "any material known to contain, or reasonably believed to contain etiologic agents," which upon strict inter-pretation would include biomedical waste.

As of January 1995, the PHS has indicated that it has included additional changes to its original proposal which may place primary reliance on biohazardous waste transportation regulations with the U.S. Department of Transportation. A final rule is projected for publication sometime in 1995 (22).

DEPARTMENT OF TRANSPORTATION

Standards regulating the interstate transport of hazardous materials were first issued by the Interstate Commerce Commission early in the twentieth century under au-thority of the Explosives and Combustibles Act of 1908. Today's standards, which derive from the Hazardous Materials Transportation Act of 1974, are administered by the Department of Transportation. The Hazardous Materials Regulations (HMR) are published in the *Code of Federal Regulations*, title 49, subtitle B, chapter 1, subchapter C, parts 171 through 180 and pertain to the interstate, and in some cases intrastate, transportation of hazardous materials in commerce. In this section we describe the development of the HMR as it applies, both directly and indirectly, to regulated (infectious) medical waste transportation. To date, the regulations as they apply to regulated medical waste have not been finalized.

Contact Information

Further information about the U.S. Department of Transportation program may be obtained from the following source:

U.S. Department of Transportation
Office of Hazardous Materials Standards
RSPA, Room 8102
400 Seventh Street, SW
Washington, DC 20590-0001
Tel: (202) 366-4488

Rule Development History

In the late 1960s, the Interstate Commerce Commission (ICC) was directed by Public Law 86-710 to adopt regulations for the safe interstate transportation of etiologic agents. These duties, as well as other duties relating to safety functions, were subsequently transferred to the Department of Transportation (DOT) following its establishment in 1967. In response to its new mandate, the Hazardous Materials Regulations Board (HMRB) of DOT took its first step toward regulating the transport of etiologic agents by adopting by reference existing classification and packaging requirements for etiologic agents that had previously been published by the Public Health Service of the Department of Health, Education, and Welfare (HEW), now the Department of Health and Human Services (HHS) (23). These first DOT standards relating to etiologic agents were adopted under Docket Number HM-96. At that time the HMRB also proposed additional standards relating to performance test criteria for etiologic agent packaging and a new hazard warning label for packages containing etiologic agents. Notification requirements were also subsequently proposed (24).

On September 30, 1972, requirements for the shipment of etiologic agents were published by the HMRB as an amendment to the Hazardous Materials Regulations (HMR) (25). An etiologic agent, now added to the DOT's Hazardous Materials List, was defined at 49 CFR Part 173.386(a)(1) to mean: "a viable microorganism, or its toxin, which causes or may cause human disease, and is limited to those agents listed in 42 CFR Part 72.25(c) of the regulations of the Department of Health, Education, and Welfare." The regulation applied to shippers of any material containing an etiologic agent by requiring such packages to be packaged, labeled, and shipped in accordance with 49 CFR Part 173.24 and other applicable parts of the regulation. Exempted from the regulation at 49 CFR Part 173.386(d) were "diagnostic specimens" and "biological products," which were regulated by HEW at 42 CFR Part 72. No distinction was made or discussed for packages containing etiologic agents, whether transported as a laboratory specimen or as a component of biohazardous waste. However, it is evident that the standards were not developed to address biohazardous waste transportation as it is conducted today.

Following publication of its etiologic agent standard, the HMRB received two petitions for reconsideration and several comments regarding the regulation of cultures of etiologic agents of less than 50 milliliters (mL) in one package. The petitioners requested that etiologic agents transported in packages of less than 50 mL be exempted from the DOT regulations to allow small packages from rural areas to be transported to laboratories via passenger-carrying aircraft rather than by the slower surface transportation. Based on the reasoning behind the petitions, the HMRB proposed an exemption for cultures of etiologic agents of less than 50 mL in one package (the 50 mL exemption) on November 29, 1972 (26). The proposal was adopted as a final rule on March 29, 1973 (27). However, HMRB noted in the preamble to the final rule that a small package containing an etiologic agent would still be subject to the HEW labeling and packaging requirements published at 42 CFR Part 72.25.

In the early 1980s, a new docket, Docket HM-181, was created to address revisions to the entire HMR. This proposed overhaul of the HMR was to be conducted by the Research and Special Programs Administration (RSPA) of the DOT. On April 15, 1982, RSPA published an Advance Notice of Proposed Rulemaking (ANPR) that addressed performance-oriented packaging standards (28). By publishing the ANPR, DOT began its formal process of exploring the possibility of simplifying the existing HMR by bringing it into alignment with the performance-based recommendations of the United Nations Committee of Experts on the Transport of Dangerous Goods (U.N. Recommendations). The U.N. Recommendations are not regulations but serve as the basis for other international hazardous materials standards. Several international organizations had developed regulations for hazardous materials transportation based on the U.N. Recommendations. These included (1) the International Maritime Organization (IMO) with its International Maritime Dangerous Waste Code, and (2) the International Civil Aviation Organization (ICAO) with its ICAO Technical Instructions for the Safe Transport of Dangerous Goods by Air Shipments of Hazardous Waste. RSPA recognized the need to align the U.S. system with those of the international community.

Five years following the publication of its ANPR, RSPA published a Notice of Proposed Rulemaking (NPRM) on May 5, 1987 (29). The proposal was to align requirements relating to classification, packaging, and hazard communication in the HMR with those of the U.N. Recommendations. The transport of etiologic agents was addressed by the NPRM by proposing a definition for infectious substances that would refer to the regulations of the Department of Health and Human Services pertaining to etiologic agents (42 CFR Part 72) but would otherwise be generally consistent with the U.N. Recommendations. The term *etiologic agent* was proposed to be synonymous with *infectious substance*. Again, however, the proposed rule did not specifically refer to or discuss definitions or requirements for regulated medical waste (biohazardous waste) as a subcategory of infectious substances.

On November 10, 1988, RSPA published an NPRM and request for comments under a new docket, Docket HM-142A, that focused specifically on the definition and criteria of an *etiologic agent* and the use of the term *infectious substance* (30).

This separate rule writing effort under a new docket was done specifically to adopt interim standards for the transportation of etiologic agents pending the adoption of final HMR rules under Docket HM-181. Due to the extensive scope of HMR revisions proposed under Docket HM-181, RSPA was concerned that it would be unable to publish a final HMR rule in a timely manner, and chose instead to develop interim standards for etiologic agents while working on final etiologic agent standards under Docket HM-181.

Under Docket HM-142A, RSPA proposed to expand the definition of etiologic agent to include, in addition to the agents listed by the Public Health Service in 42 CFR Part 72.3, any agent that poses a similar degree of hazard, such as the human immunodeficiency virus. RSPA noted that the Public Health Service had not updated its list since July 1, 1980 (31), and that the current definition was not as broad as that for infectious substances defined under Division 6.2 of the U.N. Recommendations on the Transport of Dangerous Goods and other international regulations based on the U.N. Recommendations. RSPA also proposed to eliminate the 50 mL exception, and to align the per package quantity limits aboard aircraft with the International Civil Aviation Organization Technical Instructions for the Safe Transport of Dangerous Goods by Air (ICAO Technical Instructions). The rule changes proposed under Docket HM-142A were intended to be implemented on an interim basis until a final rule could be published under Docket HM-181.

Despite predictions of long-term rule writing under Docket HM-181, RSPA published its final rule under Docket HM-181, entitled "Performance-Oriented Packaging Standards; Changes to Classification, Hazard Communication, Packaging and Handling Requirements Based on U.N. Standards and Agency Initiative," on December 21, 1990 (32). As stated in the preamble to the standard, the rule was revised to "(1) simplify and reduce the volume of the HMR; (2) enhance safety through better classification and packaging; (3) promote flexibility and technological innovation in packaging; (4) reduce the need for exemptions from the HMR; and (5) facilitate international commerce." This was accomplished through alignment with internationally based performance standards derived from the U.N. Recommendations and by other RSPA initiatives.

The final rule adopted under Docket Number HM-181 addressed the transportation of etiologic agents and infectious substances but did not specifically identify how the rule would address infectious wastes containing one or more of these agents. *Infectious substance* was defined in 49 CFR Part 173.134(a)(1) to mean "a viable microorganism, or its toxin, which causes or may cause disease in humans or animals, and includes those agents listed in 42 CFR Part 72.3 of the regulations of the Department of Health and Human Services or any other agent that has the potential to cause severe, disabling or fatal disease. The terms *infectious substance* and *etiologic agent* are synonymous."

The rule amendments were based in part on comments to Docket HM-181, comments to the proposed rule on the transport of etiologic agents issued under Docket HM-142A, and recommendations from the U.N. Subcommittee on the Transport of Dangerous Goods (U.N. Subcommittee). A purpose of the changes was to bring the regulations for etiologic agents in line with the U.N. Recommendations for infectious substances. How the rule applied to the transportation of regulated medical waste remained ambiguous.

The interim standard for etiologic agents transportation developed under Docket HM-142A was published on January 3, 1991, almost two weeks following the adoption of final standards for etiologic agents under Docket HM-181 (33). The standards of Docket HM-142A represented an abbreviated version of the standards for infectious substances in Docket HM-181. Compliance with the final rule under Docket HM-142A was effective on February 19, 1991, although compliance was authorized immediately. Compliance with the standard published under Docket HM-181 was not effective until October 1, 1991, although such compliance was authorized on or after January 1, 1991. Because of this, RSPA recommended that shippers implement the standards adopted under Docket HM-181 as soon as practicable rather than follow the interim rules of Docket HM-142A.

The interim standard adopted under Docket HM-142A (1) revised the definition of "etiologic agent", (2) removed the 50 mL exception from regulation for etiologic agents, and (3) clarified quantity limitations for etiologic agents transported aboard aircraft. The term *etiologic agent* was defined to mean "a viable microorganisms, or its toxin, which is listed in 42 CFR Part 72.3 of the regulations of the Department of Health and Human Services or which causes or may cause sever, disabling or fatal human disease." This definition was similar to that adopted under Docket HM-181 but did not define the term *etiologic agent* as synonymous with the term *infectious substance*.

The first reference to application of the hazardous materials standards to infectious waste transport was made in the preamble to the final rule under Docket HM-142A (33). One commenter requested RSPA to specifically exclude regulating infectious waste as an etiologic agent although such waste may contain human pathogenic organisms. At that time, the RSPA's HMR did not specifically refer to infectious waste, nor did the rule contain specific requirements for the transport of infectious waste. In its response, RSPA stated that it believed that most infectious waste would not contain etiologic agents and that in many cases, if medical waste is known to contain an etiologic agent, it is treated on-site to destroy the agent by a method such as steam sterilization or incineration. However, RSPA noted that if an infectious waste did contain an etiologic agent and was offered for transport, it would have to meet the HMR requirements for etiologic agents found in 49 CFR Parts 171–180.

Other considerations relating to the transport of infectious medical waste noted at the time of final rule publication under Docket HM-142A were the ongoing actions of the U.N. Subcommittee of Experts on the Transport of Dangerous Goods. At that time, one issue being addressed by the U.N. Subcommittee was the discussion of medical waste in the context of the overall U.N. recommendations rather than as an adjunct to Division 6.2, Infectious Substances.

Following the January 3, 1991 publication of the final rule under HM-142A, RSPA received a petition for reconsideration from the National Solid Waste Management Association (NSWMA), which expressed concerns regarding the impact of the rule on the waste management industry. In its petition, NSWMA recommended that RSPA revise its definition for etiologic agents to exclude solid waste or medical waste as defined in 40 CFR Part 259.10 of the Environmental Protection Agency's interim standards on regulated medical waste. NSWMA disputed RSPA's preamble discussion on medical waste which was published in the *Federal Register* on January 3, 1991 in which RSPA stated its belief that most infectious waste did not contain etiologic agents and most was treated at the site of generation before being transported for disposal (33). In a subsequent meeting with RSPA to clarify its petition for reconsideration, NSWMA also requested that RSPA reestablish the 50 mL exception for infectious substances which had been removed by the rule revision. Based on the NSWMA petition, RSPA agreed that medical waste should be treated differently than most infectious substances and on February 22, 1991, extended the effective date of its interim standard from February 19, 1991 to September 30, 1991 to allow time for RSPA to respond to the petition (34).

On September 18, 1991, RSPA incorporated Docket HM-142A into Docket HM-181 (35). This was based on RSPA's original intent to have Docket HM-181 requirements supersede the interim provisions of Docket HM-142A following its effective date. In partial response to the request by the NSWMA to reestablish the 50 mL exception for infectious substances, RSPA extended the effective date of the final rule established under Docket HM-181 for cultures of infectious substances of 50 mL or less total quantity per package from October 1, 1991 to October 1, 1992. RSPA did not extend the effective date for quantities of infectious substances exceeding 50 mL, which remained October 1, 1991.

On September 26, 1991, the NSWMA submitted a letter to RSPA requesting that the agency clarify either by letter or interim rulemaking that the rules published on January 3 and September 18, 1991 do not apply to *medical waste* as defined in 40 CFR Part 259.10(b), *regulated medical waste* as defined in 40 CFR Part 259.30(a), and *mixtures* as defined in 40 CFR Part 259.31. RSPA published a partial response to the NSWMA petition on October 1, 1991 by extending the effective date for infectious substance requirements relating to hazard communication (shipping papers, marking, and labeling) and classification to October 1, 1992 (36). RSPA extended the compliance date again to allow itself time to examine the substantive concerns of the NSWMA as the rule applies to regulated medical waste.

On December 20, 1991, RSPA published a final rule under Docket HM-181 to address more appropriately the transportation requirements for regulated medical waste in light of concerns raised by petitioners (37). In response to the NSWMA petition for reconsideration, RSPA stated in the preamble to the rule that "since the majority of these wastes are untreated and, thus, may potentially contain infectious substances, RSPA strongly believes that the public and transport personnel be protected from the hazards of these materials during transportation." In addressing standards for regulated medical waste (RMW), RSPA in effect created RMW as a subcategory of infectious substances [e.g., infectious substances that are (1) contained in medical waste, or (2) constitute medical waste]. The definition for RMW was taken from the EPA's expired interim final rule published at 40 CFR Part 259. In its preamble discussion, RSPA also noted that the packaging requirements for these wastes were consistent with those published by the EPA in its former medical waste tracking program, which expired on June 22, 1991. RSPA noted that it revised the regulations of 49 CFR Part 173.197(1991) so that the requirements for an infectious substance that is regulated medical waste would be less rigorous. As stated in the preamble to the rule, "if an infectious substance is being offered for transportation or transported, the infectious substance must be labeled, packaged, and offered for transportation in accordance with the HMR. If the infectious substance is also medical waste, or is contained in medical waste, then the shipper may use the less rigorous packaging requirements that are provided for RMW."

This final rule represented DOT's first effort to address directly the transportation of regulated medical waste. In the rule, RSPA required that after October 1, 1992, all shipments of RMW be accompanied by shipping papers. Packages were required to be marked "Regulated Medical Waste" and marked with the identification number "NA 9275" as required by 49 CFR Part 172.301. Packages were also required to be identified with the "Infectious Substances" label as required under 49 CFR Part 172.432. RSPA also required that packages containing RMW for interstate transport meet the packaging standards of 49 CFR Part 173.197. These standards called for packages to meet the requirements of 40 CFR Part 178 at the Packing Group II performance level. In general, packages were required to be rigid, leakproof, impervious to moisture, and have sufficient strength to prevent tearing or bursting under normal conditions of use and handling. Additional requirements were also established for packages containing medical sharps (e.g., hypodermic needles or scalpel blades) or volumes of liquids exceeding 20 cubic centimeters.

The final rule published on December 20, 1991 resulted in two petitions for reconsideration and a number of requests for clarification with regard to regulated medical waste and infectious substances. A call was made for a uniform federal approach to the regulation of infectious substances. Commenters and petitioners noted conflict and overlap with several other federal standards that address infectious substances. Among the most obvious overlap noted were the various package labeling requirements by each of the federal agencies regulating infectious substances.

RSPA partially responded to these concerns by extending the compliance date in 49 CFR Part 171.14(b), pertaining to hazard communication and classification requirements from October 1, 1992 to April 3, 1993 (38). RSPA also extended the compliance date for packaging requirements of infectious substances, which had been inadvertently omitted from the December 20, 1991 final rule, from October 1, 1992 to April 3, 1993. The extension was made to allow RSPA adequate time to respond to the two petitions and comments and requests for clarification of its December 20, 1991 final rule.

On March 3, 1993, RSPA announced an ANPR, a public hearing scheduled for March 17, 1993, and a request for comments under a new Docket HM-181G pertaining to the regulation of infectious substances (39). RSPA called for the public hearing to gain further information about the regulation of infectious substances in light of petitions and comments that had been received. In its request for comments, RSPA posed 29 questions that address issues involving (1) consistency with other regulations, (2) nonbulk packagings, (3) bulk packagings, and (4) scope.

Recognizing that additional time would be needed to address and respond adequately to all concerns, RSPA revised 49 CFR Part 171.14(b) applicable to infectious substances by extending the compliance date from April 1, 1993 to January 1, 1994 (40). During the interim, RSPA noted that a person may comply with either the applicable "old" requirements of the HMR (e.g., those in effect on September 30, 1991 adopted under Docket HM-142A) or the current requirements adopted under Docket HM-181. On December 20, 1993, RSPA revised 49 CFR Part 171.14(b) applicable to infectious substance classification, hazard communication, and packaging requirements by again extending the compliance date from January 1, 1994 to October 1, 1994 to allow additional time to address concerns by commenters (41). On September 22, 1994, this compliance date was extended again, to October 5, 1995 for regulated medical waste and to January 1, 1995 for cultures and stocks of substances infectious to humans. The extension was made to allow additional time to coordinate with other federal agencies with jurisdiction over infectious substances and to allow RSPA to publish a notice of proposed rulemaking, evaluate comments, and make appropriate changes to the HMR (42).

RSPA published a notice of proposed rulemaking and notice of public hearing (NPRM) on December 21, 1994 (43). This effort focused only on transportation requirements for infectious substances and RMW in the short term so that these materials may be safely transported without undue burden on the industry regulated. Other long-term issues, such as coordinating the rule with the international regulations and adopting bulk packaging standards for RMW, would be addressed in future DOT rulemaking. Of note, DOT proposed that the definition for RMW be changed from list-based (as previously proposed) to criteria-based. The proposed criteria-based definition for *regulated medical waste* is "a waste or reusable material, other than a class 7 (radioactive) material or a culture or stock of an infectious substance, which contains an infectious substance and is generated in: (a) the diagnosis, treatment or immunization of human beings or animals; or (b) research pertaining to the diagnosis, treatment or immunization of human beings or animals; or (c) the production or testing of biological products."

This definition excludes cultures and stocks regulated by the more stringent packaging standards of infectious substances. Under the proposal an *infectious substance* is defined as "a viable microorganism, or its toxin, which causes or may cause disease in humans or animals, and includes those agents listed in 42 CFR Part 72.3 of the regulations of the Department of Health and Human Services and any other agent that causes or may cause sever, disabling or fatal disease. The terms *infectious substance* and *etiologic agent* are synonymous."

Status of Regulatory Development

Until final rules for regulated medical waste transportation are promulgated, it is RSPA's position that a person may comply with either the applicable "old" requirements of the HMR (e.g., those in effect on September 30, 1991 adopted under Docket HM-142A) or the current requirements adopted under Docket HM-181. In practice, many in the biohazardous waste transportation industry are not observing any of the DOT rules relating to regulated medical waste before the final transportation rules have been published (44).

U.S. POSTAL SERVICE[1]

In 1992 the U.S. Postal Service (USPS) published a final rule entitled "Mailability of Sharps and Other Medical Devices" under Sections 8.1 through 8.10 of the USPS Domestic Mail Manual (DMM) (45). The rule, which is incorporated by reference at 39 CFR Part 111.1, was established to protect USPS employees and customers from biological hazards and physical injuries that may be associated with the mailing of medical waste through the USPS system to a destination disposal facility. In this section we present the background of regulatory development and a summary of the requirements for mailing packages containing biohazardous waste through the USPS mail system.

Contact Information

Additional information about USPS requirements may be obtained by contacting:

U.S. Postal Service
Business Mail Acceptance
Customer Service Support
475 L'Enfant Plaza, SW Room 8430
Washington, DC 20260-6808
Tel: (202) 268-5168
Fax: (202) 268-4404

Rule Development History

The USPS rule "Mailability of Sharps and Other Medical Devices" was published as a final rule on June 30, 1992 (45). The rule was amended on September 21, 1992 (46) and again on November 1992 (47). The final rule, which is published in the USPS Domestic Mail Manual (DMM) under Sections 8.1 through 8.10 (48), is incorporated by reference into 39 CFR Part 111.1 (49).

In its first proposal, the USPS noted that most packages containing either syringes or other pathogen-containing contents were usually mailed at fourth-class rates (50). Such materials were described as those sharps and unsterilized containers that had been listed by the Environmental Protection Agency as regulated medical

[1]Adapted from Turnberg WL. United States Postal Service: Mailability of sharps and other used medical devices. *Medical Waste Analyst*, 2(11):1–6, copyright 1994 by Technomic Publishing Company, Lancaster, Pennsylvania. Reprinted by permission.

waste in 40 CFR Part 259.30(a). In its supplemental discussion, the USPS stated that the fourth-class mechanical parcel sorting systems used for the bulk mailing operations occasionally caused packages to leak or break open. The USPS reasoned that removal of these packages from the mechanized handling of fourth-class mail service would limit the likelihood of damage to these packages and reduce employee exposure to potentially pathogenic microorganisms. By requiring these packages to be mailed either by registered first-class or priority mail, return receipt requested, an automatic tracking system would be established and the packages would be handled under tighter control with greater accountability. Workers would be further protected through the establishment of specific USPS packaging requirements.

On March 18, 1992, the USPS published an additional proposal that expanded on the rule first proposed (51). In this proposal, the design and construction of packages were tied to rules of the U.S. Department of Transportation under 40 CFR Part 178.604 (leakproof test), 178.606 (stacking test), 178.609 (test requirements for packaging for infectious substances), in addition to other requirements to assure package integrity. The proposal called for the integrity of packages to be tested and certified by an independent organization, and that following certification, they receive authorization for use from the USPS. USPS also proposed a four-part manifest or mail disposal service shipping record.

USPS published its final rule on June 30, 1992 establishing an immediate effective date for the requirement that packages containing sharps and other medical devices be mailed as first-class or priority mail, and a December 28, 1992 effective date for the remainder of the rule (45). On September 21, 1992, the USPS amended its final rule in an attempt to bring its shipping papers and hazards communication requirements further into alignment with those published by the U.S. Department of Transportation under 40 CFR Part 172, and to preclude such packages from rejection by airlines (46). The effective date of these changes was left unchanged from that of the original final rule published on June 30, 1992.

Responding to negative comments to these amendments as being unrealistic and unworkable, the USPS published its second and last amendment to date on November 24, 1992, which eased the shipping and hazard communication requirements published in the previous amendment (47). The USPS established a March 21, 1993 effective date for these amendments.

The USPS Domestic Mail Manual (DMM) rules pertaining to the mailability of sharps and other medical devices appears under Section 8.0, "Etiologic Agent Preparations, Clinical Specimens, and Biological Products" (48). The test of the DMM has been incorporated into 39 CFR Part 111.1 by reference. Applicable text relating to the shipment of biomedical waste through the USPS system is described in the following section.

USPS Regulations

The USPS regulation entitled "Mailability of Sharps and Other Medical Devices" is published in the USPS Domestic Mail Manual Sections 8.1 through 8.10, which is incorporated by reference at 39 CFR Part 111.1.

Wastes Regulated by the Rule. The USPS rule pertains to sharps and other medical devices. Section 8.2(e) of the DMM defines these terms as follows:

(e) *Sharps* means any item with a projecting cutting edge or fine point that was used in animal or human patient care or treatment or in medical research, or industrial laboratories, including but not limited to hypodermic needles, syringes (with or without the attached needles) pasteur pipettes, scalpel blades, blood vials, needles with attached tubing, and culture dishes (regardless of the presence of infectious agents). Also included are other types of broken or unbroken glassware that were in contact with infectious agents, such as used slides or cover slips, The term *sharps* does not include new unused medical devices such as hypodermic needles, syringes, and scalpel blades.

(f) *Other medical devices* means any devices used in animal or human patient care or treatment or in medical research that are not, or do not contain, a projecting sharp.

Sharps and Unsterilized Containers Packaging Requirements. Packaging requirements for sharps and unsterilized containers are found in Section 8.5 of the DMM as follows:

(a) A parcel containing the types of used materials defined in Section 8.2(e) is nonmailable unless it bears the international biohazard symbol on a label with either a fluorescent orange or fluorescent red background. Such parcels are mailable only as first-class or priority mail.

(b) Used sharps must be packaged in a securely sealed, leak-resistant, and puncture-resistant primary container, the total volume of liquid contents of which may not exceed 50 mL. The primary container must maintain its integrity when exposed to temperatures between 0 and 120°F.

(c) The primary container must be packaged within a watertight secondary containment system. The secondary containment system may consist of more than one component, If one of the components is a plastic bag, it must be at a minimum 3.0 mils thick and reinforced with a fiberboard sleeve. A plastic bag by itself does not satisfy the requirement for a secondary containment system. Several primary containers may be enclosed within a secondary containment system to prevent breakage during ordinary processing.

(d) The secondary containment system must be enclosed within an outer shipping container constructed of 200-pound grade corrugated fiberboard or similar material of equivalent strength. The secondary containment system must fit securely within the shipping container to prevent breakage during ordinary processing.

(e) There must be enough material within a watertight barrier to absorb and retain three times the total liquid allowed within the primary container (150 mL per primary container) in case of leakage.

(f) Each parcel must not weigh more than 35 pounds.

(g) Each package prepared for mailing must be designed and constructed so that if subjected to the environmental and test conditions in 49 CFR Part 178.604 (leakproof test), 178.606 (stacking test), 178.608 (vibration standard), 178.609 (test requirements for packaging for infectious substances (etiologic agents), in addition to a bursting test for the shipping container and an absorbency test for the absorbent material commensurate with the requirements in Section 8.5(e), there is no release of the contents to the environment and no significant reduction in the effectiveness of the packaging.

(h) All mailed packages containing used sharps must be accompanied by a four-part manifest or mail disposal service shipping record. The manifest must be placed in an envelope affixed to the outside of the shipping container. The manifest must comply with any applicable requirements imposed by the laws of the state from which the package is mailed. At a minimum, the information shown in Exhibit 8.5(h) must be on the manifest.

(i) Each distributor or manufacturer of mailing kits or packaging assemblies, including containers, cartons, and any other related material to be used to mail sharps to a storage or disposal facility, must obtain an authorization from the USPS. Before applying for this authorization, each such type of the mailing kit must be tested and certified against the standards in Section 8.5(g) by an independent company or organization. This authorization may be obtained by applying in writing to the Business Mail Acceptance Manager, USPS Headquarters. The letter of application must contain the following:

(1) Address of the headquarters or general business office of the distributor or manufacturer.

(2) Addresses of all disposal and storage sites.

(3) List of all types of mailing kits to be covered, with proof of package testing certifications by the independent testing facility that subjected the materials to the testing requirements in all of Section 8.5.

(4) Copy of the proposed manifest to be used with all mailings.

(5) Twenty-four-hour telephone numbers for emergencies.

(6) List of the types of sharps to be mailed for disposal.

(j) Each package must be mailed using merchandise return service, and each authorized manufacturer (or distributor) must provide to the Business Mail Acceptance Manager a surety bond of $50,000 or a letter of credit as proof of sufficient financial responsibility to cover disposal costs if the manufacturer (or distributor) ceases doing business before all its shipping containers are disposed of, or to cover cleanup costs if spills occur while the containers are in USPS possession. Each primary and shipping container must bear a label, which cannot be detached intact, showing:

(1) Company name of the manufacturer or the distributor.

(2) U.S. Postal Service Authorization Number.

(3) Container ID number (or unique model number) signifying that the pack-

aging material is certified and that the manufacturer or distributor obtained authorization as required by Section 8.5(i).

Other Used Medical Devices Packaging Requirements. Packaging requirements for other used medical devices are found in Section 8.6 of the DMM as follows:

(a) Other used medical devices, as defined in Section 8.2(f), must be mailed as first-class or priority mail.

(b) Other used medical devices must be packaged in a securely sealed, leak-resistant primary container, the total liquid volume of which must not exceed 50 mL, unless the devices are shipped in formalin or its equivalent. The primary container must maintain its integrity when exposed to temperatures between 0 and 120 degrees Fahrenheit.

(c) The primary container must be enclosed in an outer shipping container constructed of 200-pound grade corrugated fiberboard or similar material of equivalent strength. The primary container must fit securely within the shipping container to prevent breakage during ordinary processing.

(d) There must be enough material between the shipping container and the primary container to absorb three times the total liquid allowed within the package unless the device is mailed in a formalin solution or its equivalent.

(e) Each parcel containing other used medical devices must bear a complete return address (not a post office box).

General Marking and Labeling Requirements. General marking and labeling requirements are found in Section 8.7 of the DMM as follows:

When applicable, the outer containers must have required labels affixed [e.g., the "Etiologic Agents/Biohazard Material" label and "Clinical Specimen/Biological Products-Biohazard" label, required by 42 CFR Part 72.3(d)]; or if the material is to be transported by air, the infectious substances label specified in *International Mail Manual* 135.4, the proper shipping name and UN number as well as a shipper's declaration for dangerous goods. The UN number for etiologic agents affecting humans is 2814. See 40 CFR Part 172, *Identification Number Cross Reference Index to Proper Shipping Names*, for a description of UN numbers.

Specific Marking Requirements for Unsterilized Containers or Devices. The specific marking requirements for unsterilized containers or devices are found in Section 8.8 of the DMM as follows:

The outside container of clinical specimens, biological products, and unsterilized containers or devices must be marked to identify the contents with the proper shipping name (e.g., "Clinical Specimens," Unsterilized Medical Devices," etc.).

Specific Marking Requirements for Sharps. The specific marking requirements for sharps are found in Section 8.9 of the DMM as follows:

Each exterior package containing sharps must be marked with the words "Infectious Waste," "Medical Waste," or a label displaying the universal biohazard symbol.

Packaging Requirements When Dry Ice Is Used. Packaging requirements when dry ice is used are found in Section 8.10 of the DMM as follows:

Generally, all outside containers containing more than 5 pounds of dry ice (carbon dioxide solid) that are eligible for air transportation must have a shipper's declaration for dangerous goods attached in triplicate. Packages containing dry ice must be designed and constructed to permit the release of carbon dioxide gas to prevent a buildup of pressure that could rupture, the packaging. After fulfilling the conditions in 8.10(a) through 8.10(c), below, the marking "ORM-A UN 1845 Carbon Dioxide Solid" or "Dry Ice" is not required [see 49 CFR Part 173.615 and 175.10(a)(13)]. A shipper's declaration for dry ice is also not required if:

a. The weight of the dry ice in the package does not exceed 5 pounds and the net weight of the dry ice is marked on the package.
b. The dry ice is a refrigerant for a material being used for diagnostic or treatment purposes (e.g., frozen medical specimens) and the material is so marked on the package.
c. The package is marked "Carbon Dioxide Solid" or "Dry Ice."

Exhibit 8.5(h)   Information Required for Infectious Substance Manifest

1. General (Mailer)
   a. Name.
   b. Complete address (not a post office box)
   c. Telephone number.
   d. Description of contents of shipping container. Describe contents as "Used Medical Sharps."
   e. Date the shipping container was mailed.
   f. State permit number of the approved facility in which the contents are to be disposed.
2. Destination Facility (Disposal Site)
   Complete address (not post office box).
3. Generator's (Mailer's) Certification
   "I certify that this carton has been approved for the mailing of used medical sharps, has been prepared for mailing in accordance with the directions for that purpose, and does not contain excess liquid or nonmailable material in violation of the applicable postal regulations. I AM AWARE THAT FULL RESPONSIBILITY RESTS WITH THE GENERATOR (MAILER) FOR ANY VIOLATION OF 18 USC 1716 WHICH MAY RESULT FROM PLACING IMPROPERLY PACKAGED ITEMS IN THE MAIL. I also certify that the contents of this consignment are fully and accurately described

above by proper shipping name and are classified, packed, marked, and labeled, and in proper condition for carriage by air according to the applicable national governmental regulations."

4. Destination Facility (Storage or Disposal Site)

   a. Printed certification or receipt, treatment, and disposal stating: "I certify that the contents of this package have been received, treated, and disposed of in accordance with all local, state, and federal regulations."

   b. Printed or typewritten name of an authorized recipient at the destination facility.

   c. Signature of the authorized recipient at the destination facility.

   d. Date representative of destination facility signed manifest.

5. Transporter or Intermediate Handler Other Than U.S. Postal Service (If Different From Destination Facility)

   a. Name.

   b. Complete address (not a post office box).

   c. Printed name of transporter or intermediate handler.

   d. Signature of transporter or intermediate handler.

6. Serialized Manifests

   The manifest or mail disposal service shipping forms must be serialized.

7. Area Reserved for Comments

   The manifest must contain an area reserved for discrepancies and comments, especially if an alternative destination facility is used.

8. Completion and Distribution of Manifest

   The manifest must contain instruction for completing the manifest and distributing copies.

   a. One copy must be kept by the generator (mailer).

   b. One copy must be kept by the transporter or intermediate handler for 90 days.

   c. One copy must be kept by the destination facility for 90 days.

   d. One copy must be mailed to the generator by the destination facility.

9. Emergency Telephone Number

   The manifest must bear the following statement with appropriate information: "IN CASE OF EMERGENCY, OR THE DISCOVERY OF DAMAGE OR LEAKAGE, CALL 1-800-XXX-XXXX."

Mailing Kits and Packaging Assemblies

Sharps Package Certification Authorization Requirements. Section 8.5(i) of the DMM requires certification that packages used for mailing sharps meet the package integrity standards of Section 8.5(g) of the rule, and that once certified by an independent company, the distributors or manufacturers of the packages obtain authorization from the USPS before use. The package integrity standards of Section 8.5(g) are as follows:

> Each package prepared for mailing must be designed and constructed so that, if subjected to the environmental and test conditions in 49 CFR Part 178.604 (leakproof test), 178.606 (stacking test), 178,608 (vibration standard), 178.609 (test requirements for packaging for infectious substances (etiologic agents)), in addition to a bursting test for the shipping container and an absorbency test for the absorbent material commensurate with the requirements in Section 8.5(e), there is no release of the contents to the environment and no significant reduction in the effectiveness of the packaging.

The USPS authorization requirement of Section 8.5(i) is as follows:

> Each distributor or manufacturer of mailing kits or packaging assemblies, including containers, cartons, and any other related material to be used to mail sharps to a storage or disposal facility, must obtain an authorization from the USPS. Before applying for this authorization, each such type of the mailing kit must be tested and certified against the standards in Section 8.5(g) by an independent company or organization. This authorization may be obtained by applying in writing to the Business Mail Acceptance manager, USPS Headquarters.

USPS Authorized Mailing Kit Manufacturers. As of June 1994, the USPS had authorized sharps mailing kits or packaging assemblies to the following manufacturers/distributors:

Safe-T-MED, Inc.
2800 High Ridge Road
Boynton Beach, FL 33426

3CI Complete Compliance Corporation
9033 Knight Road
Houston, TX 77054

Sytech Medical
P.O. Box 2217
Glenview, IL 60025-6217

OnGard Systems, Inc.
1800 Fifteenth Street, Suite 100
Denver, CO 80202-1165

Browning-Ferris Industries
1350 Connecticut Avenue
Washington, DC 20036-1701

Glaxo Inc.
Five Moore Drive
Research Triangle Park, NC 27709

Enviro-American
22 Northeast 46th Street
Oklahoma City, OK 73105

MEDX (Sharps Disposal, Inc.)
7600 Northwest 69th Avenue
Medley, FL 33166

Ventura Waste Management
2715 East Main Street
Ventura, CA 93003

Compliance Management Association
P.O. Box 379
Birmingham, MI 48012

American Medical Disposal
P.O. Box 54914
Oklahoma City, OK 73154

Home Medical Waste, Inc.
P.O. Box 3240
Cleveland, TN 37320

Sage Products, Inc.
815 Tek Drive
Crystal Lake, IL 60014-9693

Stericycle of Washington
21220 87th Avenue Southeast, Suite 300
Woodinville, WA 98072

STATE REGULATIONS

Most state legislatures and governments have developed biohazardous waste laws
and regulations though the degree of regulatory variability is significant. In a 1995
survey conducted by *Waste Age/Infectious Waste News*, 44 states reported having
specific medical waste regulations in place, although a nationwide lack of uniformity between regulations was noted (52).

In many circumstances, local jurisdictions may adopt and implement biohazardous waste regulations that are equal to or more stringent than those adopted by their
state. As an example, in Washington State, biohazardous waste is not regulated
comprehensively by the state. As a result, many local health jurisdictions have
adopted and are implementing biohazardous waste regulations to meet local needs.
For instance, the Seattle–King County Health Department is implementing regulations that define infectious waste and specify specific standards for handling, packaging, labeling, transportation, treatment, and disposal. The neighboring jurisdiction

located immediately to the south, the Tacoma–Pierce County Health Department, has implemented a similar program to the Seattle–King County program but bases its implementation costs on fees collected through a generator permit program.

The reader should be aware that biohazardous waste regulations are often revised. To assist with obtaining updated copies of regulations, a listing of state agencies and contacts is presented in Appendix B. Local contacts can be obtained through the applicable state agency contacts identified in the appendix. It is essential that both state and local regulatory agencies be consulted periodically to ensure that copies are maintained of current regulations.

REFERENCES

1. U.S. Department of Labor, Occupational Safety and Health Administration. Occupational Exposure to Bloodborne Pathogens; Final Rule. *Federal Register*, 56:64004–64182. December 6, 1991.
2. Occupational Safety and Health Act of 1970. Public Law 91-596, 91st Congress, December 29, 1970 as Amended by Public Law 101-552. November 5, 1990.
3. U.S. Department of Labor, Occupational Safety and Health Administration. All About OSHA. OSHA 2056. 1994 (revised).
4. U.S. Department of Labor, Occupational Safety and Health Administration. Consultation Services for the Employer. OSHA 3047. 1991 (revised).
5. U.S. Department of Labor, Occupational Safety and Health Administration. Hepatitis B Risks in the Healthcare System. OSHA Instruction CPL 2-2.36. November 30, 1983.
6. U.S. Department of Labor, Occupational Safety and Health Administration. The Risk of Hepatitis B Infection for Workers in the Healthcare Delivery System and Suggested Method for Risk Reduction. Office of Occupational Medicine, Directorate of Technical Support. October 1993.
7. Joint Advisory Notice; Department of Labor/Department of Health and Human Services; HBV/HIV. *Federal Register* 52:41818–41824. October 30, 1987.
8. U.S. Department of Labor, Occupational Safety and Health Administration. Occupational Exposure to Hepatitis B Virus and Human Immunodeficiency Virus; Advance Notice of Proposed Rulemaking. *Federal Register* 52:45438–45440. November 27, 1987.
9. U.S. Department of Labor, Occupational Safety and Health Administration. Enforcement Procedures for the Occupational Exposure to Hepatitis B Virus (HBV) and Human Immunodeficiency Virus (HIV). OSHA Instruction CPL 2-2.44. January 19, 1988.
10. U.S. Department of Labor, Occupational Safety and Health Administration. Enforcement Procedures for the Occupational Exposure to Hepatitis B Virus (HBV) and Human Immunodeficiency Virus (HIV). OSHA Instruction CPL 2-2.44A. August 15, 1988.
11. U.S. Department of Labor, Occupational Safety and Health Administration. Enforcement Procedures for the Occupational Exposure to Hepatitis B Virus (HBV) and Human Immunodeficiency Virus (HIV). OSHA Instruction CPL 2-2.44B. February 27, 1990.
12. Centers for Disease Control. Update: Universal Precautions for Prevention of Transmission of Human Immunodeficiency Virus, Hepatitis B Virus, and Other Bloodborne Pathogens in Health-Care Settings. *MMWR*, 37(24):377–387. June 24, 1988.

13. Centers for Disease Control. Recommendations for prevention of HIV transmission in healthcare settings. *MMWR*, 36(2S). August 21, 1987.

14. U.S. Department of Labor, Occupational Safety and Health Administration. Occupational Exposure to Bloodborne Pathogens: Proposed Rule and Notice of Hearing. *Federal Register* 54:23042–23139. May 30, 1989.

15. U.S. Department of Labor, Occupational Safety and Health Administration. Occupational Exposure to Bloodborne Pathogens; Approval of Information Collection Requirements. *Federal Register* 57:12717. April 13, 1992.

16. U.S. Department of Labor, Occupational Safety and Health Administration. Occupational Exposure to Bloodborne Pathogens. *Federal Register* 57:29206. July 1, 1992.

17. U.S. Department of Labor, Occupational Safety and Health Administration. Enforcement Procedures for the Occupational Exposure to Bloodborne Pathogens Standard, 29 CFR Part 1910.1030. OSHA Instruction CPL 2-2.44C. March 6, 1992.

18. Personal communication with Richard C. Knudsen, Chief, Biological Safety, Office of Health and Safety, Centers for Disease Control. January 25, 1995.

19. Public Health Service. Statement of Organization, Functions, and Delegations of Authority. *Federal Register*, 36(94):8893. May 14, 1971.

20. Public Health Service. Interstate Shipment of Etiologic Agents; Final Rule. *Federal Register*, 45(141):48626–48629. July 21, 1980.

21. Public Health Service. Interstate Shipment of Etiologic Agents; Notice of Proposed Rulemaking. *Federal Register*, 55(42):7678–7682. March 2, 1990.

22. Personal communication with Lynn Myers, Biological Safety, Office of Health and Safety, Centers for Disease Control. December 12, 1994.

23. U.S. Department of Transportation. Proposed Requirements for Shipment of Etiologic Agents. Docket HM-96. *Federal Register*, 36:25163. December 29, 1971.

24. U.S. Department of Transportation. Proposed Requirements for Shipment of Etiologic Agents. Docket HM-96. *Federal Register*, 37:14728. July 22, 1972.

25. U.S. Department of Transportation. Shipment of Etiologic Agents—Final Rule. Docket HM-96. *Federal Register*, 37:20554. September 30, 1972.

26. U.S. Department of Transportation. Shipment of Etiologic Agents on Passenger-Carrying Aircraft—Final Rule. Docket HM-96. *Federal Register*, 37:25243. November 29, 1972.

27. U.S. Department of Transportation. Etiologic Agents. Docket HM-96. *Federal Register*, 38:8161. March 29, 1973.

28. U.S. Department of Transportation. Advance Notice of Proposed Rulemaking. Docket HM-181. *Federal Register*, 47:16268. April 15, 1982.

29. U.S. Department of Transportation. Performance-Oriented Packaging Standards; Miscellaneous Proposals. Docket HM-181. Notice of Proposed Rulemaking. *Federal Register*, 52:16482. May 5, 1987.

30. U.S. Department of Transportation. Etiologic Agents—Notice of Proposed Rulemaking. Docket HM-142A. *Federal Register*, 53:45525. November 10, 1988.

31. Public Health Service. Interstate Shipment of Etiologic Agents—Final Rule. 45 FR 48627. July 1, 1980.

32. U.S. Department of Transportation. Performance-Oriented Packaging Standards; Changes to Classification, Hazard Communication, Packaging and Handling Requirements Based

on U.N. Standards and Agency Initiative—Final Rule. Docket HM-181. *Federal Register*, 55:52402. December 21, 1990.

33. U.S. Department of Transportation. Etiologic Agents—Final Rule. Docket HM-142A. *Federal Register*, 56:197. January 3, 1991.

34. U.S. Department of Transportation. Etiologic Agents—Final Rule; Delay of Effective Date. Docket HM-142A. *Federal Register*, 36:7312. February 22, 1991.

35. U.S. Department of Transportation. Incorporation of Docket No. HM-142A into Docket No. 181. Docket 181. *Federal Register*, 47160. September 18, 1991.

36. U.S. Department of Transportation. Infectious Substances (Etiologic Agents); Revisions to Transitional Provisions; Final Rule. Docket HM-181. *Federal Register*, 56:49830. October 1, 1991.

37. U.S. Department of Transportation. Performance-Oriented Packaging Standards; Revisions and Response to Petitions for Reconsideration; Final Rule. Docket HM-181. *Federal Register*, 56:66142. December 20, 1991.

38. U.S. Department of Transportation. Infectious Substances; Correction and Extension of Compliance date. Docket HM-181. *Federal Register*, 57:45442. October 1, 1992.

39. U.S. Department of Transportation. Infectious Substances; Notice of Public Hearing and Advance Notice of Proposed Rulemaking. Docket HM-181G. *Federal Register*, 58:12207. March 3, 1993.

40. U.S. Department of Transportation. Infectious Substances; Extension of Compliance Date. Docket HM-181. *Federal Register*, 58:12182. March 3, 1993.

41. U.S. Department of Transportation. Infectious Substances; Extension of Compliance Date. Docket HM-181G. *Federal Register*, 58:66302. December 20, 1993.

42. U.S. Department of Transportation. Infectious Substances; Extension of Compliance Date. Docket HM-181G. *Federal Register*, 59:48762. September 22, 1994.

43. U.S. Department of Transportation. Notice of Proposed Rule Making and Notice of Public Meeting. Docket HM-18G. *Federal Register*, 59:65860. December 21, 1994.

44. Personal communication with Teresa Gwynn, U.S. Department of Transportation. February 8, 1995.

45. U.S. Postal Service. Mailability of Sharps and Other Medical Devices—Final Rule. *Federal Register*, 57:29028. June 30, 1992.

46. U.S. Postal Service. Mailability of Sharps and Other Medical Devices—Amendment to Final Rule. *Federal Register*, 57:43403. September 21, 1992.

47. U.S. Postal Service. Mailability of Sharps and Other Medical Devices—Amendment to Final Rule. *Federal Register*, 57:55112. November 24, 1992.

48. *Code of Federal Regulations*. 30 Part 111. Revised as of July 1, 1993.

49. U.S. Postal Service. *Domestic Mail Manual*. DMM Issue 47. April 10, 1994.

50. U.S. Postal Service. Mailability of Sharps and Unsterilized Containers and Devices—Proposed Rule. *Federal Register*, 56:36750. August 1, 1991.

51. U.S. Postal Service. Mailability of Sharps and Other Medical Devices—Proposed Rule. *Federal Register*, 57:9402. March 18, 1992.

52. Malloy MG. Medical waste in '95. *Waste Age*, 26(6):49–62. 1995.

4

GUIDELINES AND INDUSTRY STANDARDS

Guidelines and industry standards addressing biohazardous waste management have been developed by various government entities and professional organizations during the past few years. Guidelines represent recommended best management practices from the perspective of the issuing agency. Unlike regulations, guidelines are not mandatory and carry no force of law.

Industry standards addressing biohazardous waste issues have been developed by various professional organizations to address a specific need. For instance, healthcare standards developed by the Joint Commission on Accreditation of Health Care Organizations (JCAHO) must be met by those healthcare organizations seeking accreditation by that organization. Industry standards are also established by organizations such as the American Society for Testing and Materials (ASTM) to provide voluntary but industry-recognized standardization for materials, products, systems, and services.

In this chapter we present a history of biohazardous waste–related guideline development during the past three decades and specifically address biohazardous waste guidelines established by the Environmental Protection Agency, National Institutes of Health, and Centers for Disease Control. Also discussed are industry standards developed by JCAHO and ASTM.

EARLY GUIDELINES

A flurry of interest in hospital waste management in the late 1960s and early 1970s resulted in two government-developed medical waste management guidelines. A 1968 guideline published by the Public Health Service of the (then) U.S. Department of Health, Education, and Welfare defined the hospital waste stream as consisting of six principal types of solid waste: (1) garbage; (2) paper, trash, and other dry combustibles; (3) treatment room wastes; (4) surgery wastes; (5) autopsy wastes; and (6) noncombustibles such as cans and bottles (1). Although no specific biohazardous waste definitional or management criteria were described, this early guideline cautioned: "The problems inherent in safely disposing of such items as syringes, examination gloves, catheters, emesis basins, and petri dishes by means of outdoor storage and municipal collection involve special safety considerations for staff, collectors, and the public. There are legal implications as well."

A 1974 guideline published by the U.S. Environmental Protection Agency entitled "Hospital Wastes" advised that wastes resulting from the diagnosing and treatment of patients are frequently hazardous (e.g., contain biological, radioactive, and chemical components, including sharp items such as disposable needles) (2). The guideline warned that because of the difficulty of segregating these wastes from other nonhazardous hospital wastes, all hospital waste must be considered potentially contaminated and receive special handling and treatment.

ENVIRONMENTAL PROTECTION AGENCY

On October 21, 1976, the U.S. Environmental Protection Agency (EPA) was charged by Congress to establish the nation's hazardous waste management system, pursuant to Public Law 94–580, the Resource Conservation and Recovery Act of 1976 (RCRA) (3). In RCRA, Congress defined the term *hazardous waste* to mean

> a solid waste, or combination of solid waste, which because of its quantity, concentration, or physical, chemical, or infectious characteristics may: (A) cause, or significantly contribute to an increase in mortality or an increase in serious irreversible, or incapacitating reversible, illness; or (B) poses a substantial present or potential hazard to human health or the environment when improperly treated, stored, transported, or disposed of, or otherwise managed. [Section 1004(5)]

Subtitle C of RCRA required the EPA to develop and promulgate regulations for identifying the characteristics of hazardous waste and for listing hazardous waste, within the meaning of *hazardous waste* as defined by statute. (Section 3001). Under its congressional mandate, EPA was required to consider the infectiousness of waste when developing standards for hazardous waste characterization and listing.

Proposed Infectious Waste Standards

On December 18, 1978, the EPA published proposed rules that addressed the criteria for identifying and listing hazardous waste (4). In its proposed hazardous waste rule, the EPA included the infectiousness of a waste within its criteria for listing hazardous waste because the term *infectious* appeared in the definition of hazardous waste found in Section 10004(5) of the act. However, when the final hazardous waste rules were published on May 19, 1980, EPA changed its mind and omitted references to its previously proposed infectious waste standards, choosing instead to address the issue as a guideline (5).

EPA Guide for Infectious Waste Management

In September 1982, the EPA published its draft guidance document for infectious waste management in response to requests for guidance (6). EPA finalized its draft manual in May 1986 in the document "EPA Guide for Infectious Waste Management" (7). The guidance, which presented EPA's perspective on acceptable infectious waste management practices, focused solely on wastes with infective characteristics. Wastes with additional characteristics such as toxicity or radioactivity were beyond the document's scope.

In its guidance document, EPA broadly defined infectious waste as "waste capable of producing an infectious disease." This definition was made in context with the factors associated with disease transmission, which include (1) presence of a pathogen of sufficient virulence, (2) dose, (3) mode of transmission, (4) portal of entry, and (5) resistance of host.

Based on elements of disease transmission, EPA recommended six categories of medical waste for special management as infectious waste:

1. *Isolation wastes* [EPA refers the reader to the CDC Guideline for Isolation Precautions in Hospitals (8)]
2. *Cultures and stocks* of infectious agents and associated biologicals (e.g., specimens from medical and pathology laboratories; cultures and stocks of infectious agents from clinical, research, and industrial laboratories; disposable culture dishes and devices used to transfer, inoculate, and mix cultures; wastes from production of biologicals; and discarded live and attenuated vaccines)
3. *Human blood and blood products* (e.g., waste blood, serum, plasma, and blood products)
4. *Pathological wastes* (e.g., tissues, organs, body parts, blood, and body fluids removed during surgery, autopsy, and biopsy)
5. *Contaminated sharps* (e.g., contaminated hypodermic needles, syringes, scalpel blades, pasteur pipettes, and broken glass)
6. *Contaminated animal waste* (e.g., contaminated animal carcasses, body parts, and bedding of animals that were intentionally exposed to pathogens)

Additional categories of medical waste were also suggested by EPA for consideration as infectious waste:

1. *Wastes from surgery and autopsy* (e.g., soiled dressings, sponges, drapes, lavage tubes, drainage sets, underpads, and surgical gloves)

2. *Contaminated laboratory wastes* (e.g., specimen containers, slides, and coverslips; disposable gloves, lab coats, and aprons)
3. *Dialysis unit wastes* (e.g., tubing, filters, disposable sheets, towels, gloves, aprons, and lab coats)
4. *Contaminated equipment* (e.g., equipment used in patient care, medical laboratories, research, and in the production and testing of certain pharmaceuticals)

Management considerations were identified in the guidance document. The guideline recommended that a responsible person at a healthcare facility develop an infectious waste management system, incorporated into a formal plan, and that the plan contains the following elements:

- Designation of infectious waste
- Handling of infectious waste (e.g., segregation, packaging, storage, transport and handling, treatment techniques, and disposal of treated waste)
- Contingency planning
- Staff training

Several infectious waste treatment options were described, including steam sterilization, incineration, thermal inactivation, gas/vapor sterilization, chemical disinfection, and sterilization by irradiation, and factors to consider for each were presented. EPA recommended treatment options for each category of infectious waste. EPA also recommended that only treated infectious waste be landfilled, unless disposal of untreated waste takes place in a controlled sanitary landfill, and that blood and blood products be either steam sterilized, incinerated, or discharged directly to the sanitary sewer for treatment in a municipal sewage treatment plant only if secondary treatment is available.

CENTERS FOR DISEASE CONTROL

The Department of Health and Human Services, Public Health Service, Centers for Disease Control (CDC) has addressed infectious waste management through recommendations published in various publications during the past few years. Next, we address guideline recommendations presented in three CDC publications since 1983.

Isolation Precautions in Hospitals

In 1983, the Centers for Disease Control published a manual for isolation precautions in hospitals entitled "CDC Guideline for Isolation Precautions in Hospitals" (8). In the manual the CDC addresses the management of infectious waste generated in the course of patient care on isolation precautions, although only in general terms. The recommendations regarding needles and syringes and dressings and tissues are discussed below.

Needles and Syringes Handling. Here the CDC recommended that needles be placed in a prominently labeled, puncture-resistant container designated specifically for this purpose. CDC recommends against the practice of recapping needles, or bending or breaking them by hand to prevent injuries. CDC warned that needle-cutting devices may cause blood to spatter onto environmental surfaces, but advised that it did not have information showing the environmental effect of this in terms of disease transmission, and made no recommendation regarding the practice.

Dressings and Tissues Handling. CDC advised that all dressings, paper tissues, and other disposable items contaminated with infective material (respiratory, oral, or wound secretions) be bagged, labeled, and disposed of as required by the hospital's infectious waste disposal policy.

Handwashing and Hospital Environmental Control

In 1985, the CDC published "Guideline for Handwashing and Hospital Environmental Control," which replaced all previously published handwashing and environmental control statements issued by CDC or the Hospital Infectious Program, Center for Infectious Diseases, CDC (9). The guideline presented a three-category ranking scheme for its recommendations:

Category I. These measures "are strongly supported by well-designed and controlled clinical studies that show their effectiveness in reducing the risk of nosocomial infections, or are viewed as effective by a majority of expert reviewers. Measures in this category are viewed as applicable for most hospitals, regardless of size, patient population, or endemic nosocomial infection rates."

Category II. These measures "are supported by highly suggestive clinical studies in general hospitals or by definitive studies in specialty hospitals that might not be representative of general hospitals. Measures that have not been adequately studied but have a logical or strong theoretical rational indicating probable effectiveness are included in this category. Category II recommendations are viewed as practical to implement in most hospitals."

Category III. These measures "have been proposed by some investigators, authorities, or organizations, but to date, lack supporting data, a strong theoretical rationale, or an indication that the benefits expected from them are cost-effective. Thus they are considered important issues to be studied. They might be considered by some hospitals for implementation, especially if the hospitals have specific nosocomial infection problems, but they are not generally recommended for widespread adoption."

Infective Material Defined. In its section on infective waste management, CDC begins with the statement:

> There is no epidemiologic evidence to suggest that most hospital waste is any more infective than residential waste. Moreover, there is no epidemiologic evidence that hospital waste disposal practices have caused disease in the community. Therefore, identifying wastes for which special precautions are indicated is largely a matter of judgement about the relative risk of disease transmission. Aesthetic and emotional considerations may override the actual risk of disease transmission, particularly for pathology wastes.

In its recommendation, CDC acknowledged the difficulty and inappropriateness of determining that a waste should be handled as infective waste based on quantity and types of etiologic agents present. It therefore based its recommendations on identifying wastes that represent a sufficient potential infection risk during handling and disposal, and for which special handling and disposal precautions appear prudent. These wastes identified by the CDC include:

1. Microbiology laboratory wastes
2. Pathology wastes
3. Blood specimens or blood products
4. Certain sharp items (e.g., needles and scalpel blades)

CDC designated a category II status in its ranking scheme to these wastes and recommended that the wastes be considered as potentially infective and handled and disposed of with special precautions. With regard to isolation wastes, CDC referred to its recommendations for disposal of isolation wastes published in 1983 in "Guideline for Isolation Precautions in Hospitals" (8), but noted that the recommendations were not categorized because the guideline recommendations for isolation precautions are not categorized. CDC also noted that although certain wastes may be contaminated with potentially infective blood, sucretions, excretions, or exudates, that it is not necessary to treat such waste as infective. However, CDC advises that special waste-handling precautions may be necessary for wastes contaminated with certain rare disease-causing agents, such as the Lassa virus.

Handling, Transport, and Storage Recommendations. For safe handling, transport, and storage of infective waste, CDC recommends that (1) personnel who are involved with infective waste handling and disposal be informed of potential safety and health risks and trained in appropriate handling and disposal methods (ranked at category II); (2) if processing or disposal facilities are not available at the site of generation, infective waste should be transported in sealed impervious containers for treatment at another hospital area (ranked at category II); and (3) infective waste awaiting treatment should be stored in a restricted area, accessible only to personnel involved in the disposal process (ranked at category III).

Processing and Disposal Recommendations. CDC recommends that (1) infective waste, in general, be either incinerated or autoclaved before disposal in a sanitary landfill (ranked at category III); (2) disposable sharps be placed intact into puncture-resistant containers located as close to the area in which they were used as practical, and that needles not be recapped, broken, bent, or otherwise manipulated by hand (ranked at category I); and (3) bulk blood, secretion fluids, excretions and secretions be carefully poured down a drain connected to a sanitary sewer, that other infectious wastes capable of being ground and flushed may be poured to a sanitary sewer, and that special precautions may be necessary for waste contaminated with the agents of certain rare diseases such as Lassa fever (ranked at category II).

The CDC also refers to the EPA's infectious waste guideline, discussed earlier in this Chapter, for additional infectious waste management options.

Prevention of HIV and HBV Transmission

In 1987, the CDC published "Recommendations for Prevention of HIV Transmission in Health-Care Settings" to consolidate and update previously published CDC recommendations for preventing HIV transmission in the healthcare setting (10). In this recommendation, the CDC established the concept of *Universal Precautions*, which emphasizes the need to treat the blood and body fluids from all patients as potentially infective. CDC's recommendation for infective waste management was unchanged from the recommendation presented in its 1985 document, "Guideline for Handwashing and Hospital Environmental Control" (9).

In 1989 CDC published "Guidelines for Prevention of Transmission of Human Immunodeficiency Virus and Hepatitis B Virus to Health-Care and Public-Safety Workers," which presented additional recommendations for infective waste disposal (11). In its recommendation, CDC stated that "the only documented occupational risks of HIV and HBV infection are associated with parenteral (including open wound) and mucous membrane exposure to blood and other potentially infectious body fluids. Nevertheless, the precautions described below should be routinely followed." CDC's recommendations for the disposal of needles, sharps, and other infective waste were unchanged from those published previously. For infective waste disposal, CDC emphasized "In all cases, local regulations should be consulted prior to disposal procedures and followed."

NATIONAL INSTITUTES OF HEALTH AND CENTERS FOR DISEASE CONTROL

The National Institutes of Health and the Centers for Disease Control jointly developed a guidance manual to protect laboratory workers and the public from infectious disease transmission in the microbiological and biomedical laboratory setting. "Biosafety in Microbiological and Biomedical Laboratories" was amended most recently in May 1993 (12). The guideline describes four classes (biosafety levels) of laboratory operation when working with various human infectious microbial agents. Elements of safe laboratory practices and techniques, safety equipment (primary barriers), and facility design (secondary barriers) necessary to prevent transmission of these agents to workers are described for each biosafety level. Recommendations are also set forth for four biosafety levels involving the use of animals in the biomedical or microbiological laboratory setting. Many elements of the guideline have been incorporated into the bloodborne pathogen standards published by the Department of Labor, Occupational Safety and Health Administration, at 29 CFR 1910.1030.

NIH/CDC Biohazardous Waste Management

The guideline recommends handling and disposal practices for managing infectious waste. These are described within the context of each of the four biosafety levels. Each must be viewed in light of other safety requirements specified for each biosafety level, including the need for a plan of operation, hazard communication, and training requirements.

Biosafety Level 1. Practices, safety equipment, and facility design are described in the guideline for facilities in which "work is done with defined and characterized strains of viable microorganisms not known to cause disease in healthy adult humans." The guideline recommends that all cultures, stocks, and regulated wastes be decontaminated before disposal by an approved method such as autoclaving. For infectious wastes requiring decontamination outside the immediate laboratory, the guideline recommends that they be placed in a durable, leakproof container and closed for transport from the laboratory. The guideline further recommends that infectious wastes destined for off-site decontamination meet all applicable packaging requirements of local, state, and federal governments before they are removed from the facility.

Biosafety Level 2. Practices, safety equipment, and facility design are recommended for facilities in which "work is done with the broad spectrum of indigenous moderate-risk agents present in the community and associated with human disease of varying severity" and described as being appropriate when "work is done with any human derived blood, body fluids, or tissues where the presence of an infectious agent may be unknown." Recommendations for the packaging, transport, and disposal of cultures, stocks, and regulated wastes do not differ from those recommended under biosafety level 1. In addition, the guideline places emphasis on the proper and safe management of any contaminated sharp item, such as needles and syringes, slides, pipettes, capillary tubes, and scalpels. With specific reference to waste disposal, the guideline recommends that used disposable needles not be bent, sheared, broken, recapped, removed from disposable syringes, or otherwise manipulated by hand before disposal. Used needles must be carefully placed in puncture-resistant containers used for sharps disposal, and these containers must be located conveniently. The guideline recommends that containers of contaminated needles, sharp equipment, and broken glass be decontaminated before disposal as required under applicable local, state, and federal regulations. The guideline also recommends that a method for decontaminating infectious wastes be available, such as autoclaving, chemical disinfection, incineration, or other approved decontamination system, but does not specify the need for this method to be located on-site.

Biosafety Level 3. Practices, safety equipment, and facility design are recommended for facilities in which "work is done with indigenous or exotic agents with a potential for respiratory transmission, and which may cause serious and potentially lethal infection." Recommendations for the packaging, transport, and disposal of cultures, stocks, and regulated wastes do not differ from those recommended under biosafety level 1. Recommendations for the handling and disposal of used sharps, including needles and syringes, do not differ from those recommended under biosafety level 2. The guideline specifies further that all potentially contaminated waste materials from laboratories or animal rooms be decontaminated before disposal. As under biosafety level 2, the guideline recommends that a method for decontaminating infectious waste is available, such as autoclaving, chemical disinfection, incineration, or other approved decontamination system, but does not specify the need for this method to be located on-site.

Biosafety Level 4. Practices, safety equipment, and facility design are recommended for facilities in which "work is done with dangerous and exotic agents which pose a high individual risk of life-threatening disease, which may be transmitted via the aerosol route, and for which there is no available vaccine or therapy." Recommendations for the handling and disposal of sharps do not differ from those of biosafety level 2. However, unlike the other biosafety levels, biosafety level 4 emphasizes that "no materials, except for biological materials that are to remain in a viable or intact state, are removed from the biosafety level 4 laboratory unless they have been autoclaved or decontaminated before they leave the facility." Biosafety level 4 also requires that "a double-doored autoclave is provided for decontaminating materials passing out of the facility. The autoclave door which opens to the area external to the facility is sealed to the outer wall, and automatically controlled so that the outside door can only be opened after the autoclave 'sterilization' cycle has been completed." Furthermore, liquid effluents from sinks, biological safety cabinets, floor drains, and autoclave chambers must be decontaminated by heat treatment before discharge to the sanitary sewer. This process must be validated through a continual temperature recorder and indicator microorganisms with a defined heat susceptibility profile.

JOINT COMMISSION ON ACCREDITATION OF HEALTHCARE ORGANIZATIONS

The Joint Commission on Accreditation of Healthcare Organizations (JCAHO) is an organization that evaluates and accredits more than 5200 hospitals and more than 6000 other healthcare organizations, and each year establishes standards that must be met for accreditation to be granted. Accreditation is used for various purposes, such as meeting certain Medicare requirements, expediting third-party payment, fulfilling state licensure requirements, favorably influencing liability insurance premiums, or enhancing community confidence (13). JCAHO accreditation is available for:

- General hospitals, as well as psychiatric, children's, and other specialty hospitals
- Home care agencies and organization, including those that provide home health, personal care and support, home infusion, and/or durable medical equipment services
- Nursing homes and other long-term care facilities
- Mental health, chemical dependency, and mental retardation and developmental disabilities services for patients of all ages in all organized service settings
- Ambulatory care providers, including outpatient surgery facilities, rehabilitation centers, infusion centers, group practices, and managed care providers such as group and staff model HMOs
- Healthcare networks

Accreditation manuals have been established for the various healthcare organizations identified above. For hospitals, these accreditation standards are published in the *Accreditation Manual for Hospitals* (14). Standards for waste management are included among the standards that must be met by a healthcare organization seeking accreditation. For hospitals seeking accreditation in 1995 for example, the standards are described in Section 2 (Management of the Environment of Care) of the *Accreditation Manual for Hospitals*. The waste management standards are written in general terms and must be documented in a written plan that is in accordance with local, state, and federal laws and regulations (e.g., OSHA's Regulations for Bloodborne Pathogens). The plan must include policies and procedures, performance standards, written criteria, and goals and objectives of the waste management program.

With regard to biohazardous waste management, in its 1995 standard, JCAHO states:

The Joint Commission recognizes that there are different definitions of hazardous waste. At this time, federal regulations do not define infectious waste or medical waste as hazardous waste. However, the intent of this standard is to implement a management process that includes all materials and wastes that require special handling in order to address identified occupational and environmental hazards. Infectious waste and medical waste fall into the special handling category since there are recognized occupational exposure issues that must be dealt with properly.

Further information about the JCAHO accreditation program may be obtained by contacting:

JCAHO
Joint Commission Customer Service
1 Renaissance Boulevard
Oakbrook Terrace, IL 60181
Tel: (708) 916-5800

AMERICAN SOCIETY FOR TESTING AND MATERIALS

The American Society for Testing and Materials (ASTM) is a not-for-profit organization that develops voluntary industry standards for materials, products, systems, and services. ASTM publishes standard test methods, specifications, practices, guides, classifications, and terminology from the work of its 134 standards-writing committees. Each year ASTM publishes more than 8500 standards in the 68 volumes of the *Annual Book of ASTM Standards*. Standards relevant to biohazardous waste management are those developed to test the strength of plastics, which include:

- ASTM Standard, Designation: D1709-91: "Standard Test Methods for Impact Resistance of Plastic Film by the Free-Falling Dart Method" (15)
- ASTM Standard, Designation: D4272-90: "Standard Test Method for Total Energy Impact of Plastic Films by Dart Drop" (16)

ASTM standards are used by the state of California as minimum-strength criteria for plastic bags used to contain biohazardous waste.

Additional information about ASTM may be obtained by contacting:

ASTM
Member and Committee Services
1916 Race Street
Philadelphia, PA 19103
Tel: (215) 299-5454

REFERENCES

1. U.S. Department of Health, Education and Welfare, Public Health Service. Solid Wastes Handling in Hospitals. Reprinted from Public Health Service Publication 930-C-16. 1968.
2. U.S. Environmental Protection Agency. Hospital Wastes. SW-129. 1974.
3. Public Law 94-580. The Resource Conservation and Recovery Act. 94th Congress. October 21, 1976.
4. U.S. Environmental Protection Agency. Hazardous Waste Regulations and Guidelines—Proposed Rules. *Federal Register*, 43:58946. December 18, 1978.

5. U.S. Environmental Protection Agency. Hazardous Waste Management System: Identification and Listing of Hazardous Waste—Final Rule, Interim Final Rule and Request for Comments. *Federal Register*, 45:33084. May 19, 1980.

6. U.S. Environmental Protection Agency. Draft Manual for Infectious Waste Management. SW-957. September 1982.

7. U.S. Environmental Protection Agency. EPA Guide for Infectious Waste Management. EPA/530-SW-86-014. May 1986.

8. Garner JS and Simmons BP. CDC Guideline for Isolation Precautions in Hospitals. HHS Publication (CDC) 83-8314. U.S. Department of Health and Human Services, Public Health Service, Centers for Disease Control, Atlanta, Georgia. July 1983.

9. Garner JS and Favero MS. Guideline for Handwashing and Hospital Environmental Control, 1985. HHS Publication 99-1117. Public Health Service, Centers for Disease Control, Atlanta, Georgia. 1985.

10. Centers for Disease Control. Recommendations for prevention of HIV transmission in health-care settings. *MMWR*, 36(2S):1S–18S. August 21, 1987.

11. Centers for Disease Control. Guidelines for prevention of transmission of human immunodeficiency virus and hepatitis B virus to health-care and public-safety workers. *MMWR*, 38(S-6):111–155. June 23, 1989.

12. U.S. Department of Health and Human Services, Public Health Service, Centers for Disease Control and National Institutes of Health. Biosafety in Microbiological and Biomedical Laboratories, 3rd ed. HHS Publication (CDC) 93-8395. May 1993.

13. Joint Commission on Accreditation of Health Care Organizations. *Joint Commission Customer Service Directory—Blue Book*. JCAHO, Chicago, 1994.

14. Joint Commission on Accreditation of Health Care Organizations. *Accreditation Manual for Hospitals*. JCAHO, Chicago. 1995.

15. American Society for Testing and Materials. Standard Test Methods for Impact Resistance of Plastic Film by the Free-Falling Dart Method, Designation: D 1709-91. 1994 Annual Book of ASTM Standards, Section 8, Volume 8.01:393–400. PCN: 01–080194–19. 1994.

16. American Society for Testing and Materials. Standard Test Method for Total Energy Impact of Plastic Films by Dart Drop, Designation: D 4272-90. 1994 Annual Book of ASTM Standards, Section 8, Volume 8.02:606–609. PCN: 01–080294–19. 1994.

5

EPA DEMONSTRATION MEDICAL WASTE TRACKING PROGRAM

This discussion summarizes the U.S. Environmental Protection Agency's (EPA) two-year medical waste tracking program as mandated under the Medical Waste Tracking Act of 1988 (MWTA). Although the program had many critics, it did serve to provide the nation with information regarding how biohazardous wastes are generated, managed, treated, and ultimately disposed of as based on the experiences of the four participating states and one participating territory. The EPA's interim final rule sunsetted on June 22, 1991 and the program was not reauthorized by Congress. At this writing, EPA's final report to Congress has exceeded its statutory deadline by over four years and there is no indication when it will be released. Nevertheless, two interim reports have been published that have provided important information to this field. Although the program has expired, it remains as a model for states developing or amending biohazardous waste management programs.

MEDICAL WASTE TRACKING ACT OF 1988

The Medical Waste Tracking Act of 1988 was signed into law on November 2, 1988 in response to public concern over beach washups of medical waste, particularly along the shores of New York, New Jersey, Connecticut, and Rhode Island and the Great Lakes states (1). The washups, which consisted primarily of garbage and other solid waste, also included small amounts of waste recognized as medical in origin. The perception of medical waste on the beaches raised public fears about the spread of the human immunodeficiency virus (HIV) through contact with such waste. The presence of waste along the beaches resulted in beach closures, dwindling use of those that remained open, and heavy impacts to local shoreline economies.

In the heat of public concern, Congress enacted quick passage of House Bill 3515, the Medical Waste Tracking Act of 1988 (MWTA), decrying the appearance of such waste on the beaches to be "repugnant, intolerable and unacceptable," which was being attributed by Congress to medical mismanagement and "midnight dumping." The MWTA, codified at 42 U.S.C. 6992, amended the Resource Conservation and Recovery Act by adding a new subtitle, Subtitle J. The sponsor of the bill, Representative Tom Luken of Ohio, declared, "medical waste pollution is lethal and grotesque. It is sickening and frightening the public, fouling our coastlines, and crippling local tourist economies. We have a critically ill medical waste disposal system—and the hypodermic needles and vials of infected blood washing ashore from Massachusetts to Florida and New York to Ohio are the leading symptoms of this illness (2)." In response to this crisis, House Bill 3515 was passed quickly by the House of Representatives on October 4 and by the Senate with amendments on October 7. Senate amendments were accepted by the House of Representatives on October 12, and the bill was signed into law by the president on November 2, 1988.

The MWTA required that the EPA develop and implement a two-year demonstration program for the tracking and management of medical waste. Only the states of New York, New Jersey, and Connecticut were required to participate in the program, although the program was open to any state desiring to petition the EPA for inclusion. Only the state of Rhode Island and the Commonwealth of Puerto Rico petitioned the EPA for voluntary inclusion in the program. Although the states contiguous to the Great Lakes were specifically named by the MWTA to participate, these states were given by the act the option to opt out of the program if desired, and all did.

The MWTA required the EPA to publish regulations within six months of the act's enactment that would list the types of waste to be tracked in the demonstration project. Eleven categories of potentially biohazardous solid waste were specifically named to be included by the EPA in its rule: (1) cultures and stocks, (2) pathological wastes, (3) waste human blood and blood products, (4) sharps, (5) contaminated animal carcasses, (6) wastes from surgery or autopsy, (7) laboratory wastes, (8) dialysis wastes, (9) discarded medical equipment, (10) biological waste and discarded contaminated materials, and (11) other wastes found to pose a threat to human health or the environment. The EPA was given the option to exclude from the list any items from categories 6 through 10 if it could be determined that these wastes do not pose a human health or environmental hazard. The EPA was also required to establish a cradle-to-grave waste tracking system and to develop rules for segregating, containerizing, and labeling the regulated medical waste.

Under the MWTA, the EPA or its authorized representative (e.g., participating states and territories) was provided with inspection authorities that would allow entry to a facility at any reasonable time to inspect or obtain evidence for compliance purposes. Enforcement provisions were steep, giving the EPA or its representative authority to assess civil penalties of up to $25,000 per day of noncompliance with the program and to assess criminal penalties for knowing violators of the regulation, with penalties of up to $50,000 per day of each violation or imprisonment for up to five years, both penalties that could be doubled for repeat offenders. Penalties of up to $250,000 or imprisonment for up to 15 years could be levied against a person who knowingly violated the regulation which resulted in knowing endangerment, and an organization could be fined up to $1,000,000 for such an offense.

The EPA was required by the MWTA to submit two interim reports and one final report to Congress. Although the final report was due by congressional mandate in September 1991, it has not yet been released by the EPA, nor has the EPA indicated when it might be released. The EPA was also required to submit a health impacts report to Congress by November 1991, which to date has not been submitted. Nevertheless, the EPA issued an interim final rule on March 24, 1989, which has resulted in much knowledge about medical waste management.

EPA'S INTERIM MEDICAL WASTE STANDARDS

The EPA "Standards for the Tracking and Management of Medical Waste," published at 40 CFR Part 259, was published on March 24, 1989 as an interim final rule (2). Although the program was established to clean up the beaches, it soon became apparent that most of the problems of medical waste on the beaches were not due to "midnight dumping" as implied in the congressional record, but rather to solid waste and wastewater management systems that were overtaxed and underdesigned. This in combination with unusual tides and weather conditions that occurred during the summer of 1988 led to the presence of garbage, which included a small amount of medical waste, on the beaches. The preamble to the EPA rule quoted from the findings of a study conducted by the New York Department of Environmental Conservation on beach washups in 1988. This study concluded:

- With few exceptions, floatable debris cannot be traced to any specific source; Most of these wastes are likely to come from:
  - The improper transport and handling of solid waste destined for disposal at the Fresh Kills landfill.
  - Inadequate handling procedures, supervision, and maintenance at the marine transfer stations.
  - Combined sewer overflows.
  - Raw sewage discharges caused by occasional breakdowns at one or more of New York City's sewage treatment plants.
  - Storm water outlets.
- Other activities that were judged less likely to contribute include:
  - Litter deposited by beach users.

- Recreational boating.
- Commercial shipping.
- Floatables that become stranded on sandbars are sometimes refloated by the tides and washed ashore.
- Weather conditions contributed to the volume and persistence of washups.
- Illegal disposal appears to account for some of the wastes (i.e., the blood vials).

Before the project began, the EPA was skeptical about the success the program might have on medical waste beach washups. Nevertheless, the EPA recognized that the program would test a system for the safe management of medical waste and would allow for the collection of information that could be used by Congress and other state legislative bodies.

Regulated Medical Waste Defined

The universe of regulated medical waste in the EPA's tracking program is graphically described in Figure 5.1. The EPA began by broadly defining the universe of medical waste regulated under the program to include:

Medical waste means any solid waste which is generated in the diagnosis, treatment (e.g., provision of medical services), or immunization of human beings or animals, in research pertaining thereto, or in the production or testing of biologicals. The term does not include any hazardous waste identified or listed under Part 261 of this chapter or any household waste as defined in 261.4(b)(1) of this chapter.

Note to this definition: Mixtures of hazardous waste and medical waste are subject to this part except as provided in 259.31. [40 CFR 259.1]

Regulated medical waste under the program was list based (e.g., based on a list of medical waste that was "included" and a list that was "excluded" from regulation). As defined in 40 CFR 259(a) of the rule, regulated medical waste was comprised of the following list of wastes:

259.30 Definition of Regulated Medical Waste

(a) A regulated medical waste is any solid waste, defined in 259.10(a) of this part, generated in the diagnosis, treatment, (e.g., provision of medical services), or immunization of human beings or animals, in research pertaining thereto, or in the production or testing of biologicals, that is not excluded or exempted under paragraph (b) of this section, and that is listed in the following table:

Note to paragraph (A): The term "solid waste" includes solid, semisolid, or liquid materials, but does not include domestic sewage materials identified in 261.4(a)(1) of this subchapter.

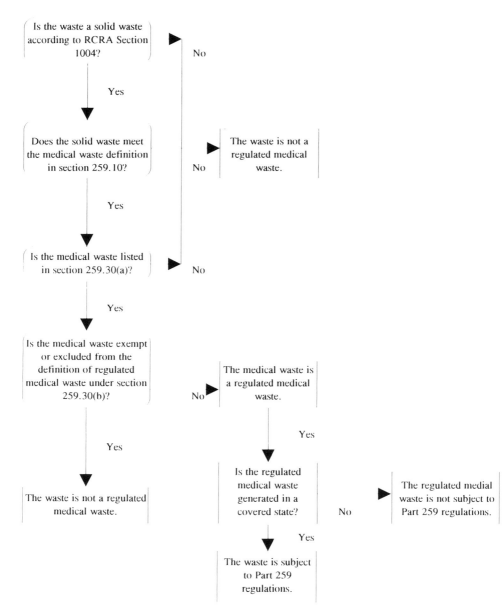

Figure 5.1   Regulated medical waste covered by Part 259. (Adapted from reference 2, page 12338.)

(1) *Cultures and Stocks.* Cultures and stocks of infectious agents and associated biologicals, including: cultures from medical and pathological laboratories; cultures and stocks of infectious agents from research and industrial laboratories; wastes from the production of biologicals; discarded live and attenuated vaccines; and culture dishes and devices used to transfer, inoculate, and mix cultures.

(2) *Pathological Wastes.* Human pathological wastes, including tissues, organs, and body parts and body fluids that are removed during surgery or autopsy, or other medical procedures, and specimens of body fluids and their containers.

(3) *Human Blood and Blood Products.* (1) Liquid waste human blood; (2) products of blood; (3) items saturated and/or dripping with human blood; or (4) items that were saturated and/or dripping with human blood that are now caked with dried human blood; including serum, plasma, and other blood components, and their containers, which were used or intended for use in either patient care, testing and laboratory analysis or the development of pharmaceuticals. Intravenous bags are also included in this category.

(4) *Sharps.* Sharps that have been used in animal or human patient care or treatment or in medical, research, or industrial laboratories, including hypodermic needles, syringes (with or without the attached needle), pasteur pipettes, scalpel blades, blood vials, needles with attached tubing, and culture dishes (regardless of presence of infectious agents). Also included are other types of broken or unbroken glassware that were in contact with infectious agents, such as used slides and cover slips.

(5) *Animal Waste.* Contaminated animal carcasses, body parts, and bedding of animals that were known to have been exposed to infectious agents during research (including research in veterinary hospitals), production of biologicals, or testing of pharmaceuticals.

(6) *Isolation Wastes.* Biological waste and discarded materials contaminated with blood, excretion, exudates, or secretions from humans who are isolated to protect others from certain highly communicable diseases, or isolated animals known to be infected with highly communicable diseases.

(7) *Unused Sharps.* The following unused, discarded sharps: hypodermic needles, suture needles, syringes, and scalpel blades.

Those wastes that were specifically excluded from regulation under the rule were identified in 40 CFR 259.3(b)(1) as follows:

(b)(1) Exclusions

    (i) Hazardous waste identified or listed under the regulations in Part 261 of this chapter is not regulated medical waste.

Note to paragraph (b)(1)(i): Mixtures of regulated medical waste and hazardous waste are subject to Part 259, except as provided in 259.31(b) of this subpart.

(ii) Household waste, as defined in 261.4(b)(1) of this chapter, is not regulated medical waste.

(iii) Ash from incineration of regulated medical waste is not regulated medical waste once the incineration process has been completed.

(iv) Residues from treatment and destruction processes are no longer regulated medical waste once the waste has been both treated and destroyed.

(v) Human corpses, remains, and anatomical parts that are intended for interment or cremation are not regulated medical waste.

Those wastes that were specifically exempted from regulation by the rule [40 CFR 259.3(b)(2)] were comprised of the following waste types:

(b)(2) Exemptions

(i) Etiologic agents being transported interstate pursuant to the requirements of the U.S. Department of Transportation, U.S. Department of Health and Human Services, and all other applicable shipping requirements are exempt from the requirements of this part.

(ii) Samples of regulated medical waste transported off-site by EPA- or State-designated enforcement personnel for enforcement purposes are exempt from the requirements of this Part during the enforcement proceeding.

The regulation of mixtures of hazardous waste with regulated medical waste were addressed in 40 CFR 259.31 as follows:

259.31 Mixtures

(a) Except as provided in paragraph (b) of this section, mixtures of solid waste and regulated medical waste listed in 259.30(a) of this subpart are a regulated medical waste.

(b) Mixtures of hazardous waste identified or listed in Part 261 of this chapter and regulated medical waste listed in 259.30(a) of this subpart are subject to the requirements in this part, unless the mixture is subject to the hazardous waste manifest requirements in Part 262 or Part 266 of this chapter.

Note to paragraph (b): Mixtures of regulated medical waste with hazardous waste that is exempt from the hazardous waste manifest requirements (e.g., under 40 CFR 261.5) remain subject to this Part.

Although upon first review, this list-based definition appears to be easily interpreted, in practice it was not. As a result, during the course of the two-year project several definitional interpretive statements were published by the EPA at the request of the participating states and the regulated community to clarify further the *regulated medical waste* definition.

Participating States and Territory

As required under EPA's hazardous waste rule, all generators of regulated medical waste (RMW) in the four covered states and one covered territory (New York, New Jersey, Connecticut, Rhode Island, and Puerto Rico) were required to identify the monthly quantity generated by each. If generating more than 50 pounds per month, the facility was subject to the full tracking and management standard. Generators of less than 50 pounds per month were not required to participate in the tracking or full transporter standards but were required to maintain records of waste generation and handling.

Medical Waste Management Requirements

For waste destined for off-site shipment to a regional treatment facility, specific segregation and packaging standards were established. The standard required that RMW be segregated into a container that is rigid, leak resistant, impervious to moisture, sufficiently strong to prevent tearing or bursting under normal conditions of use and handling, and sealed to prevent leakage during transport. Additional standards were established for sharps, waste containing fluids, and oversized RMW [259.41].

The EPA established standards for on-site storage of RMW, which required generators to

> (a) store the regulated medical waste in a manner and location that maintains the integrity of the packaging and provides protection from water, rain and wind; (b) maintain the regulated medical waste in a nonputrescent state, using refrigeration when necessary; (c) lock the outdoor storage areas containing regulated medical waste (e.g., dumpsters, sheds, tractor trailers, or other storage areas) to prevent unauthorized access; (d) limit access to on-site storage areas to authorized employees; and (e) store the regulated medical waste in a manner that affords protection from animals and does not provide a breeding place or a food source for insects and rodents [259.42].

In addition, the rule specified decontamination procedures for reusable containers [259.43], labeling requirements [259.44], and marking (identification) requirements [259.45].

Tracking Requirements. The medical waste tracking system involved a tracking form that required signatures of waste release or receipt from the generator, the destination facility, and the transporter. An adaptation of the tracking form is presented in Figure 5.2. In its most straightforward application, the generator would prepare the tracking form, which would include enough copies to provide the

State Information Block   (Name, Address, Contacts, Phone Numbers, etc.)

State Tracking Form Number (if applicable)

Medical Waste Tracking Form

| 1. Generator's Name and Mailing Address | 2. Tracking Form Number |
| | 4. State Permit or ID No. |

3. Telephone Number

| 5. Transporter's Name and Mailing Address | 6. Telephone Number ( ) |
| | 7. State Transporter Permit or ID No. |

EPA Med. Waste ID No.

| 8. Destination Facility Name and Address | 9. Telephone Number ( ) |
| | 10. State Permit or ID No. |

11. US EPA Waste Description	12. Total No. Containers	13. Total Weight or Volume
a. Regulated Medical Waste (Untreated)		
b. Regulated Medical Waste (Treated)		
c. State Regulated Medical Waste		

14. Special Handling Instructions and Additional Information

15. Generator's Certification

Under penalty of criminal and civil prosecution for the making or submission of false statements, representations, or omissions, I declare, on behalf of the generator that the contents of this consignment are fully and accurately described above and are classified, packaged, marked, and labeled in accordance with all applicable State and Federal laws and regulations, and that I have been authorized, in writing, to make such declarations by the person in charge of the generator's operation.

Printed/Typed Name   Signature   Date

INSTRUCTIONS FOR COMPLETING MEDICAL WASTE TRACKING FORM

Copy 1 - GENERATOR COPY: Mailed by Destination Facility
Copy 2 - DESTINATION FACILITY COPY: Retained by Destination Facility
Copy 3 - TRANSPORTER COPY: Retained by Transporter
Copy 4 - GENERATOR COPY: Retained by Generator

As required under 40 CFR Part 259:

1. This multi-copy (4-page) shipping document must accompany each shipment of regulated medical waste generated in a Covered State.

2. Items numbered 1-14 must be completed before the generator can sign the certification. Items 4,7,10,11c, & 19 are optional unless required by the State. Item 22 must be completed by the destination facility.

For assistance in completing this form, contact your nearest State office, Regional EPA office, or call (800) 424-9346.

16. Transporter 1 (Certification of Receipt of Medical Waste as described in items 11,12 & 13.

Printed/Typed Name   Signature   Date

| 17. Transporter 2 or Intermediate Handler (name and address) | 18. Telephone Number |
| | 19. State Transporter Permit or ID Number |

EPA Med Waste ID No. | | | | | | | | | | | | |

20. Transporter 2 or Intermediate Handler (Certification of Receipt of Medical Waste as described in items 11, 12 & 13)

Printed/Typed Name   Signature   Date

21. New Tracking Form Number (for consolidated or remanifested waste)

22. Destination Facility (Certification of Receipt of Medical Waste as described in items 11,12 & 13.

☐ Received in accordance with 11, 12 &13

Printed/Typed Name   Signature   Date

23. Discrepancy Box (Any discrepancies should be noted by item number and initials.

generator, each of the transporters (if more than one is involved), and each intermediate handler (if involved) with one copy, and the owner or operator of the destination facility with two copies. The destination facility was required to keep one copy for their records and return the second copy to the generator. In addition to the tracking requirement, the EPA established an elaborate record-keeping system to account for the regulated medical waste that was generated. The system included procedures to follow for tracking forms that were unaccounted for that involved filing "exception reports" with the participating state and the EPA. Specific record-keeping requirements were also established for those medical facilities that burned their waste in on-site incinerators.

Transporter Requirements. Much of the information generated in the course of the EPA's medical waste tracking program came from the transporters of regulated medical waste. The EPA demonstration project defined a series of specific requirements for transporters to follow that restricted waste acceptance from generators unless the waste was properly labeled, marked, and accompanied by a proper tracking form. Transporters were required to notify, as prescribed by the rule, both the EPA and the covered state to register that they are conducting such activities. In turn, the EPA issued a unique EPA Medical Waste Identification Number to transporters for each covered state in which they operated. Transporter vehicles had to meet specific physical requirements, such as having a fully enclosed, leak-resistant cargo-carrying body, and specific identification on two sides and the back. The EPA rule identified specific tracking and record-keeping requirements for transporters to follow to maintain a cradle-to-grave accounting of the RMW.

Perhaps the most important transporter responsibility in terms of overall project findings was the requirement for transporters to submit tracking reports to both the EPA and each of the covered states in which they conducted business. Information contained in these reports served as the basis for much of the information collected during the project, which included:

- Total number of generators from whom the transporter accepted RMW
- The name, address, and type of each generator
- The amount by weight and waste category (untreated or treated) of RMW accepted from each generator
- The total RMW from all generators in the Covered State that the transporter delivered to a second transporter or to a transfer facility
- The total, by weight and waste category, of RMW that the transporter delivered to an intermediate handler or to a destination facility
- The name and address of each intermediate handler and destination facility to which waste was delivered and the amount, by waste category that was delivered
- The total number of intermediate handlers and destination facilities to which waste was delivered

Treatment/Destruction/Disposal Facility Requirements. Treatment/destruction/ disposal facilities are those facilities that render RMW to either incinerator ash or residues from other treatment processes that have rendered the waste both treated and destroyed. These facilities included *destination facilities*, in which the waste is rendered treated and destroyed and no longer subject to regulation, and *intermediate handlers*, in which the waste is either only treated (e.g., steam sterilized but not shredded) or only destroyed (e.g., shredded but not sterilized) but the residue remains regulated under the standard. Both were required to meet specific tracking and recordkeeping obligations to complete the cradle-to-grave accounting of the RMW.

Final Report Status

The EPA was mandated under the Medical Waste Tracking Act of 1988 to submit two interim reports and a final report to Congress by September 1991 identifying project findings. As of February 1995, only two interim reports have been published and serve as the basis of information presented in this section. The final report and much of its supporting documentation, now several years overdue, is still awaiting release by the EPA.

REFERENCES

1. Medical Waste Tracking Act of 1988. Public Law 100–582. November 1, 1988.
2. 40 CFR Parts 22 and 259. Standards for the Tracking and Management of Medical Waste; Interim Final Rule and Request for Comments. *Federal Register*, 54(56):12326–12395. March 24, 1989.

PART III

MANAGEMENT

6

FACILITY MANAGEMENT

In this chapter we address the management of biohazardous waste when generated by medical facilities such as hospitals, laboratories, physicians' offices, dental clinics, veterinary clinics, long-term healthcare facilities, clinics, blood establishments, and funeral homes. Although biohazardous waste types and volumes vary significantly among the various potential generators, common basic principles of management can be applied. The management approach presented in this chapter is provided primarily as an overview of commonly observed practices which can be incorporated into any facility's biohazardous waste management program. Nevertheless, the reader should be aware that waste management programs are based principally on the federal, state, and local requirements established by government agencies with which the biohazardous waste generator must comply.

As a first step, the biohazardous waste management system developed by each generating facility should address those persons in the facility responsible for developing the program and ensuring that it is followed. Those responsible may consist of a waste management team comprised of key department heads (e.g., directors of safety, environmental services, engineering, infection control) for larger generators such as hospitals, or of a single person for small generators such as physicians' offices. Whatever approach is taken, those given this responsibility to develop and implement the program must also have the full backing by the facility management to carry out their duties.

Biohazardous waste management requires an understanding of all aspects of the waste stream within a given medical facility, from the point of initial generation through final treatment and disposal. Proper management would take into account mechanisms to protect the health of the healthcare provider, housekeeping and custodial staff within a facility, and waste collectors/workers and the public once the waste has left a facility for final disposal.

119

The program should address protocols to meet all applicable federal, state, and local management requirements. For a program to be effective, protocols for each element must be recorded in a written biohazardous waste management plan that is maintained at a location where it can easily be accessed by facility users and government or accreditation agencies when requested. Protocols should be established to meet all applicable federal, state, and local requirements and accreditation agency standards if applicable. When developing protocols, the health and safety of facility staff, visitors, waste industry workers, and the public should also be taken into account.

In general, the key elements to an effective biohazardous waste management program should include:

- Management plan
- Definition
- Identification
- Segregation, containment, and labeling
- Storage
- Treatment
- Transportation (on-site and off-site)
- Disposal
- Contingency planning
- Monitoring and record keeping
- Staff training

Each of these elements is discussed in the following sections.

MANAGEMENT PLAN

Central to a facility's biohazardous waste management program is the biohazardous waste management plan (1). In the plan should be documented information describing all aspects of the waste management program from the point of waste generation to final disposal. Written plans should be prepared by both larger generators such as hospitals and smaller generators such as dentists, although the plans of smaller generators may only need to be one or two pages in length. Nevertheless, the exercise of writing a plan, large or small, focuses attention on how waste is managed by a facility and documents step-by-step how waste management is to be carried out. A written plan also provides a written record should one ever become necessary.

Once written, the plan should be reviewed periodically and updated as necessary, and should be made readily available to all employees involved in waste management activities. The plan should be written in a format that is easily understood, maintained as a working document by the facility, and be available for inspection by government regulators and the public if requested. The plan should address:

- Compliance with applicable regulations
- Responsibilities of all involved staff (e.g., infection control members, environmental control and housekeeping personnel, department and individual responsibilities)
- Definition of all biohazardous wastes handled within a facility
- Procedures for biohazardous waste management, including:
  - Identification
  - Segregation
  - Containment
  - Labeling
  - Storage
  - Treatment
  - Transport
  - Disposal
  - Monitoring and record keeping
  - Contingency planning
- Training for all involved staff (professional, medical, environmental services, housekeeping, etc.)

Planning Under Federal OSHA

Under the Occupational Safety and Health Administration's (OSHA) bloodborne pathogen standard, each employer with occupational exposure is required to establish a written Exposure Control Plan [CFR 1910.1030(c)(1)]. The plan must document the exposure determination for employees occupationally exposed to blood or other potentially infectious materials. For personnel involved with waste management activities, this would require identifying the frequency of exposure and the tasks and procedures in which exposure may occur. The plan must also address methods of compliance, such as engineering controls and personal protective equipment, hepatitis B vaccination and postexposure follow-up, communication of hazards to employees, recordkeeping, and procedures for evaluating exposure incidents. Under the OSHA rule, employers are required to keep the plan in a location accessible to employees and OSHA inspectors. The plan must be reviewed and updated at least annually and whenever necessary to reflect new or modified tasks and procedures or revised employee positions within the facility.

DEFINITION

This element requires knowledge of the specific definition for the wastes within a facility to be categorized as *biohazardous wastes*. Because there is no national standard defining what wastes comprise the biohazardous waste stream, each facility's waste management team must develop its in-house definition based on definitions established in all applicable federal, state, and local biohazardous waste–related regulations and based on the need to protect health and safety. Regulatory definitional inconsistencies are common, and the waste management team will be expected to deal with these. Inconsistencies can often be resolved through discussions (which should always be documented in writing) with the regulatory agencies, although a facility may ultimately have to establish an expanded in-house definition that includes all of the inconsistencies established by the various governmental bodies to ensure compliance with each.

Biohazardous Waste Characterization

When biohazardous waste streams are being defined by either governmental regulatory agencies or facility operators, such definitions should be made recognizing the elements necessary for infectious disease transmission to occur and the fact that infectious agents are commonly found in the general waste stream whether the waste originates from residential or medical facility sources. The relative risk of disease causation from human pathogens is based on the type and concentration of pathogen, the susceptibility of the host, a method of transmitting the organism to the host, an infective dose of the pathogen, and a portal of entry into the host.

There is no uniformly accepted definition of what constitutes a biohazardous waste (1). Furthermore, biohazardous waste has never been linked epidemiologically to infectious disease transmission in the community and has been linked to only one case of disease transmission within a medical facility that involved a needle-stick injury by housekeeping staff (2). As a result, accurately defining waste streams due to their potential infectivity is difficult and often leads to inconsistent definitions between regulatory authorities.

In its infectious waste guideline, the EPA defined biohazardous waste as waste capable of producing an infectious disease. However, based on the extreme variability associated with biohazardous disease causation, no method exists to determine the "infectivity" that a waste may possess. Therefore, biohazardous waste definitions rely on judgments that should be (but not always are) based on certain high-risk factors such as organism concentration of the waste, the ability for the waste to penetrate the skin, and other considerations.

The Agency for Toxic Substances and Disease Registry was required under the Medical Waste Tracking Act of 1988 (MWTA) to analyze risk factors associated with various categories of biohazardous waste. A discussion of the ATSDR findings is presented in Chapter 2. In addition to that discussion, the following is presented based on infectious waste categories identified by the EPA in 40 CFR Part 259 and on additional miscellaneous categories of medical waste.

Category 1: Cultures and Stocks. Microbiology cultures and stocks and associated biologicals may include:

> Cultures and stocks of infectious agents and associated biologicals, including: cultures from medical and pathological laboratories; cultures and stocks of infectious agents

from research and industrial laboratories; wastes from the production of biologicals; discarded live and attenuated vaccines; and culture dishes and devices used to transfer, inoculate, and mix cultures. [40 CFR 259(a)(1)]

Discussion. Laboratories grow microbiological agents on artificial media for purposes of propagation (e.g., vaccine production), identification (e.g., clinical analysis) and research (e.g., antibiotic testing). Human biohazardous disease-causing agents (etiologic agents) must be grown in high concentrations to achieve these purposes. For this reason, laboratory workers must use extreme caution to avoid laboratory-acquired infections. Nevertheless, occupational infection among laboratory workers is well documented from inadvertent exposure to these pathogenic agents, although infections have not been attributed to contact with waste per se (3–11). The Environmental Protection Agency (1) and the Centers for Disease Control (12) have recommended in their guidelines that these wastes be rendered noninfectious prior to disposal into the general waste stream.

Category 2: Pathological Wastes. Pathological waste may include:

Human pathological wastes, including tissues, organs, and body parts and body fluids that are removed during surgery or autopsy, or other medical procedures, and specimens of body fluids and their containers. [40 CFR 259(a)(2)]

Discussion. Pathological waste has not been implicated in disease transmission in the scientific literature. Nevertheless, both the EPA (1) and the CDC (12) designate pathological waste as having potentially infective qualities and recommend special treatment of this waste category prior to disposal. Improper disposal of pathological waste is also aesthetically unacceptable. It is often recommended that human pathological waste be destroyed by incineration, cremation, or internment.

Category 3: Waste Human Blood and Blood Components. Waste human blood and blood components may include:

(1) Liquid waste human blood; (2) products of blood; (3) items saturated and/or dripping with human blood; or (4) items that were saturated and/or dripping with human blood that are now caked with dried human blood; including serum, plasma, and other blood components, and their containers, which were used or intended for use in either patient care, testing and laboratory analysis or the development of pharmaceuticals. Intravenous bags are also included in this category. [40 CFR 259(a)(3)]

Discussion. Blood and blood components are recognized by the CDC as posing a threat of disease spread due to the potential presence of hepatitis B virus and human immunodeficiency virus. Certain body fluids (cerebrospinal fluid, synovial fluid, pleural fluid, peritoneal fluid, pericardial fluid, and amniotic fluid) are also associated with such hazards (13). Infectious disease transmission has been implicated among healthcare workers due to inadvertent contact with human blood through cuts or mucous membranes (e.g., eyes, nose) (14), although never in association with the waste stream. Disease transmission risk resulting from contact with human blood or body fluids may be increased when these body substances are in liquid form and capable of splashing in the eyes or other mucous membranes. Therefore, it is prudent to dispose of liquid blood and body fluids carefully via the sanitary sewer, a practice recommended by both the CDC and EPA (12,15). Risk due to handling and potential spills increases when liquid body substances are bottled and transported off-site for special treatment. Liquid body substances should never be discharged directly to the general waste stream. Although the potential for disease transmission from high quantities of dried blood caked on gauze or dressings is less than when in liquid form, many regulatory agencies include caked blood in their definition of biohazardous waste.

Category 4: Sharps. Sharps waste may include:

> Sharps that have been used in animal or human patient care or treatment or in medical, research, or industrial laboratories, including hypodermic needles, syringes (with or without the attached needle), pasteur pipettes, scalpel blades, blood vials, needles with attached tubing, and culture dishes (regardless of presence of infectious agents). Also included are other types of broken or unbroken glassware that were in contact with infectious agents, such as used slides and cover slips. [40 CFR 259(a)(4)]

In its interim rule, EPA also included a category for unused sharps (category 7) which includes "unused, discarded sharps, hypodermic needles, suture needles, syringes, and scalpel blades."

Discussion. When improperly handled or disposed of, sharps present a safety and potential disease transmission hazard because of their ability to create a portal of entry through the skin. The role of infectious disease causation from needle-stick injuries during the course of medical care administration has been well documented in the medical literature (14,16–26), although only one case of disease transmission has been identified that involved contact with a hypodermic needle associated with waste (2). Injuries have also been observed among nonmedical occupational groups. A survey of waste collectors in Washington state observed 10 percent of respondents reporting having sustained a needle-stick injury in the year preceding the survey, although none was apparently associated with infection (27).

As reported by ATSDR, improperly disposed sharps waste does present an injury hazard and a potential infectious disease transmission hazard. Therefore, proper packaging, treatment, and disposal methods should be identified to prevent physical injury and potential biohazardous disease transmission from improperly disposed sharps waste. It may also be prudent to include unused sharps within the biohazardous waste definition because they still present an injury hazard, and should an injury occur, it would not be known whether the sharp had or had not been used.

Category 5: Animal Waste. Animal waste may include:

> Contaminated animal carcasses, body parts, and bedding of animals that were known to have been exposed to infectious agents during research (including research in

veterinary hospitals), production of biologicals, or testing of pharmaceuticals. [40 CFR 259(a)(5)]

Discussion. Animal waste has not been implicated in disease transmission in the scientific literature, nor does it present portal-of-entry hazards that are greater than would be found in the general waste stream. Nevertheless, this waste by definition is the result of artificially inoculating animals with infectious agents and warrants special disposal considerations.

Category 6: Isolation Waste. Historically, isolation waste has been viewed as waste generated by hospitalized patients who are isolated to protect others from communicable diseases (1). The EPA, in its interim demonstration tracking program, modified the common definition of *isolation waste* by including only those wastes generated from patients known to be infected with certain highly communicable diseases. These were defined as class 4 agents in the Centers for Disease Control's publication, "The Classification of Etiologic Agents on the Basis of Hazard," which included such diseases as Crimean hemorrhagic fever, Venezuelan equine encephalitis virus, and the Machupo, Lassa, and Marburg viruses (28). The EPA's demonstration program definition is as follows:

> Biological waste and discarded materials contaminated with blood, excretion, exudates, or secretions from humans who are isolated to protect others from certain highly communicable diseases, or isolated animals known to be infected with highly communicable diseases. [40 CFR 259(a)(6)]

Discussion. In general, isolation wastes are described as those wastes that are generated in the course of treating a patient on isolation precautions; a system specifically developed by the CDC to protect healthcare workers and patients from potential infection risks (29). A weakness with this system is the need for an infectious disease diagnosis, which is often difficult to determine. People with specific infections may shed the causative microorganism whether diagnosed or not, or whether at home or in a medical facility. Many people never exhibit signs of overt infectious disease, yet may be chronic carriers of an infectious disease agent. For these reasons, various systems have been designed to protect healthcare professionals by treating every patient as a potential source of infection. The CDC, in its system of *universal precautions*, which applies directly to contact with blood and certain body fluids, also cautions that the system was developed specifically for healthcare worker protection and should not be applied to waste disposal. Furthermore, there is no evidence to suggest that general isolation waste poses any more infection hazard than that of general waste. However, it may be prudent to provide special handling and disposal for wastes that have come into contact with patients or animals that are known to be infected with certain highly communicable disease agents. Such decisions should be made in consultation with the medical facility's infection control staff.

Miscellaneous Waste Categories. Miscellaneous wastes may include wastes from surgery or patient care, such as soiled dressings, sponges, drapes, drainage sets, or underpads. Other miscellaneous wastes include dialysis unit wastes that were in contact with the blood of patients undergoing hemodialysis treatment. In its 1986 guideline, the EPA has categorized these wastes under an "optional" biohazardous waste category, recommending that the final decision on whether to include such waste within the defined biohazardous waste category be made by the medical facility infection control decision makers (1). The CDC does not view these wastes as presenting an increased infection hazard compared to waste found in the general solid waste stream (12), and as a result they are usually, but not always, excluded from regulatory definitions.

Federal Definitions

Table 6.1 summarizes how various federal agencies define biohazardous waste in both regulation and guideline. Specific definitions for each are presented in Chapters 3 to 5.

IDENTIFICATION

Once the biohazardous waste stream has been defined, the next step is to conduct an audit of the waste stream. Before a management strategy can be developed, the biohazardous waste team must identify:

- Location of waste generated
- Types and amounts (volume or weight) of waste generated

Lists identifying types and amounts (either in volume or weight, depending on how disposal costs are based) of wastes generated should be developed for each location where biohazardous waste is generated. By documenting location of waste generation, types, and amounts generated, informed management decisions can be made.

SEGREGATION, CONTAINMENT, AND LABELING

In general, biohazardous waste should be segregated from the general waste stream at the point of origin by the generator of the waste into clearly marked containers that take into consideration the waste type (e.g., liquid wastes, nonsharp/nonliquid wastes, sharp wastes). To contain costs, facility staff should be trained to segregate only that which has been specifically defined as biohazardous waste by the facility. Casual disposal of nonbiohazardous waste materials can dramatically increase a facility's waste disposal costs.

Nonsharp biohazardous wastes (solid/semisolid wastes) should be segregated into disposable leakproof containers or plastic bags that meet specific performance standards. A suggested performance standard would be the ASTM "Standard Test

TABLE 6.1 Comparison of Biohazardous Waste Definitions: Federal Regulations and Guidelines

	OSHA Regulation	PHS Regulation	USPS Regulation	Proposed DOT Regulation	EPA Guideline	CDC Guideline	EPA Tracking Program
Cultures and stocks		Yes[a]		Yes[b]	Yes	Yes	Yes
Pathological waste	Yes	Yes[a]		Yes[c]	Yes	Yes	Yes
Blood and blood products	Yes	Yes[a]		Yes[c]	Yes	Yes	Yes
Contaminated sharps	Yes	Yes[a]	Yes	Yes[c]	Yes	Yes	Yes
Contaminated animal waste		Yes[a]		Yes[c]	Yes		Yes
Isolation waste		Yes[a]		Yes[c]	Yes		Yes
Other			Yes[e]				Yes[d]

[a] Any material known or reasonably believed to contain an etiologic agent listed in 42 CFR 72.3.
[b] Regulated as an infectious substance.
[c] Any medical waste other than cultures and stocks containing an etiologic agent.
[d] Only wastes involving certain highly communicable infectious agents.
[e] Any device used in medical practice that does not contain a sharp.

Methods for Impact Resistance of Plastic Film by the Free Falling Dart Method" (30), Designation D1709-91, which has been used in California's biohazardous waste rule. The bags should also be constructed so as to preclude ripping, tearing, or bursting under normal use. These bags should be tagged or effectively marked by the generator as containing biohazardous waste. The bags should be secured to prevent expulsion of contents during handling, storage, or transport. Liquid biohazardous wastes should be prevented from entering these bags.

Sharps waste (e.g., hypodermic needles) should be contained in rigid, leakproof, puncture-resistant, break-resistant containers that can be tightly lidded during storage, handling, or transport. The most suitable container material would be plastic.

ECRI, a nonprofit health services research agency, recently published its evaluation and summary of 21 plastic and 2 high density cardboard disposable sharps containers from 12 manufacturers (31,32). The ECRI rating system was based on "a sharps container's ability to safely facility the collection of medical waste sharps and to retain them under all circumstances of use and handling with minimal training or need for special procedures on the part of those who use or handle them." Each of the study sharps containers was evaluated for (1) overfill warning (contents visibility, fill line); (2) impact resistance; (3) puncture resistance at room temperature; (4) inlet characteristics (finger access, ease of inserting sharps); (5) closure characteristics (closing safety, ease of closing, protection against reopening); (6) leak resistance; (7) freestanding ability; (8) handling; (9) identification (color—base/top, labeling; (10) construction quality; (11) assembly; (12) instructions; and (13) capacity (volume to fill line in liter and in percent of manufacturer-specified volume).

All evaluated units, which ECRI noted represented only a small part of most manufacturer's product line, were rated "Conditionally Acceptable" due to "various shortcomings that might expose people to injury during use or handling." An assessment check list for sharps containers was provided for use by hospital staff responsible for purchasing sharps containers.

Liquid biohazardous wastes should be segregated into leakproof containers that are capable of transporting the waste without spillage. When storing, handling, or transporting wastes that have been packaged as described, additional protection should be provided to the original containers (plastic bags, sharps containers, liquid containers) by placing them into other durable containers, such as disposable or reusable pails, cartons, boxes, drums, or portable bins.

If containers are to be reused for biohazardous waste storage, handling, or transport, they should be thoroughly washed and decontaminated by an approved method each time they are emptied unless the surfaces of the containers had been protected by disposable liners, plastic bags, or other means. The process of cleaning should include agitation (scrubbing) to remove any visible solid residue, followed by disinfection. Disinfection could be accomplished using chemical disinfectants. A minimum standard should require that the disinfectant be used in accordance with manufacturer's recommendations for tuberculocidal and viricidal killing capacities or by a comparable process. These containers should not be used for any other purpose unless they have been properly disinfected and have had biohazardous waste symbols and labels removed.

Specific segregation, containment, and labeling requirements are specified in rules published by OSHA for occupational waste handling, by PHS when wastes are transported over the highways, and by the USPS when wastes are mailed through the USPS mail system. Regulations have also been proposed by the DOT. These requirements are described in Chapter 3. Additionally, state and local rules must be consulted to ensure compliance with all requirements.

STORAGE

Facilities that store biohazardous wastes should have a specific storage area for that purpose. The storage area should be inaccessible to unauthorized entry. The area should offer protection from animals, the elements (e.g., rain and wind), and should not provide a breeding place or a food source for insects or rodents. Storage time and temperature should be considered due to putrefaction of the waste with time. Microorganisms will grow and decompose the waste in storage, creating the unpleasant odors associated with putrefaction or rotting garbage.

TREATMENT

Biohazardous waste treatment options are described in detail in Chapters 8 to 11. In general, treatment options are available either on- or off-site, depending on the facility's needs and budget. Options include incineration, steam sterilization, and various alternative treatment technologies, including microwave irradiation, electrothermal deactivation, chemical treatment, or ionizing radiation.

TRANSPORTATION

Transport of biohazardous waste must be considered as the waste moves through the facility to storage areas, and if the waste is to be transported off-site to a treatment/disposal facility. Carts used to transport waste within a facility from the point of generation to the storage site should be used only for that purpose and not for other purposes (e.g., food carts and miscellaneous equipment transfers). Carts should be cleaned and disinfected routinely.

Off-site transportation must meet requirements established by the Public Health Service, the Department of Transportation (see Chapter 3), and all state and local transportation requirements relating to inter- and intrastate transport of biohazardous waste. In general, the waste destined for off-site transport should be transported only in leakproof and fully enclosed containers or vehicle compartments. Biohazardous waste should not be transported in the same vehicle with other waste or medical specimens unless separately contained. Biohazardous waste spills should be decontaminated promptly. The waste should only be transported to a treatment facility that meets all local, state, and federal environmental regulations.

CONTINGENCY PLANNING

Generators of biohazardous waste must be prepared for unexpected events such as equipment failure, spills, exposures, or any other event that would interrupt the day-to-day management of the waste stream. Unexpected events must be anticipated and procedures developed to address them should it ever become necessary. Procedures to follow should be documented and maintained in the facility's biohazardous waste management plan and updated periodically as necessary (33).

Equipment Failure

Biohazardous waste generators should have an alternative treatment plan in place when the primary treatment option becomes unavailable due either to a malfunction or routine equipment maintenance. If a facility relies on an in-house waste treatment technology that is temporarily out of service, another option must be identified and be available immediately. Alternative arrangements should include how the wastes are to be stored and transported to an alternative treatment location. It is a common practice for medical facilities to contract with off-site treatment providers to cover this contingency need.

Spills

Biohazardous waste generators must be prepared to decontaminate and clean up a biohazardous waste spill in their facility. Exposures can occur during the spill event or during the cleanup process through direct contact with the spilled material or through inhalation or skin contact with airborne particles and aerosols. To avoid further chance of human exposure, spills must be reacted to quickly, effectively, and in an organized manner.

Spills of biohazardous materials are quite common in medical institutions, and procedures to follow during a spill emergency should be identified and documented in the facility's management plan. Most spills involving a biohazard require only that the contaminated area be cleaned and properly disinfected. However, there can be a wide difference in the severity of spills between hospitals and microbiological research laboratories, for instance, due to the type, concentration, and virulence of infectious agents that may be involved.

Most spills can be decontaminated using an inorganic chlorine solution of 0.5 percent free chlorine, which represents a 1:10 dilution of common household bleach. Solutions of 0.5 percent free chlorine have been shown to exhibit sporicidal activity, are tuberculocidal, inactivate vegetative bacteria and are fungicidal and virucidal (34). The infection control staff or an infection control consultant should be contacted to determine the most suitable disinfectant, concentration, and contact time to use.

Evacuation would not be necessary for most spills that may commonly occur in a medical facility, although the need to secure the spill area from access is important to preclude human exposures. Spills in research facilities that may involve highly communicable disease agents, for example, would require more stringent cleanup procedures to prevent exposure of cleanup personnel and other laboratory workers. In these instances, evacuation may be mandatory. Next, we provide information about what should be included in a spill containment cleanup kit, steps that should be taken during a cleanup operation, and what to do should an exposure occur.[1,2]

Spill Containment and Cleanup Kit. Both large-scale and small-quantity bio-hazardous waste generators should maintain a spill containment and cleanup kit in a location accessible to where the wastes are being managed (33,35). Kits should consist of at least the following items:

Liquid Absorbent Material. Enough absorbant should be on hand to absorb 1 gallon of liquid for every cubic foot of biohazardous waste that is normally managed within the vicinity of the spill kit, or 10 gallons, whichever is less.

Disinfectant. The kit should contain at least 1 gallon of hospital-grade disinfectant which is effective against mycobacteria. The disinfectant should be contained in a sprayer that can dispense its contents in either a mist or a stream. The dispenser should be capable of discharging at a sufficient distance to protect the worker from direct exposure with the spill or exposure due to splashing from the sprayer.

Containment Supplies. A sufficient quantity of durable red plastic bags should be available to double enclose 150 percent of the maximum accumulated waste. The bags should be accompanied by sealing tape or devices, and warning labels or tags. Bags should be available to overpack containers and boxes normally used for biohazardous waste containment by the facility. The kit should also contain forceps and a dustpan and brush for solids pickup.

Safety Equipment. The kit should be equipped with safety equipment for at least two workers. Included should be liquid-impermeable and disposable overalls, boots or shoe coverings, gloves, goggles, tape for sealing ankles and wrists, cap and protective breathing devices. Boots should be of thick rubber and gloves of neo-prene or other appropriate polymer. A first-aid kit and boundary marking tape should also be included with the safety equipment package.

[1]Adapted from the Council of State Governments Center for Environment. Model Guidelines for State Medical Waste Management, pp. 13–14, copyright 1992 by the Council of State Governments, Lexington, Kentucky. Reprinted by permission.
[2]Adapted from Reinhardt PA and Gordon JG. *Infectious and Medical Waste Management*, p. 216, copyright 1991 by Lewis Publishers, Inc., an imprint of CRC Press, Boca Raton, Florida. Reprinted by permission.

Containment and Cleanup Procedures. During the containment and cleanup of a biohazardous waste spill, the following steps should be observed (33,35):

1. Leave the area immediately to prevent human exposure and to allow airborne particles and aerosols to settle.
2. Determine if an exposure occurred. If so, begin immediate medical action as outlined in the management plan.
3. Identify what has been spilled and the nature and extent of the spill.
4. Restrict access to the spill area.
5. Don the cleanup outfits (overalls, boots or shoe coverings, gloves, cap; if necessary, goggles or respirator).
6. Spray the broken biohazardous waste containers and materials with the disinfectant.
7. Remove the spilled material. Contact with the spilled waste should be minimized.
   a. *Solid Wastes.* Solids should be gathered using forceps or pan and brush and placed into labeled disposal containers and disposed of as biohazardous waste. Nonsharp solids should be placed into labeled plastic bags or other suitable disposal container. Broken glass should be handled, contained, and disposed of as sharps waste;
   b. *Liquid Wastes.* Liquid absorbant material should be used to remove spilled liquids, which should then be treated and disposed of as biohazardous waste;
8. Disinfect, rinse, and clean the area, absorbing after each step. Take additional cleanup steps determined to be necessary;
9. Dispose of collected spill material as biohazardous waste.
10. Remove personal protective equipment, discarding reusable items as biohazardous waste and disinfecting reusable items.
11. Thoroughly wash hands and all exposed skin.
12. Replace all items used in the spill containment kit, restoring the kit to readiness.

As the containment and cleanup procedures are being established by a facility, all applicable federal, state, and local regulations must be understood and incorporated into the protocol to ensure regulatory compliance. For instance, worker exposure protection requirements found in OSHA's bloodborne pathogen regulation (29 CFR 1910.1030) must be observed.

Exposure Incidents

If a worker is exposed to an infectious agent during a spill event or during the cleanup of a spill, procedures should be followed to ensure that the person receives the essential healthcare necessary to minimize complications due to the exposure.

Developing strategies upfront rather than in response to a crisis provides for a much more directed approach to spill exposures. Postexposure follow-up requirements are mandated under OSHA's bloodborne pathogen requirements [CFR 1910.1030 (f)].

RECORD KEEPING

Accurate record keeping provides an essential history of a facility's waste management practices. Waste management records represent a document of practices that can be used by the facility to make informed waste management decisions. Records are essential for demonstrating compliance with environmental and public health requirements. Accurate records are also essential in terms of potential liability protection should it ever become necessary. To see the importance of this, one only has to look at the plight of many waste generators of two decades ago who are now faced with Superfund cleanup cost settlements under the Comprehensive Environmental Response, Compensation and Liability Act of 1980, due to past largely undocumented waste management practices.

STAFF TRAINING

The biohazardous waste management plan should address specific training and educational needs for professional staff and housekeeping/custodial staff. Training should include (33):

1. An explanation of the waste management plan
2. Assignment of roles, responsibilities, and expectations
3. Risks associated with the waste management work environment
4. The location and proper use of personal protective equipment
5. Components of the waste management system (waste identification, segregation, containerization, labeling, transport, treatment, and disposal)
6. Regulations and the consequences of failing to comply (regulatory enforcement consequences)
7. Procedures to follow should a needle stick or other exposure occur

Training should be conducted following development and implementation of the management plan, when new employees are hired, whenever management practices change, and as a periodic refresher.

BIOHAZARDOUS WASTE MINIMIZATION

Minimizing biohazardous waste generated by a medical facility, waste that requires costly handling, treatment, and disposal, is an important cost-containment step. For example, biohazardous waste haulage is at least five times more expensive than general waste haulage (36). This discussion is limited to minimizing only the biohazardous waste component of all waste generated by a medical facility. Techniques for minimizing biohazardous waste include (1) proper segregation, (2) waste reduction, (3) product reuse, and (4) recycling.

Segregation

The primary mechanism for reducing the volume of biohazardous waste generated by a medical facility involves proper waste segregation. For example, in an effort to reduce the volume of regulated medical waste (biohazardous waste) in its facility after the implementation of EPA's Medical Waste Tracking Program, the Beth Israel Medical Center located on Manhattan's Lower East Side implemented a program of "regulation interpretation, education, training, placement of appropriate waste receptacles, and monitoring" that resulted in a significant decrease in the volumes of biohazardous waste generated by the facility (36). In 1990 the medical center generated 2,294,000 pounds of regulated medical waste, with an annual cost of $858,550 for disposal. By 1992, this volume had been reduced to 1,201,600 pounds with a total disposal cost of $311,087 as a result of implementing its biohazardous waste reduction program. Figure 6.1 presents regulated waste generation rates and disposal costs for Beth Israel Medical Center from January 1989, before implementation of the Medical Waste Tracking Act rules, and April 1993.

A similar program of "aggressive waste segregation and behavior modification" was conducted by the Mount Sinai Medical Center in New York City in response to skyrocketing biohazardous waste disposal costs resulting from increased regulatory policies and the Medical Waste Tracking Act of 1988 (37). During this period the facility observed a red bag waste reduction of from 12,000 pounds per day, which occurred immediately after implementation of the Medical Waste Tracking Act rules, to 4000 pounds per day one year later. This translated to a disposal cost reduction of from $110,000 per month to $40,000 per month.

The percentage of a medical facility's waste that is designated as "biohazardous" is largely determined by how the waste stream is defined by prevailing regulations. Because of the high costs associated with biohazardous waste handling, treatment, and disposal, it is important to ensure that staff involved with segregating biohazardous waste from general waste have been adequately educated and trained to make this determination. Far too often, waste audits in hospitals have revealed general wastes such as soda cans or office wastes being disposed into biohazardous waste red bags. Such practices result in significant but completely unnecessary waste disposal cost increases to the facility.

Techniques to improve segregation efficiency begin with the waste audit. Audits may be conducted by observing staff waste segregation practices and by examining contents of biohazardous waste red bags and containers. One purpose of the audit is to identify the consistency by staff of segregating only wastes that meet the biohazardous waste definition, and excluding all other nonbiohazardous waste types. Another purpose is to identify the locations of red bags that may be used for general waste disposal.

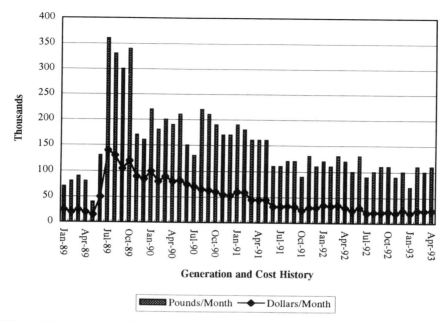

Figure 6.1 Beth Israel Medical Center regulated medical waste. From Brown J. Hospital waste management that saves money—and helps the environment and improves safety. *Medical Waste: Regulatory Analyst*, 1(10):4, copyright 1993 by Technomic Publishing Company, Inc., Lancaster, Pennsylvania. Reprinted by permission.

The audit should include a survey of biohazardous waste disposal container locations. Clearly marked biohazardous waste disposal containers should be conveniently located within all areas in which biohazardous waste is generated. But experience has found that improperly located red bags will result in their improper and costly use for general waste disposal (36,37). Red bags in patient rooms should either be placed in locations to minimize use by visitors for general waste disposal or removed from patient rooms altogether. A clear bag (general) waste container should be located next to all red bag containers for the biohazardous waste segregation program to be successful (36). In addition, each facility should ensure that enough clear bags have been ordered and that they are conveniently stocked in all supply closets so that red bags wont be substituted by housekeeping staff for general waste collection (36).

Each facility should develop a plan for biohazardous waste segregation and ensure that all staff involved with waste segregation have received adequate training and education and are committed to following the plan. The Mount Sinai Medical Center experience noted that the single most important factor in achieving success in reducing the volume of red bag waste is cooperation by the nursing department (37). For this reason, the bulk of early educational efforts focused on nursing.

Reduction

Once proper segregation techniques have been implemented, the biohazardous waste stream can be reduced further by using less of a product that will require special handling as biohazardous waste when disposed. In a 1991 report by the Minnesota Hospital Association entitled "The MHA Recycling and Conservation Guide," one hospital estimated that up to 10 tons of plastic per year would be removed from the biohazardous waste stream by requiring that biohazardous waste be only single bagged instead of double or triple bagged as had been the previous policy (38). Another hospital identified in the report estimated a plastic disposal cost savings of $7740 through an 86-ton reduction in plastic waste by converting to lighter, stronger plastic bags for containing infectious wastes. That hospital also projected a plastic purchase cost savings of $90,000 a year by changing plastic bags. "It is extremely important that a hospital make clear to the vendor its specifications for durable, lightweight bags, and make sure the bags provide protection." As stated by a spokesperson for the hospital, "We had our vendor perform the ASTM-D1709 dart test on the bags to test durability, and we weighed the cases they were shipped to us in to make sure they met our specifications." The Beth Israel Medical Center switched from disposable to reusable sharps containers throughout its facility (36) by contracting with a company offering a full-service sharps collection system. Since the switch, that medical center reports that the incidence of overfilled containers has dropped significantly, needle-sticks associated with disposal are rare, and the facility has eliminated the need to incinerate 2700 disposable containers each month at an off-site location.

Product Reuse

Another biohazardous waste reduction technique involves replacing disposable with reusable materials when feasible. During the past two decades, disposable products have come into wide use by medical institutions due to their infection control benefits. However, in many cases the benefits of reusable over disposable products are being reexamined in part due to mounting costs of waste disposal. For example, the Methodist Hospital in St. Louis Park, Minnesota switched to a reusable, cost-effective isolation gown that meets OSHA requirements to reduce the volume of biohazardous waste requiring incineration by the facility (38).

Decisions regarding the return to reusable medical items should be made on a case-by-case basis, addressing potential effects on staff safety and patient care. In its 1990 Report to Congress, the Office of Technology Assessment states (39): "The use (and reuse) of disposables can be considered on an item by item basis, in light of how they will be used, including consideration of infection risks and other factors associated with those risks."

Recycling

Recycling involves the diversion of a used product to a future use rather than to disposal. Options for recycling untreated biohazardous waste are extremely limited due to the infectious nature and inherent hazards associated with this waste stream. However, options for recycling biohazardous waste that has been treated and rendered noninfectious are available, although from a technical standpoint, treated biohazardous waste would no longer be considered part of the biohazardous waste stream. For example, plastics or metals that have been rendered noninfectious may be removed for future processing. Plastics recycling is being conducted commercially by Stericycle, Inc. of Deerfield, Illinois, a company that offers a regional biohazardous waste treatment service based on an electrothermal deactivation treatment process (see Chapter 10). As part of the process, treated plastics are diverted for recycling into medical sharps containers and other products by the Sage Company. Other plastics are used for their high Btu (British thermal unit) value as refuse derived fuel, which is packaged into bales and burned as fuel in cement kilns.

REFERENCES

1. U.S. Environmental Protection Agency. EPA Guide for Infectious Waste Management. EPA/530-SW-86-014. May 1986.
2. U.S. Department of Health and Human Services, Agency for Toxic Substances and Disease Registry, Public Health Service. The Public Health Implication of Medical Waste: A Report to Congress. 1990.
3. Barkley WE, Wedum AG, and McKinney RW. The hazard of infectious agents in microbiological laboratories. In: Block SS, Ed., *Disinfection, Sterilization and Preservation*, 3rd ed., pp. 566–576. Lea & Febiger, Philadelphia. 1983.
4. Dimmick RL, Vogl WF, and Chatigny MA. Potential for accidental microbial aerosol transmission in the biological laboratory. In: Hellman A, Oxman MN, and Pollack R, Eds., *Biohazards in Biological Research*, pp. 246–266. Cold Spring Harbor Laboratories, Cold Spring Harbor, New York. 1973.
5. Pike RM. Laboratory-associated infections: Summary and analysis of 3921 cases. *Health Lab Sci* 13(2):105–114. 1976.
6. Pike RM. Laboratory-associated infections: Incidence, fatalities, causes, and prevention. *Annu Rev Microbiol*, 33:41–66. 1979.
7. Sulkin SE. Laboratory-acquired infections. *Bacteriol Rev*, 25:203–211. 1961.
8. Skinhoj P. Occupational risks in Danish clinical chemical laboratories. *Scand J Clin Lab Invest*, 33:27–29. 1974.
9. Harrington JM and Shannon HS. Incidence of tuberculosis, hepatitis, brucellosis and shigellosis in British laboratory workers. *Br Med J*, 1:759–762. 1976.
10. Centers for Disease Control. Occupationally acquired human immunodeficiency virus infections in laboratories producing virus concentrates in large quantities. *MMWR*, 37(S-4):19–22. April 1, 1988.
11. Centers for Disease Control. Acquired immune deficiency syndrome (AIDS): Precautions for clinical and laboratory staffs. *MMWR*, 31(43):577–580. November 5, 1982.
12. Garner JS and Favero MS. Guideline for Handwashing and Hospital Environmental Control, 1985. HHS Publication 99-1117. Public Health Service, Centers for Disease Control, Atlanta, Georgia. 1985.
13. Centers for Disease Control. Update: Universal precautions for prevention of transmis-

sion of human immunodeficiency virus, hepatitis B virus, and other bloodborne pathogens in healthcare settings. *MMWR*, 37(24):377. June 24, 1988.

14. Centers for Disease Control Update. Human immunodeficiency virus infections in healthcare workers exposed to blood of infected patients. *MMWR*, 36(19):285–289. May 22, 1987.

15. Allbee S. Update on Discharge of AIDS Contaminated Waste to Municipal Wastewater Treatment Facilities. USEPA Memorandum, February 11, 1988.

16. Bock KB, Tong MJ, and Bernstein S. The risk of accidental exposure to hepatitis B virus via blood contamination in medical students. *J Infect Dis*, 144:604. 1981.

17. Adams B. Sharps management. In: *A Compendium of the Proper Management of Toxic and Hazardous Materials in Health Care Facilities*, pp. 239–248. North Carolina State University, Raleigh, North Carolina. April 1987.

18. Crow S and Weinstein SA (Eds.). Disposable needle and syringe containers. *Infect Control*, 6(1):41–42. 1985.

19. Krasinski K, LaCouture R, and Holzman RS. Effect of changing needle disposal systems on needle puncture injuries. *Infect Control*, 8(2):59–62. 1987.

20. McCormick RD and Maki DG. Epidemiology of needle stick injuries in hospital personnel. *Am J Med*, 70:928–932. 1981.

21. Ribner BS, Landry MN, Gholson GL and Linden LA. Impact of a rigid, puncture resistant container system upon needlestick injuries. *Infect Control*, 8(2):63–66. 1987.

22. Jagger F, Hunt EH, Brand-Elnaggar, J, and Pearson, RD. Rates of needle-stick injury caused by various devices in a university hospital. *N Engl J Med*, 319(5):284–288. August 4, 1988.

23. Crossle KB. Needlesticks in nursing homes and physicians' offices. Presented at the 4th Conference on Occupational Hazards to Health Care Workers, Seattle, Washington. May 15–17, 1989.

24. Neuberger JS, Harris J, Kundin DW, Bischone A, and Chin TDY. Incidence of needlestick injuries in hospital personnel: Implications for prevention. *Am J Infect Control*, 12(3):171–176. 1984.

25. Walsh SS, Pierce AM, and Hart CA. Drug abuse: A new problem. *Br Med J*, 295:526–527. 1987.

26. Alexander WD, Corrigan C, Todd P, and Wells M. Disposal of plastic insulin syringes and needles. *Br Med J*, 295:527. 1987.

27. Turnberg WL and Frost F. Survey of occupational exposure of waste industry workers to infectious waste in Washington State. *Am J Public Health*, 80(10):1262–1264. October 1990.

28. U.S. Public Health Service. Classification of etiologic agents on the basis of hazard. U.S. Public Health Service Ad Hoc Committee on the Safe Shipment and Handling of Etiologic Agents. 1974.

29. Garner JS and Simmons BP. CDC Guideline for Isolation Precautions in Hospitals. HHS Publication (CDC) 83-8314. U.S. Department of Health and Human Services, Public Health Service, Centers for Disease Control, Atlanta, Georgia. July 1983.

30. American Society for Testing and Materials. Standard Test Methods for Impact Resistance of Plastic Film by the Free-Falling Dart Method, Designation: D 1709-91. 1994 Annual Book of ASTM Standards, Section 8, Volume 8.01:393–400. PCN: 01-080194-19. 1994.

31. ECRI. Sharps Disposal Containers. Health Devices, 22(8–9):359–412. August–September 1993.

32. ECRI. Sharps Disposal Containers—Summary. Health Devices, 23(12):464–467. December 1994.

33. The Council of State Governments. Model Guidelines for State Medical Waste Management. Lexington, Kentucky. 1992.

34. Favero MS and Bond WW. Chemical disinfection of medical and surgical materials. In: Block S, Ed., *Disinfection, Sterilization, and Preservation*, 4th ed., p. 633. Lea & Febiger, Philadelphia. 1991.

35. Reinhardt PA and Gordon JG. *Infectious and Medical Waste Management*. Lewis Publishers, Inc., Chelsea, Michigan. 1990.

36. Brown J. Hospital waste management that saves money—and helps the environment and improves safety. *Med Waste: Regul Anal*, 1(10):1–8. July 1993.

37. Connelly MJ. Waste management in an urban medical center: The Mount Sinai program. *Med Waste Anal*, 2(10):1–7. July 1994.

38. Paprock J. The MHA Recycling and Conservation Guide. Minnesota Hospital Association, Minneapolis, Minnesota. 1991.

39. U.S. Congress, Office of Technology Assessment. Finding the Rx for Managing Medical Wastes. OTA-O-459. U.S. Government Printing Office, Washington, DC. September 1990.

7

HOME HEALTHCARE WASTE MANAGEMENT

This chapter addresses the management of biohazardous waste generated from homes in the course of home healthcare (acute or chronic healthcare) and self-care (e.g., diabetes care). Managing biohazardous waste generated in the course of home healthcare or self-care activities brings additional challenges to regulatory agencies, home healthcare providers, and healthcare educators. Although the home care biohazardous waste stream shares many similarities to its institutional counterpart, it is handled outside the institutional context often by untrained, unskilled individuals or family members whose primary interest is on the administration of healthcare and not on the disposal of waste. Home care agencies are often unclear about their role in waste management and how they should approach waste management decisions (1).

In this chapter we first address biohazardous waste management in the course of home healthcare. We then address the management of home sharps waste generated in the course of self-care activities such as diabetes care.

HOME HEALTHCARE

Home healthcare has been expanding rapidly during the past decade as a means of reducing healthcare costs in response to changes in Medicare reimbursement (2). Home healthcare is broadly defined as consisting of medical, nursing, pharmaceutical, and other care provided to patients in their place of residence (3). The burgeoning home healthcare industry is evidenced by the increasing number of home healthcare agencies offering client services. An example of the diversity in ownership and affiliation is seen in the results of a survey of 52 responding home healthcare agencies in northern California which appears in Table 7.1. The diversity of direct-

TABLE 7.1   Ownership/Affiliation and Number of Direct-Care Employees Among 52 Home Care Agencies in Northern California, 1991

	Agency Ownership		Direct-Care Employees		
	Number	Percent	Range	Means	Total
Private	12	23.1	9–173	78.2	938
Affiliated with hospital	27	51.9	8–350	49.0	1322
Corporation or association	10	19.2	6–236	87.4	874
Other	3	5.8	10–53	29.6	89
TOTAL	52	100.0	6–350	62.8	3223

Source: From Smith WA and White MC. Home Health Care Occupational Health Issues. *American Association of Occupational Health Nurses Journal,* 41(4):181, copyright 1993 by the American Association of Occupational Health Nurses, Atlanta, Georgia. Reprinted by permission.

care employee categories for 49 of those facilities is presented in Table 7.2. Often unaccounted for is the patient care provided by family and friends, who also play a significant role in the wastes that are placed into the trash.

Biohazardous waste generated in the course of home healthcare activities bears similarities to institutionally generated biohazardous waste. In a separate report on home healthcare agencies in northern California, White et al. identified the number and percentage of respondents to a survey performing various medical procedures in the course of care activities, which are presented in Table 7.3. Such procedures are like those conducted in the institutional setting and generate similar biohazardous waste.

TABLE 7.2   Categories of Direct-Care Employees Among 49 Home Care Agencies in Northern California, 1991

Category	Number	Percent
Registered nurses	1012	36.5
Home health aides/attendants	769	27.8
Physical therapists	241	8.7
Licensed vocational nurses	133	4.8
Homemakers	114	4.1
Occupational therapists	104	3.8
Speech therapists	92	3.3
Others	304	11.0
TOTAL	2770	110.0

Source: From Smith WA and White MC. Home Health Care Occupational Health Issues. *American Association of Occupational Health Nurses Journal,* 41(4):181, copyright 1993 by American Association of Occupational Health Nurses, Atlanta, Georgia. Reprinted by permission.

**TABLE 7.3   Number and Percentage of Agencies
Performing Certain Procedures Among 58 Home Health
Care Agencies in Northern California, 1991**

Procedure	Perform Procedure	
	Number	Percent
Urinary drainage		
Condom catheter	55	94.8
Intermittent catheter	55	94.8
Indwelling Foley catheter	55	94.8
Indweling suprapubic catheter	47	81.0
Home enteral therapy		
Nasogastric or other enteral tube	57	98.3
Airway and pulmonary		
Tracheostomy care	52	89.7
Suctioning	54	93.1
Respiratory therapy	36	62.1
Wound and stomal care		
Dressing changes	56	96.6
Decubitus care	55	94.8
Enterostomal care	54	93.1
Parenteral therapy		
Insertion		
Peripheral	46	79.3
Central	18	31.0
Site care		
Peripheral	55	94.8
Central	54	93.1
Infusion pump management	54	93.1
Hyperalimentation	48	82.8
Medications	54	93.1
Home peritoneal dialysis	8	13.8

Source: From White MC and Smith W. Infection control in home care
agencies. *American Journal of Infection Control,* 21(3):149, copyright
1993 by Mosby-Year Book, Inc., St. Louis, Missouri. Reprinted by permission.

Developing a Waste Management Strategy

Biohazardous waste management in the home healthcare setting presents challenges
because such activities are less standardized and controlled than in their institutional
counterparts. Nevertheless, poor waste management practices increase the risk of
injury or infection to others both inside and outside the home (e.g., waste industry
workers). It is the responsibility of the home healthcare agency to develop proper
waste management systems and ensure that those responsible for following them,
including family and friend caregivers, have been adequately educated.

When developing systems for managing home care biohazardous waste programs, home care agencies should:

1. Identify all applicable regulations, accreditation standards, and guidelines and the agencies and contacts responsible for their enforcement or implementation.

2. Based on regulations, accreditation standards, and guidelines, develop protocols for biohazardous waste management that address the following components only as applicable:
   • Identification
   • Segregation
   • Containerization
   • Labeling (when necessary)
   • Storage
   • Transportation (only as necessary)
   • Disposal
   • Education
   • Record keeping

3. Develop and conduct an educational program for both the skilled agency home care providers and the family and friends day-to-day care providers.

4. Implement the program through agency staff and the team of family and friend home caregivers.

These components are presented in the following discussion of regulations, standards, and guidelines and management protocol development strategies.

Regulations, Standards, and Guidelines. The first step in developing the biohazardous waste management strategy is to identify all applicable regulations, accreditation standards, and guidelines and the agencies and contacts responsible for their enforcement or implementation. Once identified, copies should be obtained, read, and understood. If questions of interpretation arise, clarification should be obtained from the agency responsible for administering the program.

Regulations. Some state and local governments have developed laws and regulations that pertain to biohazardous waste generated from homes. Contact should be made with those government agencies that may have jurisdiction over the home waste stream. A state agency contact list is provided in Appendix B to assist in making this determination. Again, if regulations do apply, copies should be obtained and kept on file, read, and understood.

The Occupational Safety and Health Administration's (OSHA) "Occupational Exposure to Bloodborne Pathogens Standard" and how it applies to biohazardous waste generators is described in Chapter 3. As described, worker protection must be provided to all occupations exposed to blood or other potentially infectious materials as defined by the rule. Such worker protection would extend to home health-care providers who are employed by "businesses affecting commerce" as defined by the Occupational Safety and Health Act, but not necessarily to those conducting day-to-day healthcare activities. The bloodborne pathogen standard's waste management requirements (e.g., containment, handling, and labeling) would apply to those regulated wastes transported by the professional care provider back to the parent agency for treatment and disposal. An OSHA representative should be contacted for case-specific information. A list of OSHA contacts is presented in Chapter 3 and Appendix A.

It is the responsibility of the regulated community to understand how applicable regulations are interpreted and implemented. Questions regarding interpretation should be directed to the enforcing agency.

Accreditation Standards. Although not required by law, accreditation standards must be met by home healthcare agencies seeking formal accreditation to provide a service. Prior to 1989, accreditation of a home healthcare agency was usually only included in the overall accreditation of an affiliated organization such as a hospital (5). However, in 1989 the Joint Commission on Accreditation of Health Care Organizations (JCAHO) expanded its accreditation from hospitals to other health-care organizations. As a result, JCAHO accreditation standards were developed for home healthcare agencies which provided the opportunity for private, nonaffiliated agencies to receive accreditation from a recognized organization.

The JCAHO addresses its waste management compliance requirements in "Standards for the Accreditation of Home Care, Safety Management and Infection Control," which states: "The instruction of appropriate staff, patients, and caregivers includes the identification, handling, and disposal of hazardous or infectious materials and wastes in a safe and sanitary manner and in accordance with law and regulation" (6).

Guidelines. Guidelines are written as recommendations, based on preferred practices from the perspective of the guideline developer. Although guidelines do not carry behind them the force of law, they serve as valuable tools for those responsible for program development decisions. The Environmental Protection Agency addressed home care activities in a guidance pamphlet entitled "Disposal Tips for Home Health Care" with recommendations for disposing of sharp medical waste that include needles, syringes, lancets, and other sharp objects, and other contaminated wastes, such as soiled bandages, disposable sheets, and medical gloves (7). EPA recommended that sharps waste be placed "in a hard-plastic or metal container with a screw-on or tightly secured lid. A coffee can will do, but you should be sure to reinforce the plastic lid with heavy-duty tape. Do not put sharp objects in any container that will be recycled or returned to a store. Do not use glass or clear plastic containers. Finally, make sure that you keep containers with sharp objects out of the reach of young children." It was recommended that nonsharp contaminated waste "be placed in securely fastened plastic bags before you put them in the garbage can with your other trash."

Other guidelines have been developed by state governments. Later in this chapter a home sharps disposal program developed and implemented by the Washington Department of Ecology is described.

Management Protocol Development. Once the applicable regulations, guidelines, and accreditation standards have been assembled, read, and understood, the next step is to develop the biohazardous waste management protocol. Requirements for home healthcare waste are largely unstandardized and written in general terms, which often allows for flexibility when developing agency programs. Whenever possible, common sense should prevail. If the waste management procedures are prohibitively costly or difficult to implement, they will in all likelihood not be followed.

Following are suggestions to consider when developing home healthcare bio-hazardous waste management protocols. These suggestions are presented without context to any state or local regulations that may apply. They are, however, based on the principles of preventing infectious disease transmission. OSHA's bloodborne pathogen standards are interjected as they would apply to the home healthcare agency employee.

Home healthcare agency personnel should not be involved with transporting biohazardous wastes, except possibly for properly contained sharps waste. When visiting a client, most use personal vehicles that are not equipped to handle spills or any of the hazards associated with transporting biohazardous wastes. Transporting biohazardous waste in a personal vehicle places the driver and possibly the driver's family at risk of injury or infection from improperly contained or spilled biohazardous waste.

Waste Identification. Home healthcare biohazardous waste can be grouped into three general categories: (1) sharps waste (e.g., hypodermic needles, syringes with needles attached), (2) liquid waste (e.g., liquid blood, body secretions), and (3) nonsharp waste (e.g., contaminated dressings or underpads, tubing).

For reference, the OSHA bloodborne pathogen standard [29 CFR 1910.1030, paragraph (b)] applies to:

1. Liquid or semiliquid blood or other potentially infectious materials *(category 2: liquid waste)*
2. Contaminated items that would release blood or other potentially infectious materials in a liquid or semiliquid state if compressed *(category 2: liquid waste; and category 3: nonsharp waste)*
3. Items that are caked with dried blood or other potentially infectious materials and are capable of releasing these materials during handling *(category 2: liquid waste)*
4. Contaminated sharps *(category 1: sharps waste)*
5. Pathological and microbiological wastes containing blood or other potentially infectious materials *(rarely encountered in home healthcare)*

Waste Handling and Disposal.

1. *Sharps Waste (Category 1).* Uncontained sharps waste presents the greatest potential for injury and infection to an exposed individual. Therefore, proper management of sharps waste represents the most important aspect of home healthcare waste management. If waste sharps are to be transported by the agency caregiver back to the agency for treatment and disposal, they should be placed immediately into a commercial sharps container that meets OSHA standards (i.e., that is closable, puncture resistant, leakproof on sides, and properly color coded). If sharps waste are to be managed by the family caregivers, they should be placed immediately into a plastic or metal container with a screw-on or tightly secured lid. Sharps containers should be maintained at the site of patient care. When full, they should be stored in a safe place until disposed. Options for disposal are described in the section "Self-Care Sharps Waste Management" of this chapter.

2. *Liquid Waste (Category 2).* All liquid waste should be poured carefully into a toilet. Nondisposable containers used to transport the waste should be washed and disinfected with a common disinfectant (e.g., 10 percent bleach solution). Disposable containers should be rinsed with water and disposed of with the nonsharp biohazardous waste. Under no circumstances should liquid waste be stored in containers on-site any longer than necessary or transported away from the home by the caregiver for treatment at a distant location.

3. *Nonsharp Waste (Category 3).* Nonsharp waste should be bagged without labeling in a common plastic garbage bag. Infection risks associated with nonsharp waste have been judged to be extremely low (8), especially if properly contained in a sealed plastic bag.

The presence of liquids in nonsharp waste should be minimized. If nonsharp waste is saturated and dripping with body fluids, it should be allowed to gravity drain into an adequately sized container before being placed into the plastic bag, and the liquid collected in the container should be poured into a toilet. Bags should never be overfilled. When ready to be disposed of, the bag should be securely fastened, placed for storage, and disposed of with the household trash.

Education. Each waste management program should include an educational element to ensure that the agency care providers, the patient and patient's caregiver family and friends are fully aware of how to approach waste management. Agency caregivers should be instructed on how to educate the patient and family and why special procedures are necessary. Educational tools such as brochures or single-page handouts should be developed and used.

Recordkeeping. A copy of the biohazardous waste management protocol should be kept on file at the agency and distributed to all agency caregivers for their personal records. The agency should also maintain a record on file of when each staff member and each client family received waste management training. Local and state regulations may restrict many of the recommendations suggested in this section. If biohazardous waste is restricted from disposal in the trash under any circumstances, other disposal options should be identified. These may include contracting with a biohazardous waste management firm, or mailing certain wastes via the U.S. Postal Service to treatment and destruction facilities.

SELF-CARE SHARPS WASTE MANAGEMENT[1]

As identified earlier in this chapter, waste syringes are generated outside the formal medical setting in the course of diabetes care, allergy treatment, and other home healthcare activities. This case study, parts of which have been adapted from work cowritten by the author (9,10), describes the home syringe disposal educational program conducted by the Washington Department of Ecology.

When disposed of improperly by home users, used syringes pose a physical injury risk and a potential infection risk to waste industry workers and possibly to the public, although the infection risk appears to be extremely low (8). To date, only one incident of infectious disease transmission has been identified that may have involved an improperly disposed syringe. This incident involved a hospital housekeeper who developed staphylococcal bacteremia and endocarditis following a needle-stick injury (11). No other cases of infectious disease transmission have been reported in the scientific literature involving exposure to a syringe in the waste stream (8).

During 1989, the Washington Department of Ecology conducted a survey of 940 waste industry workers in Washington State to determine occupational exposures to medical waste in the municipal waste stream (12). Of the 241 respondents identifying themselves as curbside waste collectors (representing 55 percent of all respondents), 10 percent reported that they had sustained a needle-stick injury on the job during the year preceding the survey. Many of these injuries were reported to have resulted from improperly disposed syringes in residential waste.

Such improper disposal of home-generated syringes has also been reported in the scientific literature. Proper syringe disposal techniques and options are often either unknown or misunderstood as identified in the literature. Based on a 1989 survey of 100 patients in New York City, 83 percent of diabetes outpatients reported disposing of their home used syringes directly into the trash, but only 14 percent reported placing the used syringes in a puncture resistant container prior to disposal (13).

A similar study of 100 injectable insulin–using individuals conducted at an inner-city hospital in Atlanta observed 93 percent disposing of their syringes directly into the trash and 3 percent into the toilet (14). Only 4 percent reported that they contained their used syringes in a puncture-resistant container prior to disposal into the general waste stream. Of those disposing of their syringes directly into the trash, 35 percent reported that they made no attempt to contain the needle. Fifty-four

[1]Adapted from Turnberg WL and Lowen LL. Home syringe disposal: Practice and policy in Washington state. *The Diabetes Educator*, 20(6):489–492, copyright 1994 by the American Association of Diabetes Educators, Chicago, Illinois. Reprinted by permission.

percent indicated that they bent or broke their needle before disposal into the trash. Eight respondents volunteered that housemates had sustained needle-stick injuries from handling their trash.

At an April 1990 meeting of the Greater Atlanta Association of Diabetes Educators, 20 diabetes educators were surveyed to measure the perception of their clients' needle disposal practices. Survey estimates indicated that 68 percent of their clients disposed of syringes directly into the trash, 4 percent placed the syringes into a puncture-resistant container before disposing into the trash, and 2 percent disposed of their syringes into the toilet (15).

Swislocki et al. reported syringe disposal practices for a group of 104 injectable-insulin users (16). Of those surveyed, only 10 reported placing their syringes into a puncture-resistant container prior to disposal. Sixty-eight reported that they either bend or break their syringes, 34 reported that they recap their syringes, and 9 reported that they throw their syringes directly into the trash. Seventeen respondents reported that they dispose of their syringes by other means, including incineration, returning them to a hospital for disposal, or disposing into hazardous waste bins at work.

The need to identify and promote safe and practical syringe disposal options to protect waste industry workers, the home syringe user, other household members, and the public is evident. To address this need, the Washington Department of Ecology (Ecology) applied for and received grant funding from the U.S. Environmental Protection Agency (EPA) in 1990 to develop and implement a biomedical waste public education program in Washington state, focusing on proper disposal of home-generated syringes. This case study identifies the methods, results, outcome, and repercussions of that public education effort and subsequent home syringe disposal laws developed by the Washington state legislature in 1994.

Public Education Project

The Washington State Home Syringe Public Education Project was conducted by Ecology in 1990 and 1991 to identify and promote safe and practical disposal options for waste hypodermic needles (syringes) when generated outside the formal medical setting in the course of home healthcare or maintenance. The project's goal was to educate home syringe users about disposal options. The project's objectives were (1) to identify practical options for syringe disposal, (2) to develop educational materials describing the disposal options, and (3) to disseminate this information to home syringe users.

Disposal Options Identification. Ecology formed a multidisciplinary advisory committee to assist with developing solutions to meet the project goal and objectives. Based on information from the project advisory committee and other sources, six reasonable home syringe disposal options were identified and targeted for promotion.

The Hand-Held Needle Clipper. The American Diabetes Association's position statement on syringe disposal includes the statement "Unless the syringe will be reused, it should be placed in a . . . needle-clipping device, which retains the clipped needle in an inaccessible compartment" (17). The hand-held needle clipper was designed to clip the needle from the plastic syringe and contain the needle within the clipper's needle collection chamber. Once done, the needleless plastic syringe plunger, without its ability to puncture, is then safely disposed in the trash, as is the clipper chamber containing needle points when full.

Only one hand-held needle clipper was identified as being commercially available. This product, the Safe-Clip Insulin Syringe Needle Clipper, is manufactured and distributed by the Becton Dickinson Corporation. The manufacturer reports that the clipper can hold as many as 2000 needle points within its small plastic depository.

Syringe Containment. The American Diabetes Association's position statement on syringe disposal also states: "Unless the syringe will be reused, it should be placed in a puncture-resistant disposal container . . . " (17). The purpose of placing syringes in puncture-resistant containers prior to disposal into the trash is to minimize the needle-point exposure to individuals involved with waste collection and disposal, and to others, such as household members or the syringe user.

In 1990, the U.S. Environmental Protection Agency published home syringe disposal guidelines based on containing the needle point (7). In its educational brochure, EPA recommended that (1) used needles, syringes, lancets, and other sharp objects be placed in a hard plastic or metal container with a screw-on or tightly secured lid; (2) sharp objects not be placed in any container that will be recycled or returned; (3) glass or clear plastic containers not be used; and (4) containers with sharp objects be kept out of the reach of young children.

Although the EPA recommendation addressed the immediate exposure hazard to the syringe generator, the generator's family, and the curbside waste collector, it did not address the possibility of the container breaking and spilling its contents due to mechanical stresses found in the waste stream, which could potentially expose workers at transfer stations or landfills. In a 1989 survey conducted in Washington State, 5 of 191 (2.6 percent) responding waste industry workers who were not involved with curbside waste collection (e.g., landfill attendants, heavy equipment operators, truck drivers, laborers) reported that they had sustained an occupational needle-stick injury within the year preceding the survey (12). This finding indicates the need for needles to be contained even after disposal into the waste stream.

A container stress study was conducted as part of this project to examine the ability of common containers found around the home and of selected commercial containers to contain syringes when subjected to compaction and other physical stresses associated with the municipal waste stream. Several common containers (bleach bottles, water bottles, milk bottles, 2-liter plastic soft drink bottles, detergent bottles, and metal coffee cans) and two commercial sharps container models (the Sage product 8920 and the Becton Dickinson product 5488 plastic commercial sharps containers) were subjected to the compactions and stresses created during solid waste disposal from the point of garbage pickup from the home through the transfer station to final disposal at the landfill.

Of the containers tested, the only container that was found capable of remaining intact and containing its contents was the common 2-liter soda bottle, which is manufactured from a polyethylene terephthalate plastic resin (PET). Commercial sharps containers commonly used by medical facilities, as well as other household containers commonly recommended for home syringe containment by diabetes educators and government officials, were observed to shatter in the waste stream, spilling their contents.

Based on its compaction studies, the Department of Ecology recommended that the common 2-liter PET soda bottle be used by home generators choosing the "syringe containment" disposal option. A bright orange warning sticker was developed by Ecology to be affixed to the PET soda bottle to provide instructions for use, warn of associated hazards, and emphasize that the filled containers should never be recycled.

Syringe Disposal with a Pharmacy. This system would ensure that syringes are properly contained and returned to a participating pharmacy for safe disposal. Although this option has merit, its availability remains extremely limited in most parts of Washington State.

Syringe Disposal with a Physician. The purpose of returning used, containerized syringes to a physician would be to ensure a safe disposal method that entirely precludes entry of the used syringe into the general waste stream. However, it has generally been observed that few physicians are willing to take on this additional responsibility.

Syringe Disposal at a Municipal Syringe Disposal Site. This option involves establishing formal used syringe dropoff centers. In Washington State, only the city of Seattle was found to provide dropoff centers at two solid waste transfer stations. To date, the Seattle program has had very limited participation.

Syringe Disposal with a Garbage Hauler. Since the 1990 adoption of biomedical waste transportation requirements in Washington State, several residential solid waste collection companies have begun offering home-generated syringe pickup services during the course of routine solid waste collection. In areas where the service is offered, the user is required to cover the costs of the special handling and disposal. Participation with the existing programs has been extremely low.

Syringe Mail-Back Programs. In these programs, syringes (or other medical waste) are packaged and mailed to a medical waste disposal company. If syringes are to be mailed through the U.S. Postal Service (USPS), they must meet all packaging and mailing requirements established in the USPS "Domestic Mail Manual" (18) which increases cost to the user. Because the USPS rules had not been adopted at the time of this project, the mail-back option was not included among the disposal options in the 1990–1991 educational program.

Educational Program

Two pieces of educational material were developed for the project: an educational brochure describing the various syringe disposal options identified above, and a bright orange fluorescent warning sticker to be affixed to a PET 2-liter soda bottle used for syringe containment. The sticker, measuring $8\frac{1}{2}$ by $3\frac{5}{8}$ inches, contained the international biohazard symbol, the Department of Ecology logo, and stated:

AVOID INJURY FROM NEEDLES! BE SAFE WITH SYRINGES!

WARNING: SYRINGES
DO NOT RECYCLE

Keep out of the reach of children!

STICKER: Remove paper backing from sticker. Apply to an empty 2-liter plastic pop bottle.

STORE: Carefully put each of your used syringes into the bottle.

SEAL: Put tape over the closed bottle cap when bottle is full.

DISPOSE: Dispose of the bottle in your household trash.

Contact your local environmental health office for regional restrictions on this disposal option.

Approximately 60,000 each of the project educational materials were distributed by Ecology to home syringe user populations, primarily through direct mailings to key groups (e.g., American Diabetes Association Washington Affiliate membership, home healthcare organizations, health departments, waste collection companies and pharmacies) and request mailings to interested parties. Information was also disseminated through the news media.

Project Impacts and Implications

User Disposal Practices. In a 1993 follow-up project, Ecology distributed surveys to 395 persons listed as members of the American Diabetes Association Washington Affiliate within six zip code areas of the state in which waste hauler syringe collection programs were not offered (19). The survey was conducted in part to determine the effect of the Washington State Home Syringe Public Education Project on home syringe disposal practices. Multiple responses to questions were allowed. Of the 188 survey respondents (response rate 48 percent), 142 (representing 75 percent of respondents) reported that either they or someone in their household used syringes. Of those 142 households, 25 (18 percent) reported that they recap the syringe and dispose in the trash; 30 (21 percent) reported that they clip the needle either with a hand-held needle clipper or with another tool; and 30 (21 percent) reported breaking their needles prior to disposing in the trash. Eighty-three (representing 58 percent of the 142 needle using respondents) reported placing their used syringe into a household container and disposing in the trash. Of those, 35 (representing 25 percent of the 142 syringe-using respondents) reported using the PET soda bottle; 11 (8 percent) used a bleach bottle; 10 (7 percent) used a coffee can; 11 (8 percent) reported using a commercial sharps container; and 11 (8 percent) reported using other containers.

Several syringe dropoff disposal alternatives were identified in the survey. However, only 11 respondents (representing 8 percent of the 142 syringe-using respondents) reported returning used syringes to a medical facility; 3 (2 percent) to a municipally operated syringe dropoff station; 6 (4 percent) to a participating syringe return pharmacy; 3 (2 percent) to a commercial medical waste collection company; and 5 (4 percent) to other dropoff alternatives. Only one respondent acknowledged placing clipped needles in a self-labeled household container into a recycling bin.

A separate study was conducted by Ecology in 1993 to evaluate participation of home syringe users in syringe pickup programs offered by two garbage hauling companies in Washington State (19). Collection company A, located in southwestern Washington, exclusively conducts waste hauling services for a community of about 65,650 homes with an estimated population of 1800 injectable insulin–using individuals. It has been offering home syringe collection since the spring of 1991. In this community, local requirements prohibit syringe disposal into the trash. Syringe collection from homes is done by a dedicated medical waste collection vehicle. At the time of the study, the company would deliver a 1-gallon commercial sharps container, capable of containing approximately 150 syringes, directly to the participant's home for a cost of $3 for the initial delivery. Filled containers were either collected directly from the participant at home, or concealed in a plastic bag at a back door for pickup by the collector at a cost of $20 per pickup. Pickup included delivery of a new 1-gallon container for no additional charge. By January 1993 (about 21 months into the program), 76 customers had enrolled in the program, which represented about 4 percent of the injectable insulin–using population within the service area. During that time, only 14 customers (18 percent) had requested pickup of a filled container.

Collection company B, located in central Washington, exclusively conducts waste hauling services for a community of approximately 30,000 homes, with an estimated injectable insulin–using population of 909. There are no restrictions on syringe disposal into the household trash in this service area. Company B began its commercial medical waste collection and disposal services in 1991 and began actively to pursue customers for home syringe pickup during the spring of 1992. At the time of the study, 1-gallon sharps container delivery and pickup was by appointment only at a cost of $15 per pickup. As of December 1993, company B had no home syringe customers.

Plastics Industry Concerns. The project recommendation for disposing of syringes in PET plastic soda bottles eventually came to the attention of the national plastic container recycling industry. The plastics industry had been developing systems to recycle PET plastics in what they described as a *closed-loop system.* In this system, recycled plastics are returned directly to the manufacture of food-grade containers such as soda bottles. The industry believed that the success and continuation of the closed-loop approach relied on consumer confidence in the recycled product.

The industry was concerned that the public image of recycled plastics would be compromised if the public believed that the recycled plastic may have been formerly used to contain used syringes. The industry was also concerned that PETs containing syringes would find their way into the recycling streams, no matter what efforts are made to prevent such occurrences, placing recycling workers at increased risk of needle-stick injury and potentially causing some degree of contamination of the recycled plastic. Additionally, several syringe-containing PET containers had been reported to enter recycling centers in Washington State, although none with the project's red warning label.

Home Syringe Legislation. The 1994 Washington state legislature addressed the concerns of the plastics recycling industry by passing legislation that redirected home syringe disposal recommendations established by the Department of Ecology. The new law indirectly restricts, under certain circumstances, the use of two practical and popular home syringe disposal option recommendations: the hand-held needle clipper and the labeled PET soda bottle containerization option. As of July 1, 1995, the new Washington state law prohibits home syringe users from disposing of their home-generated syringes into public or privately operated recycling containers or recycling sites at any time. The law also prohibits syringes from being disposed of into the general waste stream, but only in those communities in the state in which a home syringe collection service is available. In those communities where such a service is offered, neither the PET soda bottle container nor needle clipper disposal options would be allowed because those options ultimately call for a contained syringe needlepoint to be disposed in the trash. Under the law, a home syringe collection service may be conducted by a public or private solid waste collection company as part of its general waste collection responsibilities. Any company choosing to offer this service must notify the public in writing that the service is available, where users can obtain sharps containers, how much the service costs, other options that are available to a community, and the legal requirements of home syringe disposal. Syringes collected for disposal must be contained in a leakproof, rigid, puncture-resistant red container that has been taped closed or tightly lidded (e.g., a commercial sharps container). However, if a collection service is not offered in a community, home users may dispose of syringes directly into the general waste without restriction.

Other disposal options identified, but not required, by the law include municipally operated sharps waste container dropoff sites and pharmacy return programs. Pharmacies offering a return program must register, at no cost, with the Department of Ecology, and Ecology must share its registration list with local health departments and other local solid waste management officials. Local health departments must enforce the law, first through education, then using civil infraction penalties for repeat violators when necessary.

The real effects of Washington State's new law in terms of protecting waste handlers from exposure to syringes in the waste stream are unknown. The program will have to be evaluated during the next few years to determine whether the syringe disposal restrictions established in law and the effective prohibition by the law of two practical, cost-effective and safe disposal options will have an effect on syringe disposal practices and the occurrence of occupational needle-stick injuries among waste industry workers.

REFERENCES

1. Ralph IG. Infectious medical waste management: A home care responsibility. *Home Healthcare Nurs*, 11(3):25–33. 1993.

2. Lorenzen AN and Itkin DJ. Surveillance of infection in home care. *Am J Infect Control*, 20(6):326–329. December 1992.

3. Catania PN and Rosner MM, Eds. *Home Health Care Practice*, 2nd ed. Health Markets Research, Palo Alto, California. 1994.

4. Smith WA and White MC. Home healthcare occupational health issues. *AAOHN*, 41(4):180–185. April 1993.

5. White MC and Smith WA. Infection control in home care agencies. *Am J Infect Control*, 21(3):146–150. June 1993.

6. Joint Commission on Accreditation of Healthcare Organizations. Standards for the Accreditation of Home Care. Joint Commission on Accreditation of Healthcare Organizations, Oakbrook Terrace, Illinois, p 9. 1992.

7. U.S. Environmental Protection Agency. Disposal Tips for Home Health Care. EPA/530-SW-90-014A. January 1990.

8. U.S. Department of Health and Human Services, Agency for Toxic Substances and Disease Registry, Public Health Service. The Public Health Implication of Medical Waste: A Report to Congress. 1990.

9. Turnberg WL and Lowen LD. Home syringe disposal: Practice and policy in Washington state. *Diabetes Educ*, 20(6):489–492. November/December 1994.

10. Turnberg WL and Lowen L. Home Care Syringe Disposal. Publication 91-74. Washington Department of Ecology, Bellevue, Washington. March 1992.

11. Jacobson T, Burke JP, and Conti MT. Injuries of hospital employees from needles and sharp objects. *Infect Control*, 4:100–102. 1983.

12. Turnberg WL and Frost F. Survey of occupational exposure of waste industry workers to infectious waste in Washington state. *Am J Public Health*, 80:1262–1264. 1990.

13. Gambardella KA and Rowen SG. Disposal of insulin syringes by diabetic outpatients in the New York city metropolitan area [Abstract]. *Diabetes*, 38(suppl 2):53A. 1989.

14. Satterfield DW, Kling J, and Gallina DL. Need to change needle disposal practice in inner city to decrease HIV transmission risk [Abstract]. *Diabetes*, 39 (suppl 1):51A. 1990.

15. Satterfield D and Kling J. Diabetes educators encourage safe needle practice. *Diabetes Educ*, 17:321–325. 1991.

16. Swislocki A, Cram D, Noth R, Dowdell L, Lamothe J, and Ogi G. Insulin syringe disposal patterns at a VA hospital. *Diabetes Care*, 14:930–932. October 1991.

17. American Diabetes Association. Insulin administration [Position Paper]. *Diabetes Care*, 13(suppl 1):28–31. 1990.

18. United States Postal Service. Mailability of Sharps and Other Medical Devices—Final Rule. Federal Register, 57:29028. June 30, 1992.

19. Lowen LD and Turnberg WL. Residential Syringe Collection by Waste Haulers. Publication 93-44. Washington Department of Ecology, Bellevue, Washington. June 1993.

8

MEDICAL WASTE INCINERATION

Incineration has historically been recognized as a preferred method for disposing of medical waste because of its ability to destroy the recognizability of the waste, kill pathogens, and reduce the volume of waste by 90 to 95 percent. Although many hospitals continue to operate on-site incinerators, this waste disposal practice has come under increased scrutiny due to concerns about chemical, biological, and radiological emissions. Many states, including California, New York, and Texas, have established air pollution control regulations that apply specifically to medical waste incinerators. The costs associated with upgrading existing incinerators to meet these emissions standards can be considerable. As a result, many medical institutions in these states have shifted from their historical reliance on incineration to other nonincineration waste treatment systems. In addition, on February 27, 1995, the U.S. Environmental Protection Agency published proposed standards for medical waste incinerators that would limit air pollutant emissions from new, modified, and existing incinerators to levels based on maximum achievable control technology (MACT) (1). Under the EPA's current schedule, a final rule will be published by April 15, 1996. The EPA estimates that this will result in an 80 percent reduction in currently projected new incinerators, and an 80 percent reduction in existing incinerators as well.

In this chapter, which is adapted in part from other work published by the author (2,3), we review the Clean Air Act Amendments of 1990, EPA's proposed rule which would amend 40 CFR Part 60, incineration technologies, and what is known about the risks associated with medical waste incineration due to the release of biological, chemical, and radiological emissions.[1,2]

[1]Adapted from Turnberg WL and Kelly KE. Health effects of medical waste incineration. Proceedings of the 1994 International Incineration Conference, Houston, Texas, pp. 663–669, copyright 1994 by the University of California Regents, Irvine, California. Reprinted by permission.

[2]Adapted from Turnberg WL. The health effects of medical waste incineration: Biological, chemical,

FEDERAL CLEAN AIR ACT

The Clean Air Act Amendments of 1990 (CAAA) directed the U.S. Environmental Protection Agency to develop two programs addressing hazardous air pollutant emissions (HAPs) and new source performance standards (NSPS) for medical waste incinerators. The standards that are now being developed are based on what can be technologically achieved with consideration given to economics, energy impacts, and cross-media pollutant transfers.

Title III of the CAAA requires air pollution control for existing, new, and modified sources of hazardous air pollutants. The amendment specifically defines 189 toxic substances that have been deemed capable of causing adverse human or environmental health effects through inhalation or other exposure routes. Potential health effects cited include carcinogenicity, mutagenicity, teratogenicity, neurotoxicity, reproductive dysfunction, or acute or chronic toxicity. Of the 189 HAPs identified by the CAAA, 11 are metals and their compounds. These include antimony, arsenic (including arsine), beryllium, cadmium, chromium, cobalt, lead, manganese, mercury, nickel, and selenium.

The EPA is required under Title III to reduce emissions from sources releasing HAP to levels that will protect public health in two phases of standards development. The first phase standards involve bringing source emissions to levels that are being achieved by well-controlled sources. The second-phase standards require further emission reductions in cases where the first-phase levels are deemed as being inadequate to protect human health. The second-phase standards are due in about a decade.

Section 129 of the 1990 Amendments singled out incineration for special rule-making, including medical waste incinerators, requiring that the two-phase approach under Title III be applied to new and existing emitters. When performance standards are published under Section 111(b) of the act, states are required under Section 111(d) to submit plans that (1) establish emission standards for the designated pollutants, and (2) provide for implementing and enforcing the standards. The state plans must be adopted and submitted to the EPA within 1 year after the regulations are promulgated. The act specifies that the process for plans submission by the states is one that is similar to the process for state submission of state implementation plans (SIPs) under Section 110.

For the past few years, the Office of Air Quality Planning and Standards of the EPA has been developing regulations for air emissions from medical waste incinerators. The program includes standards to be developed under Section 129 of the CAAA of 1990, including new source performance standards under Section 111(b) and emission guidelines for existing medical waste incinerators under Section 111(d) of the Clean Air Act. The EPA's medical waste incinerator regulatory development program is comprised of three phases: (1) data gathering, (2) data analysis and control alternatives, and (3) proposal and promulgation of standards.

and radiological emissions. Proceedings of the 1994 International Incineration Conference, Bellevue, Washington, pp. 599–603, copyright 1995 by the University of California Regents, Irvine, California. Reprinted by permission.

Because the EPA failed to meet its CAAA mandated statutory deadline for publishing incineration standards, legal action was taken by the Sierra Club and the Natural Resources Defense Council (NRDC). As a result, under a consent decree filed by the EPA, the Sierra Club, and the NRDC with the U.S. District Court of the Eastern District of New York, the EPA Administrator was required to sign a notice of proposed rulemaking by February 1, 1995 and a notice of final rulemaking by April 15, 1996.

EPA-PROPOSED REGULATIONS

On February 1, 1995, the Administrator of the EPA signed a notice of proposed rulemaking under the title of "Standards of Performance for New Stationary Sources and Emission Guidelines for Existing Sources: Medical Waste Incinerators." This would amend 40 CFR Part 60 (1). In its preamble discussion, the EPA estimated that for new medical waste incinerators (MWIs), the proposed emission standards would reduce nationwide emissions of dioxins and furans by 99 percent; particulate matter (PM), carbon monoxide (CO), hydrogen chloride (HCl), lead (Pb), and cadmium (Cd) by more than 95 percent; mercury (Hg) by 92 percent; and sulfer dioxide (SO_2) and nitrogen oxides (NO_x) by 25 percent. For existing MWIs the EPA estimated emission reductions of 99 percent for dioxins and furans; 98 percent for PM, CO, and HCl; 97 percent for Cd; 94 percent for Hg; and 37 percent for SO_2 and NO_x.

Under the proposal, a *medical waste incinerator* (MWI) or *MWI unit* is defined as: "any device that combusts any amount of medical waste." In its proposal, the EPA has not included a size exemption for smaller units. Any unit that burns medical waste, whether or not with other fuels or types of waste, will be subject to the standards if adopted as proposed. The EPA has defined the term *medical waste* to mean

> any solid waste that is generated in the diagnosis, treatment, or immunization of human beings or animals; in research pertaining thereto; or in the production or testing of biologicals. Biologicals means preparations made from living organisms and their products (including vaccines, cultures, etc.), intended for use in diagnosing, immunizing, or treating humans or animals or in research pertaining thereto. The term medical waste does not include any hazardous waste identified or listed under 40 CFR 261, or any household waste as defined in 40 CFR 261.4(b)(1), or any human or animal remains not generated as medical waste.

New MWIs include those for which construction has begun after the date of publication of the proposed standard in the *Federal Register* (February 27, 1995) and MWIs that are modified after the effective date of the new, as yet unpublished standards. Existing MWIs include those that were constructed on or before February 27, 1995. Changes made to existing MWIs for the purpose of complying with the emission guidelines are not considered as modifications that would require compliance with the NSPS emission standards for new MWIs.

Although the proposed standards would establish emission limits for both new and existing MWIs, the type of air pollution control technology necessary to meet the standards is not specified. The proposed MACT emission limits are summarized in Table 8.1. The EPA has also proposed other requirements that would address (1) operator training and qualification, (2) siting requirements, (3) compliance and performance testing requirements, (4) monitoring requirements, (5) reporting and recordkeeping requirements, and (6) inspection requirements (for existing MWIs).

Section 129 of the CAAA requires that emission standards must reflect MACT, and that the degree of emissions reduction that is determined to be achievable for new MWIs cannot be less stringent than that which is achieved in practice by the best controlled similar unit. This level, termed the *MACT floor*, defines the minimum level of emissions control and is based on the top 12 percent best performing units in each category. To adopt standards of greater stringency, the administrator must consider costs, any nonair-quality health and environmental impacts, and energy requirements that are associated with the limits (1).

For new MWIs, MACT and MACT floors have been proposed for three incinerator types based on how the waste is fed into the system. These are (1) new continuous MWIs, (2) new intermittent MWIs, and (3) new batch MWIs. Continuous MWIs are those that have been designed to allow waste charging and ash removal during combustion. Intermittent MWIs allow waste charging but not ash removal, and batch units do not allow either waste charging or ash removal when operating.

For new continuous, intermittent, and batch MWIs, the level of emission control

TABLE 8.1 Proposed Emission Limits for New and Existing Medical Waste Incenerators[a]

Pollutant	Emission Limits at 7 Percent Oxygen, Dry Basis
Particular matter	30 milligrams (mg) per dry standard cubic meter (0.013 grain per dry standard cubic foot)
Carbon monoxide	50 parts per million (ppm) by volume
Dioxins/furans	80 nanograms (ng) per dry standard cubic meter total dioxins/ furans (35 grains per billion dry standard cubic feet) or 1.9 ng per dry standard cubic meter toxic equivalency (0.83 grain per billion dry standard cubic feet)
Hydrogen chloride	42 ppm by volume or 97 percent reduction 9-hour average
Sulfur dioxide	45 ppm by volume
Nitrogen oxides	210 ppm by volume
Lead	0.10 mg per dry standard cubic meter (44 grains per million dry standard cubic feet)
Cadmium	0.05 mg per dry standard cubic meter (22 grains per million dry standard cubic feet
Mercury	0.47 mg per dry standard cubic meter (210 grains per million dry standard cubic feet) or 85 percent reduction

Source: Reference 1.

[a]All pollutant values based on 12-hour average, except as noted.

**TABLE 8.2   Proposed MACT Floor Emission Levels
for Existing MWIs**

	MWI Type		
Pollutant	Continuous	Intermittent	Batch
PM, mg/dscm	46	69	69
CO, ppmv	76	90	91
CDD/CDF, ng/dscm	1,619	12,906	14,606
HCl, ppmv	43	115	911
SO_2, ppmv	284	414	1,166
NO_x, ppmv	257	216	220
Pb, mg/dscm	8.7	11.8	23.1
Cd, mg/dscm	0.56	1.8	3.4
Hg, mg/dscm	4.0	15.6	18.5

Source: Reference 1.

achieved by a unit operating with a dry sorbent injection followed by a fabric filter control system (DI/FF) with carbon injection is considered to be MACT floor. In its proposal, EPA considers the MACT floor to be the most effective level of control. Because there are no additional alternatives to consider, the EPA has determined in its proposal that the emissions control achieved by the DI/FF with carbon injection system is considered MACT.

For existing MWIs, the EPA did not have sufficient information to establish a MACT floor due to the limited test data for existing units. To determine emission limitations for the best 12 percent of units in each subcategory, the EPA examined air quality permits and state regulations. EPA estimates that there are 338 continuous, 3018 intermittent, and 336 batch MWIs in use nationwide (1). Proposed MACT floor emission levels for each pollutant were established by EPA based on averaging the emission limitations reported by the top 12 percent of units in each category that are subject to the most stringent permit or state regulation limitations. The proposed MACT floor emission levels for each category are presented in Table 8.2. However, EPA has determined in its proposal that control technologies are capable of achieving MACT emissions levels, which for many pollutants, are considerably higher than the MACT floor emissions levels. These levels are presented in Table 8.2.

MEDICAL FACILITY WASTE CHARACTERISTICS

Medical facility waste encompasses all waste generated by a healthcare or health-related facility. Depending on the facility, such waste can include a facility's general waste (e.g., office or cafeteria waste), infectious waste, radioactive waste, or toxic chemical waste. In practice, waste streams are often intermixed, which adds complexity to disposal decisions. Because the waste that is burned by a medical facility can ultimately influence stack emissions or ash residue contaminants, knowledge of the category, source, and volume of waste destined for incineration should be understood.

General Waste. General waste consists of those waste materials that are not regulated or handled as hazardous or otherwise potentially dangerous or special waste. Examples include paper goods, cardboard, plastics, food scraps, and glassware, as well as other miscellaneous general wastes. Such waste originates from any area of a healthcare facility. U.S. hospitals typically generate a median of about 15 pounds of hospital waste per patient per day (4).

Biohazardous Waste. Also referred to as infectious, biomedical, regulated medical waste, or red bag waste, biohazardous waste consists of those wastes that have been defined in federal, state, or local regulation based on their potential to contain human pathogens and transmit infectious disease. Categories of waste that are typically defined as biohazardous waste include microbiological stocks and cultures, blood and blood products, sharps waste (e.g. syringes), human or animal pathological waste, and communicable disease isolation waste. Depending on how it is defined, biohazardous waste typically makes up about 15 percent of the total hospital waste (4).

Toxic Chemical Waste. Although volumes are small relative to industrial operations, hospitals routinely use toxic substances for both patient diagnosis and treatment. These substances include such materials as antineoplastic drugs (seven of which are listed by the EPA as hazardous waste), formaldehyde, photographic chemicals, solvents, mercury, and waste anesthetic gases, as well as other toxic, corrosive, and miscellaneous chemicals (5).

Low-Level Radioactive Waste. Low-level radioactive waste is typically generated during the course of patient diagnosis or treatment from nuclear medicine or clinical testing laboratories. Waste containing radioisotopes often includes anatomical samples from humans and animals, syringes, aqueous and organic liquids, soiled clothes and paper, material used for cleanup, and scintillation vials.

In one study characterizing low-level radioactive waste (LLRW) at a 560-bed university hospital that is physically connected to the school of medicine, 67 percent of the LLRW waste generated was characterized as solid, 7 percent liquid, 8 percent scintillation vials, and 18 percent biological, for a total LLRW volume of 75.8 cubic meters collected during 1990 (6). Radioisotopes included ^{125}I (25.5 percent), ^{32}P (19.1 percent), ^{3}H (14.5 percent), ^{14}C (8.7 percent), ^{35}S (6.2 percent), ^{131}I (1.1 percent), ^{51}Cr (0.8 percent), ^{45a}C (0.4 percent), microspheres (17.6 percent: isotopes included ^{153}Gd, ^{141}Ce, ^{113}Sn, ^{103}Ru, ^{95}Nb, and ^{46}Sc, with each waste package containing five simultaneously), and other isotopes (6.1 percent).

Plastics and Metals. The plastics content of medical waste has increased dramatically during the past two decades as medical procedures have come to rely on plastic disposable products. During the early 1970s plastics accounted for approximately 10 percent of the overall medical waste stream. In the late 1980s, plastics content rose to more than 30 percent (7), about five times more than municipal solid waste (8). The amount of chlorinated plastics (e.g., PVCs) has been found to influence directly the generation of acid gases and may also influence the production of dioxins and furans. Sources of heavy metals in medical waste include needles, surgical blades, and foil wrappers as well as less obvious sources such as fillers, stabilizers, colorants, and inks that are used in the plastics production (7).

Heating Value. The heating value for medical waste can range from as low as 2000 Btu per pound for anatomical waste to more than 17,000 Btu per pound, with a typical value of about 8000 to 8500 Btu per pound, depending on plastic and moisture content (9). Although most medical waste incinerators are rated in pounds per hour, the actual throughput limiting factor is the heating value of the waste.

TECHNOLOGIES AND POLLUTION CONTROL

Most medical waste incinerators operating in the United States today fall into one of three categories: (1) excess-air (retort) systems, (2) controlled-air (starved-air) systems, and (3) rotary kilns (9). Depending on design, incinerators may be operated as continuous, intermittent, or batch units.

Excess-air systems are characteristic of the older units in use and are usually operated in a batch mode. To operate these systems, waste is placed manually in the combustion chamber, the charging door closed, and the afterburner ignited. When the stack reaches its target temperature, the primary burner ignites. Moisture and volatile organic materials vaporize, the volatile gases burn in the chamber and the stack, and the waste begins to burn. Air is supplied at a constant rate. Typically, these systems operate in a starved-air mode until the organic material is burned. At the end of the cycle, the chamber operates with excess air (10). After the waste is burned the primary burner shuts off. A diagram of an excess-air combuster is presented in Figure 8.1.

Controlled-air systems are among the most widely used. For these systems, the lower chamber is operated in a controlled (e.g., starved-air) mode. Waste is fed into a lower combustion chamber which receives only about half of the air that is stochiometrically necessary to combust the waste. Partially burned gases in the lower chamber enter the secondary chamber through a connecting duct, which is referred to as the flame port, because about 75 percent of the total combustion air is introduced through ports in this short duct (10). Conditions in the secondary chamber are sufficient for complete combustion of hydrogen and carbonaceous material from the lower chamber. Gas temperatures at the primary zone exit are typically adjusted to about 1400 to 1600°F. In the secondary chamber, gases typically have a dwell time of at least 1 second and exit at a temperature of 1800°F. Supplemental heat is derived from an auxiliary fuel-fired burner in the secondary chamber if the minimum exit temperature of the gas cannot be maintained. A diagram of a controlled air combuster is presented in Figure 8.2.

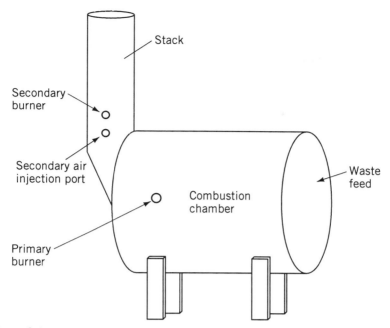

Figure 8.1 Modular excess air (batch) incinerator. (From reference 10, page 36.)

Rotary kiln incinerators consist of a slightly inclined cylinder (kiln) that rotates slowly during the combustion cycle. The rotation mixes the waste and moves it through the incinerator. During the process, the ash drops from the kiln onto a conveyor and is transported to a waste collection hopper. Flue gas exits the kiln and enters a secondary chamber that is equipped with an auxiliary fuel burner and temperature control system. The secondary chamber is typically designed for 1 or 2 seconds of gas residence time at 2000°F (10). A rotary kiln diagram is presented in Figure 8.3.

Most incinerators are operated on-site, although regional incinerators are common. For example, in California, of 146 operating medical waste incinerators identified, 94 percent (137 facilities) were smaller on-site incinerators and 6 percent (9 facilities) regional incinerators burning solely medical waste (11). The estimated number of MWIs in the United States is presented in Table 8.3. Although this information was recently published as a final document by the EPA (12), it is based on information from a draft document published in 1991 (13). These data may represent an overestimation of the number of incinerators combusting medical waste today. For example, in 1990 there were 13 hospitals conducting on-site combustion of medical waste in the Greater Seattle region (14). Due to increased air emissions standards in that region and a stepped-up enforcement program conducted by the regional air pollution control authority, as of 1995, 11 of these facilities have ceased operation rather than face the cost of upgrading older inadequate units. The two

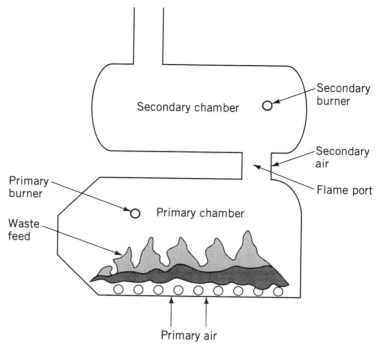

Figure 8.2   Dual-chamber, starved-air incinerator. (From reference 10, page 38.)

units that have continued operation were constructed within the past 5 years and are currently meeting regional air emissions restrictions.

Table 8.4 presents data on the estimated age of existing incinerators, which is based on information obtained from state air control agencies and hospital associa-

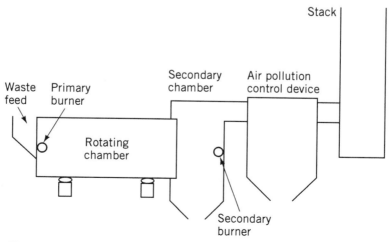

Figure 8.3   Typical rotary kiln incinerator. (From reference 10, page 38.)

TABLE 8.3 Estimated Number of Medical Waste Incinerators in the United States

Facility Type	Units	Percent of Total
Hospitals	3150	63
Laboratories	500	10
Veterinary facilities	550	11
Nursing homes	500	10
Commercial facilities	150	3
Other/unidentified facilities	150	3
Total	5000	100
Municiple combusters that cofire medical waste	31	

Source: Reference 13.

tions. Again, this information is based on data collected during the same time period and highlights the need for older poorly controlled units to either upgrade or cease operation.

Brinckman categorized pollutants of combustion into three groups: (1) those controlled by the combustion process (e.g., carbon monoxide, hydrocarbons, and organic constituents; (2) particulate matter, which includes heavy metals, possibly dioxins and furans, and entrained ash; and (3) acid gases (e.g., hydrogen chloride, hydrogen fluoride, and oxides of sulfur (15). Control of pollution from medical waste incinerators requires attention to three process periods: (1) precombustion (waste stream control), (2) combustion (facility operation and maintenance), and (3) postcombustion (emission control equipment) (16).

Precombustion practices to control stack emissions from medical waste incinerators involve identifying the source of toxic substances in the waste stream and removing these from the waste before incineration. It has been shown that the sources of emission contaminants can be identified by examining the individual

TABLE 8.4 Summary of Available Medical Waste Incinerator Age Data

Medical Facilities	Number of Units in Sample	Age (years) Range	Age (years) Average
Hospitals[a]	108	1–33	14
Laboratories[b]	2	10–21	16
Veterinary facilities[b]	10	3–21	13
Nursing homes[c]	6	18–33	24
Commercial facilities[d]	5	3–6	5
Other/unidentified facilities[c]	8	7–30	16

Source: Reference 12.

[a]Based on data from California, Rhode Island, and Washington.

[b]Based on data from Washington.

[c]Based on data from California and Washington.

[d]Based on data from California and Rhode Island.

components of the waste stream. For example, levels of lead and cadmium in medical waste emissions have been shown to be linked to paper printing inks and plastic stabilizers (17). Medical waste has been found to consist of about 30 percent plastics, of which about 15 percent has been shown to be polyvinyl chlorinated plastic (PVC). PVC, which contains about 45 percent chlorine, has been identified as the major contributor to HCl emissions in medical waste incinerators (17). Minimizing pollutants in waste via pollution prevention before the incineration process will have the effect of reducing toxic emissions.

Combustion operating practices have a direct affect on pollutant emissions. Combustion efficiency is influenced by temperature, residence time, and air–fuel mixing. For example, operational parameters were examined as part of the EPA's regulatory development program for medical waste incinerators (18). The test program collected information from seven incinerators. Parameters included combustion temperature, residence time, waste feed rate and excess air levels on carbon monoxide (CO), particulate matter (PM), and chlorinated dibenzo-p-dioxins and chlorinated dibenzofuran (CDD/CDF) emissions (18). Based on results, emissions for the seven incinerators tested exhibited greater variability between facilities than within facilities, indicating the effect of combustion controls on emissions. Emissions from older facilities were found to be most influenced by inadequate secondary chamber temperature. The results indicated that for facilities with adequate residence times and airflow control, secondary chamber temperatures had little effect on emissions in the range 1600 to 2000°F. For newer facilities, the results suggested that control of primary chamber temperature is necessary to minimize CO and CDD/CDF emissions. It is evident that medical waste incinerator operation and maintenance must be optimized to minimize emissions.

Postcombustion controls are used to capture pollutants from flue gas to minimize releases to the environment. These systems are capable of achieving a substantially high pollutant removal rate. Wet scrubbers and fabric filters (baghouses) are an emissions control option to minimize the release of pollutants from medical waste incinerators (19). Three types of wet scrubbers are used: (1) low-energy scrubbers (e.g., spray tower), (2) medium-energy scrubbers (e.g., impingement scrubber), and (3) high-energy scrubbers (e.g., venturi). Dry scrubber systems mix an alkaline reagent such as sodium bicarbonate or hydrated lime with the flue gas. The reagent mixes with the acid gas emissions in the flue gas, producing nonhazardous products such as salts of chlorine which are collected as particulate matter on a fabric filter (baghouse). A secondary acid gas neutralization occurs on the fabric filter as unreacted reagent continues to react with acid gases.

HEALTH EFFECTS

Carcinogenic Air Emissions

California Air Resources Board. In 1990, the California Air Resources Board (CARB) published a report presenting a proposed control measure designed to reduce dioxins and other emissions from medical waste incinerators (10). Following the CARBs designation of dioxins as a toxic air contaminant in 1986, the board evaluated the need and appropriate degree of control for that compound. Medical waste incinerators were given the highest priority by CARB for dioxins control measure development because (1) medical waste incinerators were designated by CARB as having the greatest potential risk of all dioxins sources that it had identified, (2) most of the incinerators identified by CARB were uncontrolled and located in residential areas, and (3) emissions testing conducted by CARB showed that medical waste incinerators are also sources of other pollutants, including cadmium, benzene, polycyclic aromatic hydrocarbons, lead, mercury, nitrogen oxides, sulfur dioxide, particulate matter, and hydrochloric acid.

A health risk assessment model was used to estimate the potential acute, chronic, and cancer health effects from combustion source pollutants. The potential cancer risk was determined by taking the lifetime average daily dose of a pollutant and multiplying it by a risk factor which was developed by the California Department of Health Services (DHS). The DHS risk factor for a lifetime of exposure to 1 picogram per cubic meter of dioxins ranged from 24 to 38 chances in a million of contracting cancer. Because dioxins emitted to ambient air can accumulate in the environment and create exposure events through other routes, a multipathway risk assessment method was developed. For the study the estimates of the total potential dioxin risk was based on an average daily dioxins dose exposure from four pathways: (1) inhalation, (2) dermal absorption, (3) soil ingestion, and (4) mother's milk ingestion.

Data were derived from emissions tests conducted on eight California medical waste incinerators. These incinerators included single-chamber units without air emission control systems that processed less than 100 pounds per hour, and multichamber units, most equipped with either a baghouse or wet scrubber, processing from 500 to about 1000 pounds of waste per hour. Of those units with emissions control equipment, none had been designed for dioxins reductions.

Dioxins emissions were expressed in terms of 2,3,7,8-tetrachlorodibenzo-p-dioxins equivalents (TCDD equivalents), based on the relative potency of the 15 dioxins or furans compared to 2,3,7,8-TCDD. The emissions rates from the eight incinerators ranged from 7 to 216 nanograms of dioxins per kilogram of waste burned (ng/kg) at the controlled facilities, and from 230 to 6200 ng/kg at the uncontrolled facilities. Dioxin emissions, and corresponding risk, were reported to vary widely depending on the waste composition, the incinerator design, and operating conditions.

The modeling showed the potential maximum individual risk ranging from 1 to about 250 chances in a million of getting cancer. Of note, the three units fitted with wet scrubbers showed a multipathway risk ranging from 1.3 to 15.7 per million cancer risk, and the one unit equipped with a baghouse exhibited a multipathway risk of 180 per million. Extrapolating these figures to all medical waste incinerators statewide, the potential statewide cancer incidence for dioxins emissions ranged from 1.2 to 1.8 cases in populations exposed. One large facility described as being equipped with a well-designed incinerator and emission control device, which met proposed control measures during the test, had risks between 1 and 10 per million.

CARB also investigated cadmium emissions rates for the eight incinerators tested, which ranged from 1.2 to 3183 micrograms of cadmium per kilogram of waste burned at facilities. A risk factor for a lifetime of continuous exposure to 1 nanogram per cubic meter of cadmium was determined to be 2 to 12 chances in a million of contracting cancer by the DHS. The health risk assessment method for cadmium was the same as for dioxin with the exclusion of the mother's milk pathway for cadmium emissions. The maximum individual lifetime risk from cadmium exposure from tested incinerators ranged from less than 1 per million from a controlled facility to 45 per million from an uncontrolled facility. Taking only the inhalation exposure into account, the risk ranged from 1 to 15 per million, with seven out of eight incinerators with a risk of less than 10 per million.

Several emission control technologies evaluated to assess the potential to reduce dioxins included wet scrubbers, fabric filters (baghouses), and dry scrubber/fabric filter combinations. Source test results reviewed by CARB on medical waste incinerators and on municipal solid waste incinerators conducted by CARB and Environment Canada found that 99 percent dioxin emission reductions are obtainable for medical waste incinerators operating under normal conditions. The dry scrubber/baghouse combination appeared to provide higher and more consistent reductions in dioxin emissions than wet scrubber systems. Based on this, CARB concluded that properly designed dry scrubber/baghouse systems would consistently meet the dioxin emission requirements of the proposed control measure, although it was also noted that such consistency may very well be achievable using the wet scrubber technology.

A 99 percent reduction efficiency or an emission rate of 10 nanograms of dioxins per kilogram of waste was identified as the best available control technology for dioxins. If such controls were applied to the eight medical waste incinerators and a 99 percent reduction achieved, the maximum individual dioxins risk per million exposed would range from 1 to 2.5 per million.

New York City Health and Hospitals Corporation. The New York City Health and Hospitals Corporation undertook a study of medical waste incinerators that examined short- and long-term health effects based on six future incineration options (9). The study facilities included three regional medical waste incinerators with capacities of 48, 165, and 330 tons per day, one coincineration facility with a capacity of 2250 tons per day, and two on-site incinerators with capacities of 4 and 13.6 tons per day. Emission factors were determined based on emission studies at other facilities, taking into consideration variables such as dispersion factors, facility operation, and pollution control equipment.

The long-term effects were judged based on the ability to cause cancer. The cancer risk by inhalation at the point of maximum impact was calculated based on the assumption of a daily 24-hour dose for a lifetime of 20 cubic meters air which contains the maximum carcinogen concentrations and 100 percent absorption by a 70-kilogram adult. The conservative cancer potency factors developed by the EPA Cancer Assessment Group for dioxins (PCDD/PCDF) and the primary trace metals of concern [arsenic, chromium(VI) and cadmium] were applied. Other organic carcinogenic emissions were assumed to contribute little to the total risk and therefore not included in the assessment.

The risk of exposure to deposited particles through all related exposure pathways was considered. In a separate risk assessment of the Brooklyn Navy Yard resource recovery facility which was characterized as having been thoroughly scrutinized in legal review, the total inhalation risk for metals and dioxins was observed to be about one-third of the total risk. Based on this finding, the upper-bound chances of cancer for a person exposed for a 70-year lifetime at the point of maximum impact was made by applying a multiplier of three times the inhalation risk.

Based on the risk assessment, the chances in a million for the most exposed person to develop cancer due to a lifetime exposure to emissions from each facility are reported in Table 8.5. The authors noted upper-bound risk estimates of less than 10 chances of cancer in a million as generally acceptable in regulatory reviews and risks of 1 in a million as negligible. Based on this assumption, only one on-site hospital incinerator presented a risk that exceeded 10 in a million. All other facilities resulted in risk within an acceptable range, and three facilities (two regional and one coincineration facility) presented a negligible risk. The risk estimates of future incineration options was much less than those observed by CARB for existing uncontrolled or inadequately controlled medical waste incinerators.

The likelihood of sensitive receptors at the point of maximum ground-level effects was discussed. Risk was considered to be insignificantly low for coincineration facilities and regional facilities due to the large areas affected and the likelihood of such facilities being located in industrial areas. On-site hospital incinerators were noted to have the greatest potential to affect sensitive receptors because many are located in residential areas that may have large numbers of sensitive people. The authors stressed the need for site-specific evaluations to make such determinations.

Noncarcinogenic Toxic Air Emissions

As part of the California Air Resources Board's 1990 risk assessment of medical waste incineration, acute and chronic noncancer health risks from exposure to other

TABLE 8.5 Chances in a Million of Developing Cancer Following Lifetime Exposure to Medical Waste Incinerator Emissions

Type of Facility	Tons/Day	Total Risk
Regional	48	1.5
Regional	165	0.14
Regional	330	0.22
Coincineration	2250	0.14
On-site	4	4.8
On-site	13.6	15.4

Source: Reference 9.

emissions from the eight incinerators were examined (10). Results of the dispersion modeling conducted for the cancer risk analysis as described above and the current recommended reference doses (RfDs) from the California Department of Health Services were used to examine potential chronic effects from lifetime average daily doses of cadmium, iron, manganese, nickel, and lead. Ambient background concentrations of these pollutants were not taken into consideration. The contribution from the incinerators did not exceed the RfDs for most pollutants, although it was noted that one facility contributed lead emissions amounting to 57 percent of the lead RfD value.

Potentially acute effects from 24-hour exposures to hydrochloric acid and particulate matter were also examined. Five of these uncontrolled or inadequately controlled facilities tested were found to be significantly above the RfD for HCl emissions. Particulate emissions were not found to contribute to a violation of the ambient particulate standard in California. Currently available emission control technologies such as the dry infection/fabric filtration technology have been shown to achieve particulate matter (including heavy metals and dioxins) and HCl emissions well within regulatory standards (11).

Radioactive Contaminants in Air and Ash

The primary regulatory authority for low-level radioactive waste (LLRW) is the Nuclear Regulatory Commission, although state governments have the authority to adopt regulations and carry on such programs. Safe levels in ambient air and water have been established by the NRC in the form of maximum permissible concentrations (MPCs). The MPC levels have been established to protect human health while considering background radiation levels. Incineration of LLRW must not be conducted in a manner that would release radioactive emissions in excess of the MPC values established by the Nuclear Regulatory Commission (18).

A study was conducted on the Mayo Foundation incinerator in Rochester, New York to identify radioactive releases from stack emissions to the ambient air (20). Prior to incinerating LLRW, the Mayo Foundation stores the waste for up to 22 months, depending on the half-life of the specific isotope and the time necessary for the isotope to decay to safe levels. Of the 26 isotopes examined, none exceeded the NRC maximum permissible concentration when measured at the top of the stack. The sum of ratios for all isotopes measured at the top of the stack compared to the NRCs maximum permissible concentration totaled 1.5×10^{-4}. Ratio sums of less than 1 are considered acceptable to the NRC and therefore safe to human health.

In 1992, Brady published a study examining the fate of tritium (^{3}H), ^{14}C, and ^{131}I in typical dual-chambered controlled-air medical waste incinerators with wet scrubber emission control systems with a 99.5 to 99.9 percent organic compound destruction and removal efficiency (21). The author reported that when these isotopes are introduced into the incinerator, between 99.9 ando 99.95 percent of ^{14}C-labeled organic is converted to carbon dioxide and traces of carbonate salts; ^{3}H burns at similar efficiencies to water vapor, HCl, and HI; and that some of the ^{131}I-labeled sodium iodide converts to hydrogen iodide and elemental iodine, with the remaining sodium iodide generally converting to submicron particulate matter, most of which is vaporized into the gas phase.

As reported by the author, because both ^{14}C-labeled compounds and ^{3}H burn with greater than 99.95 percent efficiency, the incinerator ash contains less than 0.05 percent of the ^{14}C and ^{3}H introduced into it. If quantities of either ^{14}C or ^{3}H exceed allowable concentrations in the exhaust stack, the only corrective measure is to reduce the volume of these isotopes from entering the incinerator. Between 20 and 30 percent of the sodium iodide remains as a crystalline ash material which discharges to the incinerator ash, with the remainder of the ^{131}I isotopes discharging to the exhaust gas, of which 99 percent will be removed in the scrubbing system and end up as sodium iodide in the waste liquid from the scrubber.

Incinerator ash may contain radioactive levels as long as the levels do not exceed the maximum permissible concentrations established by the NRC in 10 CFR 20. In this context, the MPC levels are used by the Mayo Foundation incinerator to calculate the amount of LLRW that can be legally incinerated at the facility during the course of a year (20).

Pathogenic Emissions in Air

The release of human pathogenic microorganisms through stack emissions is often identified as a concern by various groups, although the pathogenic hazard to health and the environment is extremely low when recognizing the elements necessary for disease transmission and the principles of microbial destruction. Infectious disease requires a source of pathogenic microorganisms, a transmission pathway from the source to the host, and an inoculation through a portal of entry of enough microbes to cause infection and disease in a susceptible host. Pathogens of high concern to the public include the human immunodeficiency virus (HIV) and the hepatitis B virus (HBV), neither of which is transmitted via the airborne route. Most other pathogens commonly found in both medical waste and municipal waste are opportunistic pathogens, requiring unusual conditions for disease transmission to occur, conditions that would not be favorable to microbial agents that may be emitted through the incinerator stack.

Although stack emission studies are limited, survival and escape of enough human pathogens to the atmosphere capable of causing disease in a susceptible host is unlikely. In 1982, Kelly reported findings from a study examining microbial emission data from a hospital incinerator under actual conditions (22). Temperatures were reported from 1380 to 1900°F in the primary chamber and from 1200 to 1950°F in the secondary chamber. Hospital waste was burned at a rate of 500 to 800 pounds per hour. Retention times were not identified. Bacterial concentrations in the stack gases were observed at concentrations of 231 colonies per cubic meter and in ambient air at 148 colonies per cubic meter. The authors of the study reported that the differences between ambient air (more than 200 percent ambient air having been added to the primary and secondary chambers) and stack emission bacterial concentrations were not statistically significant ($p > 0.05$), but noted that the ambient air did not account for all bacteria recovered from the stack gas. Because these bacterial isolates were not identified, it was not determined whether human pathogens were present.

In 1989, Allen et al. published findings on the potential for a hospital incinerator to release human pathogenic bacteria to the ambient air through stack emissions (23). The test incinerator, located in a large primary care facility, was described as a manual-load, two-chambered unit with a burn capacity of about 100 pounds per hour designed to maintain a temperature of 1400°F in both combustion chambers. Bacteria were recovered from stack emissions, although the counts were not significantly different from counts observed in ambient air. *Bacillus subtilis*, the indicator organism used for the study, was not recovered from the stack emissions, suggesting that the source of the recovered bacteria was not from the unburned waste. It was concluded that the most likely source of the bacteria in the stack emissions was the incinerator combustion air.

Bacterial emissions were examined by Blenkharn and Oakland from a newly constructed conventional oil-fired two-chamber controlled-air clinical waste incinerator, with a primary chamber designed to operate at 1470°F and the secondary chamber at 1800°F (24). The system had a minimum design loading rate of 350 kilograms per hour. Viable gram-positive and gram-negative bacteria were recovered from the exhaust flue gases during normal operation in numbers of up to 400 colony-forming units per cubic meter. Although reasons for the survival could not be explained directly, the authors postulated that survival may be attributed to rapid transit times of fine particles suspended in air within the incinerator, or to inadequate combustion due to the presence of large volumes of liquids, and that bacterial survival may be a common occurrence from older, inadequately controlled or maintained units. The authors did not feel that it was appropriate to attribute any public health hazard to low-level emissions of viable bacteria from clinical incinerator stacks. In a separate study, Segall also noted that because microorganisms typically sorb to available particles, most that escape in emissions would be expected to sorb to fly ash particles, further reducing the opportunity for release to the ambient air (25).

Pathogenic Releases to Ash

In 1986, the EPA concluded that complete and effective combustion of medical waste would have the effect of killing microorganisms in the waste (26). This was also the conclusion of the Mayo Foundation incinerator report, which quoted a declaration by the National Research Council's Committee on Hazardous Biological Agents in the Laboratory as stating that " . . . with primary combustion temperatures of at least 1600°F, secondary combustion temperatures of 1800°F with good mixing, and a gaseous retention time of approximately 2 seconds . . . all pathogens and proteinaceous materials are denatured (27)." Nevertheless, inadequate combustion practices have been shown to result in viable microorganisms present in incinerator ash. In 1969, Peterson and Stutzenberger examined municipal incinerator residues from four 1960s-era incinerators with operating temperatures ranging from 1200 to 2000°F to evaluate the ability of these units to destroy bacteria normally associated with solid waste (28). None of the incinerators were observed to produce a sterile residue. The ash from the incinerators was associated with unburned material such as vegetables, animal wastes, and newspapers with readable print, which identified that the waste had not been entirely burned. A subsequent investigation by Peterson focused on the distribution and survival patterns of pathogenic microorganisms before and after incineration (29). Ash and quench water residues from eight 1960s-era municipal incinerators with recorded operating temperatures of between 1200 and 2000°F were examined for total bacteria, total and fecal coliform bacteria, heat-resistant spores, and selected enteric pathogenic bacteria. Again, none of the incinerators were found to produce a sterile residue. The enteric pathogens *Salmonella derby* and *S. st. paul* were isolated from quench water of one incinerator. Fecal coliforms, a bacterial indicator of fecal pollution, were also isolated from seven of the eight incinerators. Although survival of potentially pathogenic microorganisms would be expected to occur only if a medical waste incinerator were improperly operated, the steps required for disease transmission to occur as identified in the preceding section minimize the health risk associated with pathogen survival in incinerator ash.

Toxic Chemical Contaminants in Ash

In 1992, Hasselriis published a report on incinerator ash from medical waste incinerators (30). Based on findings from studies of municipal solid waste incinerator ash studies and limited medical waste ash studies, the author's conclusions included the following points: (1) Ash residues contain heavy metals that were originally in the waste materials; (2) dry fly ash is a toxic material and should be conditioned with water or moist ash as soon as possible to prevent harm to humans and the environment; and (3) tests of ash residues and quench water from medical waste incinerators have shown that the leachates tested by the toxicity characteristic leaching procedure (TCLP) meet EPA limits, on average, although the data may vary widely. The human health significance of these findings was not addressed.

Conclusion

Although generally faced with public opposition, medical waste incineration remains an important option for the treatment, destruction, and disposal of medical waste. The incineration industry has changed substantially during the past decade as new regulations are instituted by state and local authorities, and the need to control pollutant releases to the environment becomes necessary. Current information indicates that medical waste incinerators can operate within acceptable ranges of risk if properly designed, operated, and maintained. Pollution prevention practices to remove contaminants from the waste stream before the waste is incinerated are recognized as an important factor in reducing the generation of pollutant emissions. The current state-of-the-art pollution control devices are capable of significantly controlling pollutant releases to the environment, again only if these systems are properly designed, operated, and maintained. Risk assessments for new facilities have become an important step in understanding the individualized pollution control needs for a proposed incinerator. Proposed EPA rules mandated by Congress as part of the federal Clean Air Act Amendments of 1990 will have the effect of closing inadequately designed facilities when adopted. Based on its proposed standard, the EPA estimates that its implementation would result in 80 percent of proposed incinerators not being built and 80 percent of existing incinerators discontinuing operation. This increased regulatory presence and focus on environmentally sound medical waste combustion practices will add further assurances to the public that new medical waste incinerators and existing units are both operating in a manner that minimizes the impact on public health and the environment.

REFERENCES

1. U.S. Environmental Protection Agency. Standards of Performance for New Stationary Sources and Emission Guidelines for Existing Sources: Medical Waste Incinerators. *Federal Register.* February 27, 1995.

2. Turnberg WL and Kelly KE. Health effects of medical waste incineration. Proceedings of the 13th International Incineration Conference, Houston, Texas. May 9–13, 1994.

3. Turnberg WL. The health effects of medical waste incineration: Biological, chemical and radiological emissions. Proceedings of the 14th International Incineration Conference, Seattle, Washington. May 8–12, 1995.

4. Rutala WA, Odette RL, and Samsa GP. Management of infectious waste by U.S. hospitals. *JAMA*, 262(12):1635–1640. September 1989.

5. U.S. Environmental Protection Agency. Guides to Pollution Prevention: Selected Hospital Waste Streams. EPA/625/7-90/009. June 1990.

6. Emery R, Marcus J, and Sprau D. Characterization of low-level radioactive waste generated by a large university/hospital complex. *Health Phys*, 62(2):183–185. February 1992.

7. Hasselriis F and Constantine L. Characterization of Today's Medical Waste. In: Green AES, Ed., *Medical Waste Incineration and Pollution Prevention*. Van Nostrand Reinhold. New York. 1992.

8. Lauber JD. New perspectives on toxic emissions from hospital incinerators. Presented at the New York State Legislative Commission on Solid Waste Management Conference on Solid Waste Management and Materials Policy, New York. February 12, 1987.

9. New York City Health and Hospitals Corporation. The New York City Medical Waste Management Study Final Report, Identification and Evaluation of Waste Management Techniques and treatment and Disposal Options, Volume 5: Task 3 Report. Prepared by Waste-Tech Waste Energy Technologies, Inc. and Konheim and Ketcham, Inc. June 1991.

10. California Air Resources Board. Survey of Medical Waste Incinerators and Emissions Control. Prepared by Energy and Environmental Research Corporation, Contract A832-155. Sacramento, California. January 1992.

11. California Air Resources Board. Proposed Dioxins Control Measure for Medical Waste Incinerators. Sacramento, California. May 1990.

12. U.S. Environmental Protection Agency. Medical Waste Incinerators—Background In-

formation for Proposed Standards and Guidelines: Industry Profile Report for New and Existing Facilities. EPA-453/R-94-042a. July 1994.

13. U.S. Environmental Protection Agency. Medical Waste Incinerators—Background Information for Proposed Standards and Guidelines: Industry Profile Report for New and Existing Facilities—Draft. Public Docket A-91-61 II-C. Office of Air and Radiation, Washington, DC. September 23, 1991.

14. Turnberg WL. An Examination and Risk Evaluation of Infectious Waste in King County, Washington. Seattle–King County Department of Public Health, Seattle, Washington. March 18, 1988.

15. Brinckman GA. The control of particulate matter, acid gas, and heavy metals emissions from medical waste incinerators. Proceedings of the Thermal Treatment of Radioactive, Hazardous, Chemical, Mixed, Energetic, Chemical Weapon, and Medical Wastes 1993 Incineration Conference, Knoxville, Tennessee. May 3–7, 1993.

16. Green AES. The future of medical waste incineration. In: Green AES, Ed., *Medical Waste Incineration and Pollution Prevention*. Van Nostrand Reinhold, New York. 1992.

17. Hasselriis F. Relationship between input and output. In: Green AES, Ed., *Medical Waste Incineration and Pollution Prevention*. Van Nostrand Reinhold. New York. 1992.

18. Wallace DD. Assessment of the effects of incinerator operating parameters on emissions from medical waste incinerators. Proceedings of the Thermal Treatment of Radioactive, Hazardous, Chemical, Mixed and Medical Wastes 1992 Incineration Conference, Albuquerque, New Mexico. May 11–15, 1992.

19. Dennis MG and Borowsky AR. Controlling emissions from medical waste incinerators. Proceedings of the Thermal Treatment of Radioactive, Hazardous, Chemical, Mixed and Medical Wastes 1992 Incineration Conference, Albuquerque, New Mexico. May 11–15, 1992.

20. Minnesota Pollution Control Agency. Air Quality Emissions and Deposition, Mayo Foundation Incinerator. Technical Work Paper 4. Prepared by Camp Dresser & McKee, Inc. 1991.

21. Brady JD. Fate of tritium, carbon[14], and iodine[131] in wet scrubber air pollution control systems on chemical and medical waste incinerators. Proceedings of the Thermal Treatment of Radioactive, Hazardous, Chemical, Mixed and Medical Wastes 1992 Incineration Conference, Albuquerque, New Mexico. May 11–15, 1992.

22. Kelly NA. Masters thesis, Health Sciences Center, University of Illinois. 1982.

23. Allen RJ, Brenniman GR, Logue RR, and Strand VA. Emission of airborne bacteria from a hospital incinerator. *JAPCA*, 39:164–168. 1989.

24. Blenkharn JI and Oakland D. Emission of viable bacteria in the exhaust flue from a waste incinerator. *J Hosp Infect*, 14:73–78. 1989.

25. Segall RR. Development and evaluation of a method to determine indicator microorganisms in air emissions and residue from medical waste incinerators. *J Air Waste Manage Assoc*, 41(11):1454–1460. 1991.

26. U.S. Environmental Protection Agency. EPA Guide for Infectious Waste Management. EPA/530-SW-86-014. May 1986.

27. Minnesota Pollution Control Agency. Ash Characterization, Handling, and Disposal, Mayo Foundation Incinerator. Technical Work Paper 6. Prepared by Camp Dresser & McKee, Inc. 1991.

28. Peterson ML and Stutzenberger FJ. Microbiological evaluation of incinerator operations. *Appl Microbiol*, 18(1):8–13. July 1969.

29. Peterson ML. Pathogens Associated with Solid Waste Processing. USEPA SW-49r. 1971.

30. Hasselriis F. Ash Disposal. In: Green AES, Ed., *Medical Waste Incineration and Pollution Prevention*. Van Nostrand Reinhold, New York. 1992.

9

STEAM STERILIZATION

Steam sterilization remains a widely used technology for treating biohazardous waste. Steam sterilizers are used to treat biohazardous waste by both small generators such as dental clinics and physicians' offices and commercial biohazardous waste treatment companies treating biohazardous waste regionwide. This discussion, which has been adapted from other work written by the author (1), addresses the process of steam sterilization and the advantages and disadvantages of this treatment technology.

PROCESS DESCRIPTION

Steam sterilizers (autoclaves) have historically been used for sterilizing biohazardous waste on location, although larger regional off-site facilities have become common. For example, Browning-Ferris Industries (BFI) has offered off-site treatment by steam sterilization of medical waste since 1976 (2). An example of a large scale BFI medical waste autoclave is presented in Figure 9.1. Other regional biohazardous waste autoclave treatment services are also commercially available (2).

Steam sterilization involves placing contaminated waste into the chamber of a steam sterilizer (autoclave, retort) and exposing the waste to pressurized steam of sufficient temperature and duration to render the waste noninfectious. Steam sterilizers operate by replacing the air within the sterilizer chamber with steam, generally by forcing the air out through a valve located at the bottom of the unit as the steam is introduced into the chamber (gravity displacement unit; see Figure 9.2) or by exhausting the air by vacuum before the steam is introduced into the chamber (preexhaust unit) (3,4). After the steam enters the chamber, the temperature will

177

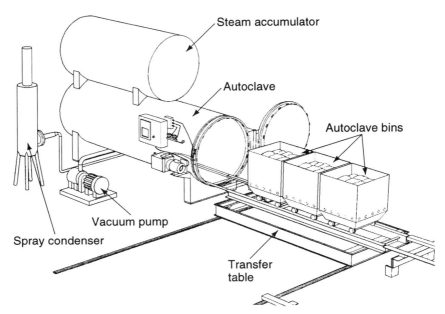

Figure 9.1   BFI medical waste autoclave. Courtesy of Browning-Ferris Industries, BFI Medical Waste Autoclave, copyright 1995 by Browning-Ferris Industries, Houston, Texas. Reprinted by permission.

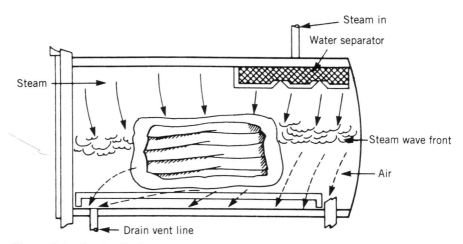

Figure 9.2   Gravity displacement steam sterilizer. (From Joslyn LJ. Sterilization by Heat. In: *Disinfection, Sterilization, and Preservation*, Fourth Edition, p. 505, Seymour S. Block, ed., copyright 1991 by Lea & Febiger. Reprinted by permission.

begin advancing to the selected setting. This is called the heat-up time (steam penetration time). The holding time begins after the entire load has reached the desired temperature that represents the minimum time–temperature requirement for achieving sterilization of a given load. The exposure time represents the entire period necessary to achieve sterilization and includes the sum of heat-up time and holding time plus a margin of error (5).

ELEMENTS OF EFFECTIVE STERILIZATION

Effective treatment requires a knowledge of many factors relating to the treatment process. These include the type of unit, type and volume of wastes to be processed, waste packaging, and placement in the chamber. Skilled, trained steam sterilizer operators are essential. Training should be provided by the employer and include proper autoclave operation as well as potential associated hazards (e.g., chemically toxic or radioactive wastes should never be autoclaved). Processing effectiveness should be monitored to ensure that treatment has been accomplished using time/ temperature charts, chemical indicators that produce color changes to correspond with the necessary time–temperature relationship to achieve sterilization, and biological indicators (e.g., spore strips of *Bacillus subtilis* or *B. stearothermophilus*) to ensure inactivation of the most resistant microorganisms. Only such monitoring will provide the operator with the knowledge regarding operational parameters that is necessary for an autoclave to treat a given waste load effectively. All units should be inspected and serviced routinely to ensure optimum operation. An adequate system of record keeping should also be in place not only for internal purposes, but for evidence of proper processing if requested by government regulators or accreditation organizations.

Effective standardization of the steam sterilization process is essential to assure treatment effectiveness. Five factors must be considered:

1. Conscientious, dependable, skilled operators
2. Correct methods of packaging to ensure steam penetration of the load
3. Proper loading of the unit
4. Approved sterilizer with demonstrated reliability
5. Adequate exposure period that provides complete steam penetration of the entire load and ensures microbial destruction with a safety margin (6)

Everall and Morris describe the ideal steam sterilizer container for treating infectious wastes as one that is reasonably inexpensive, corrosion resistant, leakproof, and capable of allowing complete steam penetration to its contents (6). Traditionally, metal containers have been used for containing laboratory wastes, but during the past few years, attention has focused on plastic bag use. The authors reported that in the course of routine laboratory autoclave surveillance using thermocouple measurements, loads contained in plastic bags frequently performed poorly. Further investigation observed that this was caused by inadequate steam penetration and removal of air from all-plastic bags. The authors suggested that when plastic bags are used to contain infective material for treatment, steps be taken to establish the necessary sterilizing time/temperature relationships and steam penetration.

Rutala et al. investigated the conditions necessary to achieve sterilization of microbiology laboratory waste by steam sterilization (7). Standardized loads (5, 10, and 15 pounds) containing contaminated petri plates in stainless steel containers or polypropylene plastic bags, with and without added water, were examined. Time/temperature profiles were determined using a digital potentiometer; sterilization effectiveness was monitored using biological monitors (*B. stearothermophilus* spores) located at the center of the test loads.

A single polypropylene bag with its sides rolled down had a significantly higher heat transfer rate than that of the same bag with its top twisted shut. Heat transfer was observed to be facilitated for smaller volumes of waste (e.g., 5-pound loads compared to 15-pound loads) and for waste contained in stainless steel containers rather than plastic bags. Heat-up time (steam penetration time) was not observed to be improved significantly by the addition of water into either the stainless steel or polypropylene containers. The addition of water was not observed to influence bacterial kills when placed into stainless steel containers, although an effect was observed when kills were placed in polypropylene bags.

Rutala et al. noted that the effective treatment of microbiological waste should be measurable. The most conservative approach would be the use of the biological indicator *B. stearothermophilus*, providing for a wide margin of error. However, the authors observed that a steam sterilizer operating cycle of 90 minutes was necessary to eliminate *B. stearothermophilus* when contained in 10 to 15 pounds of microbiological waste, which they believed would be viewed as unrealistic and unnecessary from a health perspective. Alternative measurement processes were not suggested, however.

A similar study was conducted by Lauer et al. in which time/temperature profiles of microbiology waste being autoclaved were observed with a temperature probe and sterilization effectiveness monitored using chemical and biological indicators (8). Waste was placed into stainless steel and polypropylene containers of similar dimensions. Given amounts of water (0, 100, and 1000 mL) were added to the containers. Waste placed into autoclavable bags with given amounts of water (0, 100, and 1000 mL) was also placed in stainless steel and polypropylene containers and tested by described procedures.

Lauer et al. found that adding water to the waste container prior to autoclaving significantly affected the effectiveness of the steam sterilization process. The test infectious waste was observed to be decontaminated adequately when processed in a gravity-displacement steam sterilizer at 121°C for 50 minutes when the waste was contained in a stainless steel container with the addition of 1 liter of water, or into an autoclavable bag with 1 liter of water, then placed into a stainless steel container. The use of polypropylene containers was observed to increase the necessary decontamination time significantly, whether or not water was added to the load. The height of the waste container was also observed to affect the time–temperature profile significantly. The importance was noted of overcoming cool, dry air pockets in the waste, which commonly occur in upright containers or autoclavable bags, when a gravity-displacement unit is used.

All steam sterilizer process parameters and monitors should be checked before releasing any load (9). These include time–temperature charts from a properly calibrated sterilizer, chemical indicators, which are advised for all loads, and periodic biological monitoring. An advantage of chemical indicators over biological indicators is that results are available immediately (10).

ADVANTAGES AND DISADVANTAGES

When properly operated, steam sterilizers are capable of sterilizing biohazardous waste. They are most suitable for decontaminating laboratory wastes such as stocks and cultures of infectious agents, contaminated glassware, and biologicals, but are capable of decontaminating other classes of biohazardous waste as well. For aesthetic and regulatory reasons, autoclaves are unsuitable for treating recognizable body parts.

An advantage of steam sterilization is that the technology has a long history of use by hospitals, laboratories, clinics, and other medical institutions for biohazardous waste sterilization. Steam sterilizers do not raise the public concern or involve the complex regulations such as is observed with medical waste incinerators. Another advantage is seen in their output capacity with minimal spatial requirements compared to that required for on-site incinerators (1).

Disadvantages presented by autoclaves are that the decontaminated waste retains the appearance of medical waste following treatment and the volume of the waste is not reduced. In some circumstances, landfills have refused to accept autoclaved waste, in part because of concern as to whether the waste was treated adequately and because it looks like medical waste. Some states have enacted regulations requiring not only that biohazardous waste be decontaminated but that it be rendered unrecognizable as medical waste. For these reasons, many steam sterilizers are now offered with waste shredding systems to render the decontaminated waste unrecognizable. Variations of autoclave systems are reviewed in Chapter 10 when we address alternative treatment technologies.

HEALTH CONSIDERATIONS

Health impacts attributed to autoclaving have not been documented (1). Nevertheless, operators may be exposed to hazardous constituents in steam emissions if they come into contact with the venting steam, such as commonly occurs when the autoclave door is opened at the end of the cycle. For this reason it is important to exclude any waste containing potentially toxic constituents such as hazardous chem-

TABLE 9.1 Facilities Reporting On-Site Steam
Sterilizers to Treat Biohazardous Waste in Washington
State

Facility Type	Number of Responding Facilities	On-Site Steam Sterilizers
Hospital	51	24 (47%)
Medical clinic	60	11 (18%)
Dental practice	32	11 (34%)
Physician's office	33	10 (30%)
Nursing home	43	3 (7%)
Funeral home	32	1 (3%)
Veterinary clinic	49	11 (22%)
Research laboratory	17	13 (76%)
Medical laboratory	25	11 (44%)

Source: Adapted from reference 11.

icals (e.g., antineoplastic drug waste, RCRA waste) or radiological wastes. Further
study on emissions from steam sterilization is warranted.

USE AND PRACTICES

Steam sterilizers are widely used to treat infectious waste generated by many different kinds of medical practices. In 1989 a study was conducted in Washington State
to gather information regarding the use of steam sterilization as a biohazardous
waste treatment method (11). Questionnaires were mailed to a total of 454 medical
facilities obtained by systematic sampling of statewide listings within each medical
facility category. Results of responses from nine facility categories (hospitals, medical clinics, dental practices, research laboratories, physicians' offices, funeral
homes, nursing homes, veterinary clinics, and medical laboratories) are presented
in Table 9.1.

REFERENCES

1. Turnberg WL. Human infection risks associated with infectious disease agents in the
   waste stream: A literature review. In: Washington State Infectious Waste Project—
   Report to the Legislature. Washington Department of Ecology, Olympia, Washington.
   December 1989.
2. U.S. Congress, Office of Technology Assessment. Finding the Rx for Managing Medical Wastes. OTA-O-459 U.S. Government Printing Office, Washington, DC. September 1990.
3. Joslyn LJ. Sterilization by heat. In: Block SS, Ed., *Disinfection, Sterilization, and
   Preservation*, 4th ed. Lea & Febiger, Philadelphia. 1991.

4. U.S. Environmental Protection Agency. EPA Guide for Infectious Waste Management. EPA/530-SW-86-014. May 1986.

5. Perkins JJ. *Principles and Methods of Sterilization in Health Services*, 2d ed. Charles C Thomas, Publisher, Springfield, Illinois. 1976.

6. Everall PH and Morris CA. Failure to sterilize in plastic bags. *J Clin Pathol*, 29:1132. July 1976.

7. Rutala WA, Stiegeland MM, and Sarubbi FA Jr. Decontamination of laboratory microbiological waste by steam sterilization. *Appl Environ Microbiol*, 43:1311–1316. June 1982.

8. Lauer JL, Battles DR and Vesley D. Decontaminating infectious laboratory waste by autoclaving. *Appl Environ Microbiol*, 44(3):690–694. September 1982.

9. Kotilainen HR and Gantz NM. Biological sterilization monitors: A four-year in-use evaluation of two systems. *Infect Control*, 6(11):451–455. 1985.

10. Fitzpatrick BG and Reich RR. Sterilization monitoring in vacuum steam sterilizers. *J Healthcare Mater Manage*, pp. 82–85. September/October 1986.

11. Turnberg, WL. Survey of infectious waste management practices conducted by medical facilities in Washington State. In: Washington State Infectious Waste Project—Report to the Legislature. Washington Department of Ecology, Olympia, Washington. December 1989.

10

ALTERNATIVE TREATMENT TECHNOLOGIES

In this chapter we address many of the nonincineration alternative treatment technologies proposed or currently available for treating biohazardous waste prior to disposal. Such systems are being relied upon increasingly as public and regulatory pressures continue to move the medical industry away from its historical reliance on incineration as the preferred method of treating biohazardous waste. For instance, if proposed standards for medical waste incinerators are implemented by the U.S. Environmental Protection Agency as currently written, it is estimated that up to 80 percent of existing incinerators will cease operation (see Chapter 8). There is a need to recognize and understand the alternative biohazardous waste treatment system options that are available.

In this chapter, 20 nonincineration alternative treatment technologies are reviewed. Four of these systems are based on steam sterilization principles, with variations from the classically designed autoclave systems (see Chapter 9 for a discussion of steam sterilization). This discussion includes such systems, although the technology of steam sterilization is not recognized as an alternative treatment technology in many states. The alternative treatment systems reviewed in this chapter are summarized in Table 10.1. In reading this chapter, the reader should be aware that biohazardous waste treatment systems are not limited to the systems reviewed. To date, over 60 companies have developed systems for treating biohazardous waste, although many are no longer in business.

Each treatment technology reviewed in this chapter begins with a summary of the principles behind the technology, followed by reviews of one or more commercially available systems that incorporate the given technology for treating biohazardous waste. The information presented in the reviews, as it could be obtained, includes:

1. Distributor information
2. Product overview
3. Models available
4. System specifications
5. Capital cost
6. Waste stream compatibility
7. Installation requirements
8. Operation requirements
9. System monitoring
10. Training
11. Service information
12. Treated waste characteristics
13. Environmental discharges
14. Efficacy testing
15. Approvals and installations
16. Corporate profile

Most information was derived from each vendor between the fall of 1994 and the spring of 1995 with final updates in September 1995. Additional information was derived from governmental reports and from the scientific literature when available.

Independent critical evaluations of many of these alternative treatment systems are relatively new, and much remains unknown regarding potential toxic or biological emissions or other factors that may affect human health or the environment. However, the review process by state regulatory agencies has been markedly standardized since publication of the "Technical Assistance Manual: State Regulatory Oversight of Medical Waste Treatment Technologies" by the State and Territorial Association on Alternate Treatment Technologies (STAATT) in April 1994 (see Chapter 11). The guideline was developed by the STAATT for state and local governments to evaluate new treatment technologies for approval within their respective jurisdictions. This guideline also has application to vendors developing new systems and to purchasers of new systems, by identifying critical operational, environmental and health factors that should be considered to ensure that the systems are operating as claimed without compromising the safety and health of the public, workers, or the environment.

The product information presented here represents a snapshot of each technology at the time it was collected. The field of biohazardous waste management is rapidly changing, and systems are continually being altered and improved as manufacturers and state regulators learn from their successes and mistakes. New companies are quickly formed and many are unable to survive. For this reason, the information presented in these reviews must be recognized in light of its temporal nature. However temporal, the information is important for those attempting to understand the intricacies of the many alternatives that are currently available or that will be in

TABLE 10.1   Reviewed Alternative Treatment System Companies and Models

Alternative Treatment Technology	Capacity	Capital Cost
Chemical–mechanical destruction technology		
Medical Materials & Technology		
Model MMT3000	3000 lb/hr	$395,000
Medical Safe TEC		
Model MST 300L	—	$ 85,000
Model MST 1200	700–1500 lb/hr	$385,000
Premier Medical Technology		
PMT System	750 lb/hr	Not applicable[a]
Steris Corporation		
EcoCycle 10 Processor	40 liters/hr	$ 20,000
Winfield		
CONDOR	600 lb/hr	$400,000
Alternative steam sterilization technology		
Ecolotec, Inc.		
MW-130	5 ft^3/load	$375,000
San-I-Pak Pacific Inc.		
Mark I Sterilizer	30 lb/cycle	$ 16,500
Mark X Sterilizer	600 lb/cycle	$470,000
SAS Systems, Inc.		
SAS-1	300 lb/hour	$400,000
Tempico		
Remedy-One Rotoclave	500–900 lb/hour	$395,000–$1,485,000
Steam reforming technology		
Synthetica Technologies		
Steam detoxification system	Up to 2400 lb/day	$500,000–900,000
Dry heat technology		
Disposal Sciences, Inc.		
DSI SYSTEM 2000	Up to 16 gal/cycle	$ 25,000
D.O.C.C.		
Demolizer System, Model 47	1 gal/cycle	$ 2,500
MedAway		
MedAway Waste Processor	Up to 35 lb/cycle	$ 49,400
Spintech, Inc.		
TAPS I	Up to 1.5 quarts/cycle	$ 995
TAPS II	Up to 6 quarts/cycle	$ 2,195
Microwave technology		
Sanitec, Inc.		
HGA-100S	220–350 lb/hr	$375,000
HGA-250S	550–900 lb/hr	$600,000
Rootan Medical Services Corp.		
Radlor 2 Sterilizer	50 lb/hr	$138,000

(*continued*)

TABLE 10.1 (*Continued*)

Alternative Treatment Technology	Capacity	Capital Cost
Electrothermal deactivation technology Stericycle, Inc.		
Custom-designed facilities	Up to 6000 lb/hr	Not applicable[b]
Plasma technology Plasma Energy Applied Technology		
Custom-designed systems	1000 lb/hr (example)	$2–$8 million
Electron beam sterilization technology Nutek Corporation		
Custom-designed systems	800 lb/hr (example)	$250,000–$500,000
Cobalt-60 irradiation technology Nordion International, Inc.		
Custom-designed systems	>5 tons/day[c]	$2–$5 million

[a]Available on a fee per pound basis only.
[b]All systems are owned exclusively by Stericycle, Inc. and are not for purchase.
[c]The Nordion system becomes economically feasible when treating >5 tons/day.

the future. Additional information on alternative biohazardous waste treatment systems may be obtained from:

- State environmental or public health agencies responsible for approving or evaluating alternative biohazardous waste treatment systems (see Appendix B)
- Electric Power Research Institute, P.O. Box 10412, Palo Alto, CA 94303, (415) 855-2993
- The Medical Waste Institute, National Solid Waste Management Association, 4301 Connecticut Avenue, NW, Suite 300, Washington, DC, 20008, (202) 244-4700

CHEMICAL DISINFECTION

Chemical agents have been used for disinfecting biohazardous waste for many years. Such treatment usually incorporates a mechanical destruction process that reduces the waste to a small particle size which increases contact with the chemical agent and ultimately renders the waste unrecognizable. Several different chemical agents are currently being marketed for use in various biohazardous waste treatment systems. For example, of the five chemical disinfection–based biohazardous waste treatment systems reviewed in this book, all render the waste unrecognizable by various shredding processes, and all incorporate different chemical agents to disinfect the waste. Of these treatment systems, one involves a proprietary dry calcium oxide–based disinfectant which is mixed with the waste in the presence of a small amount of water as the waste is being shredded; another disinfects the preshredded and hammermilled waste in a sodium hypochlorite solution; one disinfects using a proprietary chlorine dicxide treatment solution with corrosion inhibitors while the waste is being shredded; one system uses a peracetic acid–based disinfectant mixed with water during shredding; and one uses a proprietary dry powder disinfectant in which the active agent is not identified by the vendor.

The ability of a chemical disinfectant to kill a targeted organism effectively depends on many factors, some being (1):

1. Type of microorganism
2. Degree of contamination
3. Type of disinfectant
4. Concentration and quantity of disinfectant
5. Contact time between the antimicrobial agent and the targeted organism
6. Other relevant factors (e.g., pH, presence of electrolytes, or complex formation and adsorption such as binding to small molecules or ions, macromolecules, or soil)

Depletion of the chemical disinfectant can also occur through consumption of the chemical agent by interaction with microorganisms, loss through volatilization, chemical decomposition, or metabolism by the microorganism itself (2). Such factors should be considered when choosing a disinfectant to disinfect biohazardous waste.

FIFRA Registration. Under the Federal Insecticide, Fungicide and Rodenticide Act (FIFRA), any chemical agent used in a treatment process may require registration with the EPA Pesticide Registration Office (3). If a manufacturer makes advertised claims on a product label that a chemical agent can achieve a level of microbial inactivation (e.g., disinfectant, sterilant, sanitizer) for a specified use (e.g., use as a surface disinfectant), that agent would have to be registered with the EPA Office of Pesticide Registration. The EPA Office of Pesticide Programs is currently preparing to release its policy on such registration as it relates to medical waste treatment under the subject heading "Efficacy Data Development Guidance for Registration of Chemical Medical Waste Treatment Products." Although the policy has not yet been formally released, it is being used as an internal policy by the EPA for registration of chemical medical waste treatment products (3). The text of the draft policy is as follows:

Notice to Manufacturers, Formulators, Producers
and Registrants of Pesticide Products

Attention: Persons Responsible for Registration of Pesticides

Subject: Efficacy Data Development Guidance for Registration of Chemical Medical Waste Treatment Products

This Notice describes EPA's policy on antimicrobial claims and efficacy data requirements in EPA registered chemical medical waste treatment products/processes. This policy is effective immediately.

I. BACKGROUND

Due to an increased generation of medical waste, the search for methods of treating and disposing of medical waste has flourished. Among the more recent innovations addressing the need for treatment and disposal of medical waste are liquid solidification and on-site mechanical-chemical systems that shred the waste reducing it to small fragments. At various points during this shredding process, the medical waste is treated with a chemical germicide solution, compacted and prepared for approved disposal.

While EPA recognizes that the mechanical-chemical treatment of medical waste is useful, disinfection sterilization claims, as explained below, may not be achievable when used to treat medical waste. As a result, such claims in relation to use of an antimicrobial pesticide not specifically registered for use in the treatment of medical waste may be false or misleading, and thus unlawful under the Federal Insecticide, Fungicide and Rodenticide Act (FIFRA) Section 12.

Under FIFRA, EPA is responsible for registering all pesticidal products with antimicrobial claims (including sanitization, disinfection, or sterilization). EPA registers such products on the basis of data developed to show that the product will perform as stated, without adverse effects to man and the environment, when the product is used according to label directions. Such data must be generated in accordance with a protocol that meets the purpose of the test standards specified in EPA guidelines and provides data of suitable quality and completeness as typified by the protocols cited in those guidelines. See 40 CFR 158.70. At this time, EPA is aware of no existing protocol or study of suitable quality or completeness that is capable of demonstrating whether any particular type of treated medical waste can be disinfected or sterilized. EPA defines disinfection and sterilization, respectively, as achieving a complete kill of non-spore microbial life forms, or bacterial spores, when tested under actual (or simulated) use conditions by EPA approved protocols.

Additionally, considering the various types of medical waste that may be treated by the germicides and the state of current chemical technology, EPA believes it is highly unlikely that medical waste, in fact, can be disinfected or sterilized with chemical agents in actual use situations.

For example, disinfection and sterilization claims imply a complete kill of all vegetative pathogens and microorganisms, respectively. However, high levels of organic soil load may inactivate a germicide. Thus, because medical waste contains high levels of organic matter, such as body fluids, the chemical germicide applied to the shredded medical waste may not in fact be able to disinfect or sterilize the waste materials. Further, it is not clear that disinfectants and sterilants are even capable of penetrating the dried blood and body fluids that are on or in porous or nonporous materials, such as bandages, syringes, hypodermic needles, bottles, test tubes with coagulated blood, plastic tubing and valves, plastic coated bandages, or other shredded waste particles coated with organic materials containing human pathogens.

II. POLICY

While EPA recognizes that solidification or mechanical-chemical treatment of medical waste may be useful, disinfection or sterilization claims associated with such uses may be false or misleading. Disinfection and sterilization are defined by EPA as achieving a complete kill of the microbes in question and, as discussed above, a complete kill of all microbes may not be demonstrable using current product performance protocols and may not be achievable by chemicals when used to treat medical waste.

The EPA considers pesticide products bearing false or misleading claims to be misbranded and therefore not consistent with the requirements of FIFRA. Accordingly, such claims should not be made in association with the treatment of medical waste without data to support the claimed level of antimicrobial activity achieved by the product/system while in operation.

EPA believes that sanitization claims, rather than disinfection or sterilization claims, are the more appropriate pesticidal claim associated with the treatment of medical waste. Sanitization is defined by EPA as a level of antimicrobial activity achieving at least a 99.9 percent reduction of the microbial population. Thus, EPA suggests that registrants interested in making sanitization claims in association with the treatment of medical waste submit an efficacy testing protocol to determine if it is of suitable quality and completeness.

Such protocols should be written for testing during actual use of the germicide in the system on the type of medical waste, (e.g. shredded soiled syringes), that it is intended to treat, containing the ratio of waste/fluid to be treated during actual use. In addition, it should show that the test is sufficiently sensitive to demonstrate the required reduction in microbial population. Once EPA concludes that the protocol is acceptable, the registrant should conduct the study and submit the data to EPA's Registration Division for review.

EPA believes that the health care industry, and the affected municipalities and States, may develop acceptable microbial reduction limits (standards), and terminology, to designate the desired level of microbial destruction for different categories of medical waste, that accept and imply less than complete destruction of microbial pathogens, (disinfection) or microbial life (sterilization), yet which are sufficient to prevent disease transmission. Statements of compliance with such standards may be included on labels of EPA registered medical waste sanitizing products.

III. EFFECTIVE DATE

(*Note:* Upon formal EPA approval.)

IV. FURTHER INFORMATION

If you have questions regarding this policy you may contact Dr. Zig Vaituzis, Antimicrobial Program Branch, Registration Division (703-305-7167).

Effluent Releases.   Some medical waste treatment technologies that employ chemical disinfection and mechanical destruction involve releasing relatively large volumes of liquid effluents to the sanitary sewer. Such volumes may affect the receiving sewage treatment system's ability to handle the chemical waste load effectively as required under applicable waste discharge permits (e.g., the National Pollutant Discharge Elimination System waste discharge permit). Because of this, the receiving sewage utility should be contacted to determine whether effluent from a medical waste treatment process may be discharged to that sanitary sewage system. The sewage utility may require that the effluent be chemically tested to determine whether it may be disposed directly to the system or if pretreatment would be necessary.

Worker Exposure. Potential worker exposure to both chemical agent and airborne microorganisms should be considered when chemical disinfection/mechanical grinding systems are employed. Contact with the chemical agent may occur as the agent is added either to the equipment or to the waste, or as a result of volatilization during the mechanical destruction process. Airborne microorganisms may also be released during the grinding process. For example, one study published in 1992 observed the presence of an indicator bacterium on the external surface of a sodium hypochlorite–based chemical–mechanical medical waste treatment system which indicated that aerosols were not contained within the system during operation, although the system functioned under negative air pressure (4).

Chemical–mechanical destruction medical waste treatment systems should operate under negative air pressure, and tests should be conducted to assure that potentially harmful chemical and microbiological releases are not occurring. These test data should be made available to any potential purchaser of the medical waste treatment equipment.

System Reviews. Five chemical disinfection/mechanical destruction medical waste treatment technologies are reviewed in this section. These systems are available from the following companies:

1. Medical Materials & Technology, Inc.
2. Medical Safe-TEC, Inc.
3. Premier Medical Technology, Inc.
4. Steris Corporation
5. Winfield Environmental Corporation

Medical Materials & Technology, Inc. (5,6)

Source

Medical Materials & Technology, Inc.
5406 Rutherglenn Drive
Houston, TX 77096
Tel: (713) 723-7270
Fax: (713) 728-9357

Product Overview. The MMT3000 system decontaminates and destroys biomedi-cal waste via a dry calcium oxide–based disinfectant and physical destruction of the waste by a series of cutting blades. The system is designed for both regional and on-site use by larger medical waste generators such as hospitals. It is not designed for use by smaller facilities such as dental offices or laboratories.

Biomedical waste is loaded into a continuous waste loading hopper that consists of a scale and auger. It is then conveyed to the primary cutting and sterilant release chamber, where it is shredded by two sets of opposing cutting blades and sterilized. At this stage the sterilant and a small amount of water are added to the waste. The sterilant, which is referred to as Cold-Ster, employs calcium oxide as the active ingredient. The waste is further reduced in size and sterilized in the cutting, convey-ing, and mixing chamber. Before the sterilized waste can exit the system, it is passed over an exit screen that captures particles that exceed $\frac{3}{8}$ inch in size. Once the waste has passed through the screen, it is passed into a waste catcher, which can transfer the waste by screw conveyor into a dumpster for general solid waste disposal. The pH of the waste is between 11.7 and 12.3 as it exits the unit. No residual liquid is produced. The entire process requires 6 minutes. Following com-pletion of a cycle, the system computer will shut down the main motor. Confirma-tion that all waste has been processed is made by the operator by raising the hopper hood and looking into the unit to see if any remains. Once done, the next batch may be processed.

Models Available

• MMT3000 system (other models are also available based on customer through-put needs)

Specifications

Use:	Regional or on-site use by large generators
Size (overall):	7 ft long × 8 ft wide × 14 ft high
Weight:	1000 lb
Electrical requirements:	440–480 volts (V), 60 hertz (Hz), three-phase electrical connections as well as regular 110 to 220-ampere (A) current
Power consumption:	40 kilowatt hours (kWh) (maximum)
Water requirements:	$\frac{3}{4}$-in. water line
Drain requirements:	None
Load capacity:	3000 lb/hr
Decontamination time:	Room temperature
Decontamination temperature:	6 minutes
Total cycle time:	Continuous process
Venting requirements:	None
Filter system:	High-efficiency particulate air (HEPA) (99.97 percent efficiency), particle, and carbon filters

Capital Cost

- $395,000

Waste Stream Compatibility, Challenges, Incompatibility

Compatible Waste Types

- Cultures and stocks of infectious agents and associated biologicals
- Liquid human and animal waste, including blood and blood products and blood fluids
- Pathological waste
- Sharps
- Other biomedical waste

Challenging Waste Types

- One hundred percent organic liquids (e.g., blood). Additional sterilant would be necessary to absorb the liquid waste.

Incompatible Waste Types

- Large metal objects such as hip joints. Large metal objects will damage the cutting teeth on the unit's shredder.
- Toxic chemical (e.g., antineoplastic drug waste, RCRA waste) or radiological waste (e.g., low-level radioactive waste).

Installation. Installation is done by Medical Materials & Technology.

Operation. The treatment process is initiated only after an operator has inserted a key and entered an assigned password into the system computer. Once done, the following steps are followed:

1. Either automatic or manual mode is entered on the control panel. (The automatic mode dispenses the sterilant and water based on the weight of the waste to be processed. The manual load allows sterilant and water volumes to be selected by the operator.)
2. The operator checks the sterilant hopper and the water connection.
3. The cart containing waste is weighed on the system's weighing station.
4. The lifting mechanism is attached to the cart and the waste is loaded into the hopper.
5. Once the waste is loaded, the amount of sterilant and water is calculated either automatically by the computer or manually by the operator, and added to the hopper.

6. Operator functions are entered on the control panel, which initiates the process.

7. Once the process has completed, the sterilized waste is removed to a waste catcher, which can transfer the waste by screw conveyor into a dumpster for general solid waste disposal.

System Monitoring. The system computer monitors the pH sensor and oxidation activity to monitor that the process is operating correctly. A pH value of between 11.7 and 12.2 demonstrates the effectiveness of the sterilant. If the sterilant concentration drops below the sterilization range, the computer aborts the process.

Training. Medical Materials & Technology will train operators of the MMT system. In addition, a training film and training manual is provided with each unit.

Service. The distributor makes available 24-hour technical assistance and prompt service.

Treated Waste

Characteristics of Treated Material. The treated waste is shredded to less than $\frac{1}{2}$-inch mesh size and is slightly moist, although moisture is quickly reduced due to the drying agent in the sterilant.

Volume Reduction. Waste volume is reduced by up to 90 percent.

Disposal of Treated Material. Depending on local restrictions, the treated, shredded waste may be disposed into the general waste stream.

Environmental Discharges

By-Products and Discharges. Only the treated waste is discharged from the system. Liquids are absorbed by the waste or by additional sterilant.

By-Product Controls. Potential air releases are triple filtered.

Efficacy Testing. Real scale efficacy studies were performed in 1994 to evaluate the treatment feasibility of simulated medical waste in a treatment system following protocols required by the Illinois Environmental Protection Agency. Tests were conducted by an independent laboratory, Howard G. Gratzner and Associates. In this test, 15-pound challenge loads consisting of blood/broth cultures, fibers, metals, sharps, plastics, pathological waste, glass, nonwoven fibers, and bottles of liquids were placed into 55-gallon red biohazard bags. Challenge loads were inoculated with 40 mL of inoculum containing approximately 6.5×10^9 colony-forming units (CFU)/mL concentrations of *Bacillus subtilus* spores (ATCC 9372) and *B. stearothermophilus* spores (ATCC 12980).

Effects of dilution and loss in the system of test organisms were determined through control runs without addition of the disinfectant. Such testing observed an average 2-log loss of test organisms in the control runs. Taking this into consideration, spore reduction was observed to exceed 6 logs for the indicator microorganisms during test runs with the addition of disinfectant into the system.

Approvals and Installations. According to the vendor, the technology has been approved or meets state requirements in 31 states with 19 applications pending throughout the United States.

Medical Safe-TEC, Inc. (7–12)

Source

Medical Safe-TEC
P.O. Box 78156
Indianapolis, IN 46278
Tel: (317) 879-8080
Fax: (317) 876-2669

Product Overview. Medical Safe-TEC, Inc. has developed a biohazardous waste treatment technology that disinfects waste using a sodium hypochlorite solution and destroys its recognizability by mechanical grinding in a hammermill. A system diagram is presented in Figure 10.1. Waste is placed on a conveyor, transported to a

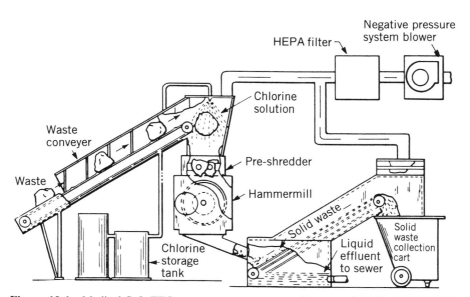

Figure 10.1 Medical Safe-TEC waste treatment system. Courtesy of Medical Safe-TEC, Inc., Medical Safe-TEC Waste Treatment System, copyright 1995 by Medical Safe-TEC, Inc., Indianapolis, Indiana. Reprinted by permission.

feed hopper, sprayed with the chlorine disinfectant in a concentration of between 1500 and 2000 parts per million (ppm), preshredded, and hammermilled in the presence of disinfectant solution. During the process, water and disinfectant are continually recirculated through the system with additional disinfectant added as needed. The slurry of ground waste and disinfectant is directed to a liquid separation. The solids are separated and conveyed to a waste collection cart where they are held for additional contact time with the residual disinfectant. Sodium biosulfite is added to the liquid separation tank at the end of the day to neutralize the sodium hypochlorite. Liquids are then discharged through a cartridge filter and an activated carbon filtration system prior to release of approximately 150 to 200 gallons of water to the sanitary sewage system. Waste is reduced in volume from 2:1 to 8:1, depending on the nature of the waste being treated. The entire process is maintained under negative pressure and air is discharged to the atmosphere through a HEPA filter.

Models Available

- MST 300L
- MST 500
- MST 1200

Specifications

Model MST 300L

Use:	Series MST 300L is designed to be placed near the source of infectious waste generation
Size (overall):	50 in. long × 58.5 in. wide × 72.5 in. high
Weight:	1000 lb
Electrical requirements:	Standard: 480 V alternating current (ac), three phase, 60-Hz, 50-A service required; 208/230 V ac
	Optional: 100-A service required, 50 Hz optional
Water requirements:	$\frac{1}{2}$-in. supply with anti-backsiphon preventer with a 6-gallon per minute (gpm) flow rate after backflow device
Drain requirements:	Floor sewer required capable of accepting a 4-gpm discharge rate; flush mounted or below grade desirable, although a pump system is optional to accommodate above-grade sewer connections
Decontamination time:	Processing takes less than 5 seconds in a hammermill with sodium hypochlorite solution
Venting requirements:	4-in. flexible duct with a vent that discharges to the atmosphere; the unit will power up to

Filter system: 25 ft of duct; a booster blower is available
                                  for longer duct runs if necessary
                                  All air is drawn through a special chlorine-
                                  resistant HEPA filter before discharge to the
                                  atmosphere

Model MST 1200

Use: Designed to centrally process the infectious
                                  waste generated by a medical facility

Weight: 10,000 lb dry weight isolated on 14 floor isola-
                                  tion mounts; maximum weight for an indi-
                                  vidual floor mount is 800 lb, with a
                                  maximum service factor of 2:1

Electrical requirements: *For machine:* Standard: 150-horsepower (hp)
                                  hammermill, 480 V ac, three-phase, 60-Hz,
                                  260-A service required
                                  Optional: 100-A hammermill, 480 V ac, three-
                                  phase, 60-Hz, 159-A service required
                                  For generator: Standard: 230 V ac, three-
                                  phase, 20-A, 60 Hz
                                  High-capacity: 480 V ac, three-phase, 20-A,
                                  60 Hz

Water requirements: $1\frac{1}{2}$-in. supply with anti-backsiphon preventer
                                  and 2 hp in-line pressure booster pump [80
                                  to 100 pounds per square inch (psi)], pre-
                                  wired with starter; control switching lines
                                  provided by Medical SafeTEC

Drain requirements: Vary with room size and configuration; 6-in.
                                  sewer main is recommended; maximum flow
                                  is 30 gpm and in some cases a 4-in. sewer
                                  will be adequate, as determined by the facili-
                                  ty engineers

Load capacity: 700 to 1500 lb/hr, depending on hypochlorite
                                  concentration used

Venting requirements: One 6-in. vent line (nonpowered); roof mount
                                  or side mount is acceptable; the hose should
                                  be flexible; no stack is necessary

Filter system: HEPA filter

Capital Cost

- MST 300L: $85,000
- MST 500: $269,000
- MST 1200: $385,000

Waste Stream Compatibility, Challenge, Incompatibility

Compatible Waste Types

- Cultures and stocks of infectious agents and associated biologicals
- Pathological waste: unrecognizable (depending on local restrictions)
- Liquid blood and blood products and blood fluids
- Sharps

Incompatible Waste Types

- Toxic chemical (e.g., antineoplastic drug waste, RCRA waste) or radiological waste (e.g., low-level radioactive waste)
- Culture slants from tuberculosis research (the manufacturer recommends autoclaving before processing)
- Large cloth items such as sheets
- Metal objects thicker than $\frac{1}{8}$ inch
- Waste size limitations (e.g., Model MST 300L): maximum size of waste to be destroyed is 5 inches by 5.5 inches by 16 inches long; size limited by the dimensions of the entry chute

Installation. Model MST 300L may be mounted against a wall and requires a floor space 6 feet long by 5 feet deep, although these dimensions may be modified. Feed hopper is at a height of 4 feet.

Operation. Operation of the MST 300L model is as follows:

1. The safety door is raised and waste is deposited (up to 15 gallons per cycle). After the lid is closed manually, the feed chute opens automatically.
2. Waste material is fed into the feed chute. The waste and disinfectant solution enter the hammermill chamber for destruction.
3. Treated solids and liquids are separated by the dewatering conveyor. Liquids are discharged to the sanitary sewer. Solid residues are deposited in collection carts.
4. Collection carts are drained further, to prepare waste for dumpster disposal as municipal solid waste.

Treated Waste

Characteristics of Treated Material. The waste is rendered unrecognizable by the process. Glass is reduced to sand or large gravel-like pieces, needles are blunted and bent, tubing and plastics are reduced to pieces of less than 2 inches, and paper is reduced to pulp.

Volume Reduction. There is 8:1 volume reduction in infectious waste, 6:1 reduction in paper type wastes, and 2:1 reduction in liquid-entrained wastes.

Disposal of Treated Material. Depending on local restrictions, the treated, destroyed waste may be disposed of as municipal solid waste.

Approvals and Installations. Medical Safe-TEC has installations in California, Michigan, Connecticut, Delaware, Georgia, Hawaii, Indiana, Kansas, Kentucky, Louisiana, Maine, Maryland, Missouri, New Jersey, South Carolina, Tennessee, Vermont, Wisconsin, Canada, Israel, Italy, Japan, Korea, Puerto Rico, and Thailand.

Corporate Profile. Medical Safe-TEC has been supplying its treatment systems since 1985.

Premier Medical Technology, Inc. (13)

Source

Premier Medical Technology, Inc.
9800 Northwest Freeway, Suite 302
Houston, TX 77092
Tel: (713) 680-8833
Fax: (713) 683-8820

Product Overview. Premier Medical Technology, Inc. (PMT) has developed a biomedical waste treatment system for on-site use that incorporates a grinding and shredding technology with waste disinfection via a proprietary disinfectant developed by the company. PMT describes the PMT 100 disinfectant as a dry powder disinfectant capable of destroying all pathogens and sporeformers in the waste. The treated waste product is reduced in volume, rendered unrecognizable, and disinfected, and may be disposed as municipal solid waste.

The PMT system is comprised of (1) a frame and support structure, (2) a waste loading subsystem, (3) two cutting assemblies, (4) two mixing augers, (5) a PMT 100 disinfectant feeding subsystem, (6) a water disbursement subsystem, (7) an air filtration subsystem, (8) process controls, and (9) PMT 100 disinfectant. The system is entirely automated and controlled by an onboard computer, which calculates all measurements, such as the disinfectant concentration, controls the sequence in which the waste is processed, and records batch and other system parameters.

Models Available

- PMT System

Specifications

Use:	On-site at a central location within the facility
Size (overall):	The machine footprint is 112 ft²; the overall size is 16 ft long × 7 ft high × 7 ft wide
Electrical requirements:	150 A at 480 V, 60 Hz, three-phase
Water requirements:	¾-in. line with a standard coupling
Drain requirements:	No sewage connection required
Load capacity:	Approximately 750 lb/hr or 6000 lb in an 8-hour shift
Venting requirements:	None
Filter system:	The air filtration subsystem provides a sealed environment using a negative pressure handling system, with prefilter and HEPA filters to prevent organic and particulates from escaping

Capital Cost

- There is no capital cost. Product is available on a fee per pound basis which is all inclusive of service, repairs and maintenance, and disinfectant. The fee per pound is based on the volume to be treated on an annual basis. Large volume generators can utilize the PMT System for $0.20 per pound or less.

Waste Stream Compatibility, Incompatibility

Compatible Waste Types

- Cultures and stocks of infectious agents and associated biologicals
- Liquid human and animal waste, including blood and blood products and blood fluids
- Pathological waste (human or animal)
- Sharps

Incompatible Waste Types

- Waste that would shorten the life of or damage the shredder, such as:
  - Large or high-grade stainless steel items
  - Implants, such as pins, rods, and joints
  - Tools and broken pieces of hospital equipment
- Toxic chemical (e.g., antineoplastic drug waste, RCRA waste) or radiological waste (e.g., low-level radioactive waste)

Installation. The manufacturer is responsible for delivery, installation, and startup testing of the system and all associated costs. The lessee is responsible for providing electrical and water lines up to the point of connection to the processor.

Operation. The PMT system is designed for control by one operator. A start button begins the process, which is totally controlled by an onboard computer. Waste is processed by individual batch and is physically destroyed, size reduced, and disinfected as it moves through the unit.

An operator coming on duty must enter an individualized operator code to access the unit. The code remains locked in until the operator goes off-shift or another operator comes on duty. After a batch of waste is staged for loading, the system computer checks the status of all conditions and notifies the operator that it is ready for processing. The operator presses the start button, and the waste is loaded and processed without further intervention from the operator.

Once the process has been initiated, if a problem is detected by the computer, the computer will notify the operator and make a decision to shut down the cycle depending on the problem. If shutdown occurs, the operator is instructed to contact the Premier service hotline so that a service technician can be dispatched.

The waste treatment process occurs in seven steps, with only the first two steps requiring input from the operator:

1. The medical waste container is placed by the operator on the scale to be weighed. Disinfectant and water volumes are automatically calculated by the computer based on the net weight of the medical waste in the container.

2. The operator will engage the start button to begin the batch process. The lift mechanism will dump the material out of the container and into the unit for processing.

3. At this stage the first phase of shredding occurs and the disinfectant (PMT 100) and water are introduced to the system.

4. The first phase of mixing of the waste with PMT 100 and water occurs.

5. The waste is further shredded to reduce volume and particle size to less than a $\frac{1}{2}$-inch mesh size.

6. Final mixing of the waste with the disinfectant and water occurs.

7. The disinfected, unrecognizable waste is discharged to a receptacle and the process is complete. Waste may be disposed of as municipal solid waste. The unit is ready for the next batch.

It is the obligation of the facility to provide a screening program to ensure that waste that could damage the system, or should otherwise not be processed by the system, be excluded from processing. Such waste includes large pieces of metal (e.g., steel rods, steel replacement joints, nonsurgical metals such as bedpans) and waste containing radioactive materials or toxic chemicals.

System Monitoring. The system is automated for record keeping. All batches being processed are recorded automatically and given a batch number, which can be used as supporting documentation for billing purposes if necessary. The system computer also records performance information about machine components and a record of maintenance activities conducted during PMT's preventive maintenance visits. A noninclusive list of information that may be recorded includes (1) batch number, (2) hospital identification number (source of waste), (3) operator identification number, (4) date, (5) weight of batch, (6) amount of PMT 100 added, and (7) processing time for each batch.

Training. PMT offers an extensive inservicing and training program for system operators and for all hospital personnel which addresses proper handling of all waste in the facility and the regulatory requirements. Training focuses on the collection and segregation of infectious waste from all areas of the facility and its movement to the PMT system. Premier staff will also work with the facility to develop a quality assurance program so that the effectiveness of the PMT system can be evaluated over time.

Service. PMT provides a comprehensive maintenance program to keep the system operational. The service includes weekly preventive maintenance visits. Service personnel are on call 24 hours, seven days a week. PMT maintains an inventory of all system parts for rapid equipment repair if necessary. PMT will pay the difference in cost for alternative waste treatment should the system not be repairable within 72 hours.

PMT also offers a system warranty for all components of any leased unit for the life of the lease under normal use. This warranty does not include repair costs associated with improper operation by the operator, such as the failure of the operator to screen large pieces of heavy metal that could damage the shredding mechanisms.

Treated Waste

Volume Reduction. Waste will be reduced by about 90 percent.

Efficacy Testing. According to PMT, Inc., efficacy testing has been performed using independent laboratories and universities such as the University of Miami School of Medicine, Cleveland State University, M.D. Anderson Cancer Center, Fox Chase Cancer Center, University of Texas Medical Branch at Galveston, Leberco Testing, Inc., Microbiology Specialists, Inc., etc. Laboratory testing followed by actual tests in the PMT System was conducted using the Illinois Efficacy Test Protocol and the testing criteria outlined in the *Technical Assistance Manual: State Regulatory Oversight of Medical Waste Treatment Technologies* published in April 1994.

STERIS Corporation (14,15)

Source

STERIS Corporation
5960 Heisly Road
Mentor, OH 44060
Tel: (216) 354-2600
Fax: (216) 639-4457

Product Overview. The STERIS EcoCycle 10 processor, used in conjunction with STERIS-SW and STERIS-LW decontaminants, is a waste decontamination and destruction treatment system designed for on-site use by large or small medical facilities such as hospitals, clinics, or laboratories. A system diagram is presented in Figure 10.2. Once the waste has been disinfected and rendered nonrecognizable by the process, it may be disposed into the general waste stream.

Waste to be treated is placed into the unit's processing chamber with a sealed, single-dose container of the disinfectant. After the chamber is sealed, a standardized volume of water is introduced into the chamber, which dilutes the disinfectant to its use concentration. The medical waste is then ground in the unit's grinding/processing chamber through a hammermill action for 3 minutes, which renders it unrecognizable as medical waste. The ground mass of material is then allowed to soak in the disinfectant liquid for an additional 7 minutes, allowing the necessary contact time to achieve a microbial destruction of greater than or equal to 99.9 percent. Following the destruction and decontamination cycle, the chamber is inverted over the system's liquid separation unit and automatically rinsed with clean water. Both the rinse water and the solids are poured into the liquid separator unit, which spins to remove water from the waste. After the liquid effluent has been discharged to a sanitary sewer, the solids are removed from the separator module and disposed of as general solid waste.

The system is recommended for two types of medical waste: (1) general medical waste (waste with a predominance of syringes, needles, scalpel blades, and other sharps and disposable medical/surgical devices), and (2) laboratory waste (waste with a predominance of blood and body fluids). The disinfectants, STERIS-SW and STERIS-LW, are based on peracetic acid [peroxyacetic acid (PA)] as the active ingredient. The diluted product is nontoxic, minimally sensitizing, and safe for direct discharge to a sanitary sewer. STERIS-SW and STERIS-LW employ the same chemical formulation as is used in STERIS 20 sterilant concentrate, which is listed as a chemical sterilant under EPA Registration No. 58779–1. On January 17, 1995, STERIS officially received EPA acceptance of amended label claims for STERIS 20 for use with EcoCycle 10 as a chemical treatment method for the decontamination of biohazardous materials. The acceptable labeling claims include the following statement:

EcoCycle 10 components

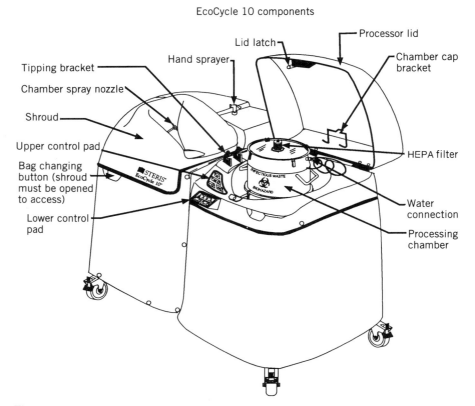

Figure 10.2 STERIS EcoCycle 10 Processor. Courtesy of STERIS Corporation, STERIS EcoCycle 10 Processor, copyright 1995 by the STERIS Corporation, Mentor, Ohio. Reprinted by permission.

Decontaminant–Santizer–Germicide

STERIS 20 is a single-use chemical decontaminant concentrate validated for use in the STERIS/ECOMED BWD3. STERIS 20 is suited to the decontamination of liquid (<1000 mL blood) and dry medical (<100 mL blood content) waste materials such as syringes, needles, sharps, glassware, laboratory wastes, and other biohazardous materials—recommended for use in hospitals, medical/surgical/dental clinics or offices, laboratories, and biomedical research facilities.

Models Available

• STERIS EcoCycle 10 Processor

The STERIS EcoCycle 10 processor is designed to work in conjunction with the Steris 20 sterilant concentrate, which is marketed in two dose packages:

1. STERIS-SW decontaminant: recommended for waste largely consisting of solids with a relatively low organic load, such as general waste generated during a surgical procedure;
2. STERIS-LW Decontaminant recommended for waste largely consisting of liquids with a high organic load, such as waste generated in a hospital laboratory;

Specifications

Use:	On-site at point of generation for large and small generators
Electrical requirements:	208 V ac, single-phase, 30-A (thermal magnetic breaker); unit is supplied with a three-wire cord, requiring the purchaser to supply a plug connector to match the receptacle
Water requirements:	40 to 100 psi, $\frac{1}{2}$-in. supply with a $\frac{3}{4}$-in. male hose connection with shutoff valve; water use: about 6 Liters per processing cycle
Drain requirements:	$1\frac{1}{4}$-in. minimum-inside-diameter standpipe, non-backpressuring, may be up to 40 in. above grade; or a $\frac{3}{4}$-in. male hose connection may be used; discharge to a sanitary sewer
Load capacity:	Approximately 2 kilograms (kg) by weight or 8 Liters (L) by volume per batch, five batches per hour
Decontamination time:	10 minutes (3 minutes grinding in sterilant, 7 minutes soaking in sterilant)
Total cycle time:	Ten to 15 minutes

Capital Cost

• Less than $20,000

Waste Stream Compatibility, Challenges, Incompatibility

Compatible Waste Types

• Cultures and stocks of infectious agents and associated biologicals
• Liquid human and animal waste, including blood and blood products and blood fluids
• Pathological waste
• Sharps
• Other biomedical waste: the system is specifically recommended for two types of medical waste:

- General medical waste (waste with a predominance of syringes, needles, scalpel blades, and other sharps and disposable medical/surgical devices)
- Laboratory waste (waste with a predominance of blood and body fluids)

Challenging Waste Types

- Cotton gauze content greater than 10 percent of waste load is difficult to process mechanically and may result in a canceled cycle condition.
- Blood content greater than 100 grams for general medical waste application (STERIS-SW decontaminant), or greater than 1000 grams for laboratory application (STERIS-LW decontaminant), may interfere with the biocidal activity of the chemical decontaminant.
- Heavy metal instruments (such as surgical staplers), orthopedic implants, and other heavy-duty orthopedic instruments may cause the grinding blades to fail.

Incompatible Waste Types

- Toxic chemical (e.g., antineoplastic drug waste, RCRA waste) or radiological waste (e.g., low-level radioactive waste)

Installation. The power and liquid separation units require a workspace of 31 by 46 inches and an operator work space of 30 by 46 inches. A 60-inch open area is required above the system. Utility connections must be within 5 feet of the treatment system location. Installation connections should be flexible to allow the units to be wheeled forward for about 3 feet for servicing. Nearby counter space and a utility sink are desirable.

Operation. The following operation steps are observed:

1. The system and system controls are checked for readiness.
2. Contaminated waste is placed in the processing chamber, which is placed into the power unit. The appropriate decontaminant package is placed in the chamber. The chamber is then secured.
3. The processor lid is closed and the start button depressed. The appropriate volume of water enters the processing chamber automatically and the hammermill destruction cycle begins. This cycle lasts for 3 minutes.
4. Following the destruction cycle, the waste is allowed to soak in the chamber for an additional 7 minutes to allow settling and to assure the necessary contact time.
5. Following the soak period, the processor lid detaches automatically, which is signaled by the cycle light turning off.
6. The chamber is inverted into the solids/liquids separation device, the chamber is rinsed automatically with clean water, liquids are filtered and dis-

charged to the sanitary sewer, and solids are retained in a bag for disposal to the general waste stream.

System Monitoring. The system is equipped with a continuous time indicator. Treatment is deemed complete based on processing time and the visual condition of the processed waste. Routine biological monitoring is not recommended as necessary by the manufacturer, although any waste load may be tested biologically.

All system parameters are sequenced by microprocessor. The system has been designed not to begin the treatment process if the "chemical added" button has not been depressed, if the water connection is not attached, or if the correct volume of water fails to flow into the chamber. If the system jams or overloads, a "canceled cycle" warning will be activated, which will include the problem that requires correction. Operating instructions will require that the load be reprocessed.

Training. An intensive operator training manual and program has been developed by STERIS to train users in the proper use and safety procedures of the EcoCycle 10 system. Training includes (1) fundamentals of the EcoCycle 10 process, (2) proper technique in performing processing cycles, (3) safe use and handling of STERIS-SW and STERIS-LW decontaminant, (4) proper technique for performing preventive maintenance procedures, (5) understanding troubleshooting information and instrument specifications, and (6) proper procedure for documenting completed cycles.

Service. STERIS provides a customer service phone number for technical assistance.

Treated Waste

Characteristics of Treated Material. The treated waste is damp, shredded, granular material, the granule size being material dependent. Glass is reduced to powder or sand-sized particles. Brittle plastics (such as styrene test tubes or petri dishes) are rendered to sand-sized particles. Semibrittle plastics (such as syringe parts) are reduced to plastic granules from $\frac{1}{4}$ to $\frac{1}{2}$ inch in size. Soft plastics such as gloves, vinyl tubing, plastic bags, and polyethylene bottles are reduced to large particles from $\frac{1}{4}$ inch to several inches in size. Paper and nonwoven materials (such as synthetic gauze, gloves, or chucks) are reduced to fiber or pulp. Metals (such as disposable instruments, needles, or scalpel blades) are blunted, bent, and/or broken. The waste is safe for subsequent handling by waste handling and landfill personnel who may come into physical contact with it.

Volume Reduction. Volume reduction is 70 to 80 percent.

Disposal of Treated Material. Liquid effluent is discharged to a sanitary sewer. The chemical active agent decomposes rapidly to oxygen and acetic acid. The hammermilled, inert solid fraction may be disposed of in the general waste stream as general solid waste.

Environmental Discharges

By-Products and Discharges. Discharges are related only to the liquid effluent and ground, inert solids that have been processed.

By-Product Controls. Once the process has been initiated, the system remains sealed until the treatment cycle has completed.

Efficacy Testing. Efficacy testing was performed by STERIS and independent laboratories on two representative waste loads: (1) a general load of approximately 1000 grams of sharps, paper, and plastic products having 7.5 to 10 percent (weight/ weight) blood, and (2) a laboratory load of approximately 2000 grams of waste having 50 percent (weight/weight) assorted sharps, plastic, paper, and other organic liquids. In all trials, samples of the ground/treated waste demonstrated a 6- to 8-log reduction in viable microbial population when subjected to a contact time of 10 minutes (a 3-minute grind and a 7-minute soak) with the STERIS decontaminant solution compared to parallel control counts of ground/untreated waste.

Efficacy testing was performed using the following test organisms:

- Bacteria
  - *Staphylococcus aureus*
  - *Enterococcus faecium*
  - *Pseudomonas aeruginosa*
- Bacteriophage
  - *MS-2*
  - *Escherichia coli* (phage host)
- Bacterial endospore
  - *Bacillus subtilis*
- Mycobacteria
  - *Mycobacterium bovis*
  - *Mycobacterium smegmatis*
- Virus
  - Poliovirus, Sabin 3 (Leon 12 a,b)
- Fungi
  - *Aspergillus fumigatus*
  - *Candida albicans*
  - *Trichophyton mentagrophytes*
- Parasite
  - *Giardia muris*

Approvals and Installations. Applications are pending.

Winfield Environmental Corporation (7,12,16,17)

Source

Winfield Environmental Corporation
9750 Distribution Avenue
San Diego, CA 92121
Tel: (800) 456-7701
Tel: (619) 271-4861
Fax: (619) 695-1299

Product Overview. The Condor mechanically disintegrates medical waste with controlled shredding and granulation. The resulting "mulch" is washed and treated automatically with an oxidant decontaminant (chlorine dioxide). After wastes have been disintegrated and treated, a dewatering process is used to recycle most of the oxidizing solution. Independent laboratory testing confirms the effectiveness of this oxidizing agent against typical infectious agents. A system diagram is presented in Figure 10.3.

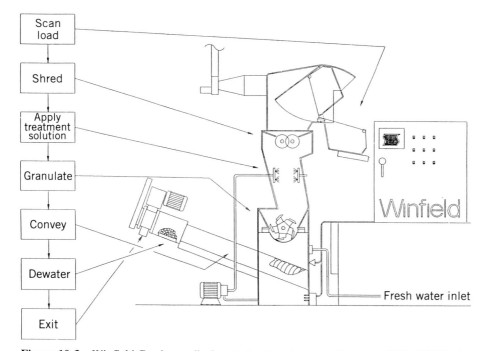

Figure 10.3 Winfield Condor medical waste treatment system. Courtesy of Winfield Environmental Corporation, Winfield Condor Medical Waste Treatment System, copyright 1995 by Winfield Environmental Corporation, San Diego, California. Reprinted by permission.

When the collection bags and their contents are disintegrated, additional oxidizer is automatically metered into the treatment solution, keeping it fresh and active. Before processing, collection bags are scanned electronically to confirm that the waste contains no detectable discarded metal instruments or radioactive isotopes, and then weighed.

Wash temperatures (77 to 140°F) and oxidizing agent concentrations (20 to 70 ppm) are monitored and controlled at effective levels automatically without operator intervention. Any deviation from the correct operating conditions is sensed electronically and indicated on the machine's control panel, preventing further processing.

Models Available

- Winfield Condor medical waste treatment system
  - Manual dock load
  - Manual floor load
  - Automatic bulk load

The Winfield Condor mobile medical waste treatment system is contained in a nominal 40-foot trailer.

Specifications

Use:	Hospitals of 300+ beds or regional processor
Size (overall):	8 ft 6 in. long × 7 ft $2\frac{1}{2}$ in. wide × 11 ft 10 in. high
Weight:	11,000 lb
Electrical requirements:	460 V ac, three-phase, 100 A
Power consumption:	25 kilowatt (kW) average
Water requirements:	$\frac{1}{2}$-in.-diameter line, 30 psi, 8 gpm; 8 to 18 gal/hr consumption requirement, depending on waste load
Drain requirements:	$1\frac{1}{2}$-in-diameter line
Load capacity:	600 lb/hr
Total cycle time:	Continuous feed
Treatment time:	3 to 5 minutes
Treatment temperature:	Wash water temperatures range from 77 to 140°F
Venting requirements:	HEPA vent, 6 in. diameter minimum to atmosphere
Filter system:	HEPA filter, flow of 325 cubic feet per minute (cfm) at 1 in. H_2O

Capital Cost

- $400,000, which includes training and installation assistance

Waste Stream Compatibility, Incompatibility

Compatible Waste Types

- Cultures and stocks of infectious agents and associated biologicals
- Liquid human and animal waste, including blood and blood products and blood fluids, if not contained in tightly closed containers and provided that liquids do not exceed 10 percent by weight of the total waste
- Sharps
- Other biomedical waste (e.g., isolation wastes)

Incompatible Waste Types

- Pathological wastes, including human and animal body parts, organs, and tissues
- Blood and body fluids that exceed 10 percent by weight of the total waste

Installation. The Condor is shipped as a self-contained unit with the hopper assembly removed prior to shipment. Upon receipt, the unit will require offloading from a flatbed trailer. It is recommended that a 12,000-pound fork truck or suitable crane be utilized for equipment placement. Upon equipment positioning, the unit is to be leveled by means of the adjustable foot pads. The hopper assembly is located over the shredder assembly and bolted into place. Electrical service is then hooked up to the control panel, water to the input solenoid, and the tank drain to a suitable hospital drain. Installation and checkout can be accomplished in approximately one day. (As described directly from information provided by the distributor.)

Operation. In general, infectious waste is fed into a hopper, from which it is shredded, sprayed with treatment solution, and then ground further in a granulator before falling into a tank filled with treatment solution. An auger screw moves the waste from the tank to a dewatering screen at a controlled rate, where the treated waste is squeezed through a constriction and then moved by conveyor to the hospital's existing trash compactor or dumpster. The following, which is adapted from the Winfield operation manual, describes operation of the Winfield system in detail.

1. *Machine Startup.* The load hopper includes a sealed entry door that is normally in its closed position until all interlocks assert correct machine safety conditions, currents, temperatures, concentration, flow rates, and fluid levels. Radiation and metal is detected by the electronic bag-reader. The control panel shows the machine status at all times and signals "machine ready."

2. *Material Entry.* Individual red bags are loaded into the hopper bin and weighed automatically. The load button closes the bin hydraulically and actuates the previously locked entry door, allowing a bag to enter the shredding chamber. If radiation or metal is detected, the operator is signaled, the hopper door will not actuate, and the shredder will stop. Any aerosols emitted from the wastes are

exhausted away from the hopper through a HEPA filter. If this exhaust system is not functioning correctly, the hopper door will not close. With the system interlocks, the load hopper and entry door act as an air lock to provide negative pressure within the equipment.

3. *Shredding Chamber.* Counterrotating cutting disks shred the bag and its contents. Cutting is accomplished at low speeds and high torques. Torque is sufficient to shred all normal medical waste, including plastic, glass, and thin-gauge metal.

4. *Granulator Transition.* Treatment solution is sprayed on the waste before it enters the granulator. A flow switch confirms that the fluid line is clear.

5. *Granulating Chamber.* Five rotating blades mesh closely with two stationary blades for fine slicing of previously shredded material. A screen at the bottom of the granulating chamber retains material until sufficient size reduction is accomplished. On falling through the screen, material enters the auger wash tank.

6. *Auger Wash Tank.* The wash tank is a reservoir of heated liquid treatment solution. Mechanical agitation of treated waste is provided by recirculation of fluid using a pump. The treatment solution temperature is maintained through heaters and associated controls.

7. *Fluid Filter.* A screen located under the auger filters the treatment solution prior to recirculation through the pump and spray manifold. The wiping action of the auger removes materials stopped by the screen and conveys them through the auger barrel together with the rest of the washed material.

8. *Auger Trough.* Material is conveyed from the wash tank by the rotation of the auger within the auger trough. The design of the auger system ensures that materials are held at nominal temperature, for nominal time, while completely wetted with treatment solution.

9. *Dewatering.* A constriction is established at the exit end of the auger. Material driven upward by the auger accumulates behind this constriction until compression is sufficient to overcome the frictional force against it. Material that is sufficiently squeezed and dewatered is then driven out and allowed to exit the machine. Water squeezed from the waste stream is returned to the wash tank via the screen filter and return line.

10. *Liquid-Level Control.* A level sensor monitors the fluid level in the wash tank and signals the solenoid to admit additional fresh water to the tank when its level drops below nominal. This automatic control compensates for the small amount of fluid not recovered from the treated waste.

11. *Chemistry Monitoring.* The wash tank's temperature and treatment concentration are monitored continuously through probes. The data logger provides hard-copy printout of all monitored system parameters for inclusion in the operator's log book.

System Monitoring. The fluid level in the wash tank is monitored and signals a solenoid to admit additional fresh water to the tank when levels drop below nominal. Parameters continually measured in the wash tank are temperature and concentration of treatment solution. The data logger proves a hard-copy printout of system parameters. The system's potency alarm level approximates a 3-log kill of spores per gram of waste.

Training. The purchase price includes two to five days of on-site training by Winfield engineering personnel. Training consists of classroom instruction as well as hands-on training on the machine.

Service. Winfield provides a nationwide service organization with a 24 to 48-hour response time. A complete spare parts inventory is maintained for same-day shipment of parts, when required.

Treated Waste

Characteristics of Treated Material. The treated waste appears as a damp granular waste with a particle size of less than $\frac{3}{4}$-inch screen. It is not recognizable as medical waste.

Volume Reduction. Volume is reduced by 70 to 80 percent.

Disposal of Treated Material. Depending on local restrictions, the treated waste may be disposed of as municipal solid waste.

Environmental Discharges

By-Products and Discharges. Aerosols are released to the air.

By-Product Controls. Aerosols emitted during the process are exhausted from the hopper through a HEPA filter and discharged to the atmosphere.

Efficacy Testing. Efficacy test results conducted on the Winfield Condor system using simulated loads of waste are summarized below.

1. Final Report: "Evaluation of the Effectiveness of the Condor Medical Waste Treatment System," April 1992 (Research Triangle Institute)
   - *Bacillus stearothermophilus* spores: kill (solids) 4.5 $\log_{10}$/gram
   - *Mycobacterium terrae:* kill (solids) 6.3 $\log_{10}$/gram
   - *Mycobacterium phlei:* kill (solids) 6,3 $\log_{10}$/gram
2. Final Report: "Evaluation on the Winfield Medical Waste Treatment System Against *Giardia* Cysts," January 29, 1992 (West Virginia University)
   - *Giardia* cysts: All killed (99.8 percent)
3. Final Report: "Antimicrobial Effectiveness of the Condor Medical Waste Treatment System," September 1992 (Research Triangle Institute)

Microbial Kill

- *Staphylococcus aureus* (4.8×10^6)
- *MS2 bacteriophage* (5.1×10^5)
- *Pseudomonas aeruginosa* (7.8×10^6)
- *Candida albicans* (3.9×10^5)
- *Penicillium chrysogenum* (1.0×10^6)
- *Bacillus stearothermophilus* (3.3×10^3)

4. Technical Report: "Evaluation of the Effectiveness of the Condor Medical Waste Treatment System," November 4, 1994. (Sorrento Biochemical Inc.)
   - *Bacillus subtilis* spores—$\log_{10}$ reduction = 4.681

Approvals and Installations. As of January 1995, there are 45 states where the Winfield system may be used or may be subject to local conditions, with applications pending in the remaining five states. Winfield system installations include three in the United States, three in Saudi Arabia, one in Mexico, two in Israel, and one in Abu Qhabi with systems pending in Oman, Italy, and Canada.

Corporate Profile. Winfield Industries, located in San Diego, California, develops, manufactures, and markets a broad range of products for the safety and protection of healthcare workers and for the handling, containment, and treatment of infectious medical waste, generated by hospitals and health care providers and clinical and research laboratories. Winfield introduced the Condor medical waste treatment system to the market in January 1992. The company's main office and western manufacturing facility, located in San Diego, California, is approximately 56,000 square feet. Winfield's eastern manufacturing and distribution center in Clarksburg, West Virginia, is approximately 50,000 square feet. In July 1991, Winfield opened a research and development facility in Escondido, California with a technical and engineering staff of 10. In August 1993, Winfield acquired Ryan Medical in Brentwood, Tennessee, a manufacturer of safety needles and safety blood collection devices. As of December 1994, the company had 180 full-time employees.

ALTERNATIVE STEAM STERILIZATION

The principles of steam sterilization are described in Chapter 9. The discussion that follows in this section focuses on steam sterilization based biohazardous waste treatment systems with alternative design features to enhance their ability to effectively manage, treat or destroy biohazardous waste. The technology of steam sterilization is not recognized as an alternative biohazardous waste treatment technology *per se* by most states or by the State and Territorial Association on Alternate Treatment Technologies. Nevertheless, autoclave systems with design variations from the classically recognized steam sterilizer are currently available and merit discussion. Such alternative design features include rotating steam sterilization chambers with mechanical destruction, continuous feed systems, heated auger supplements, waste compactors, and shredders. Examples of some commercially available alternative biohazardous waste steam sterilization processing systems are described in the following reviews.

System Reviews. Four alternative biohazardous waste steam sterilization treatment systems are reviewed in this section. These systems are available from the following companies:

1. Ecolotec, Inc.
2. San-I-Pak Pacific Inc.
3. SAS Systems, Inc.
4. Tempico Rotoclave Medical Products, Inc.

Ecolotec, Inc. (18,19)

Source

Ecolotec, Inc.
100 Springdale Road, A-3, Suite 290
Cherry Hill, NJ 08003
Tel: (609) 346-2447
Fax: (609) 346-1779

Product Overview. The technology developed by Ecolotec, Inc. (U.S. Patent 5,217,688) is capable of disinfecting or sterilizing biomedical waste using a vacuum/pressure steam autoclave that heats waste to temperatures of between 245 and 270°F and simultaneously destroys the waste so that it is no longer recognizable as medical in origin. A schematic diagram of the Ecolotec system is presented in Figure 10.4. The Ecolotec equipment is a batch unit ideal for on-site application, having the advantage of minimal space requirement and easy operation without harmful effluent, emission, or residue other than completely sterile waste suitable for disposal with other nonregulated housekeeping or kitchen waste. Due to the fact that the Ecolotec system provides true and verifiable on-site sterilization rather than mere disinfection, it is ideally suited for applications with high infection or contamination hazards, such as microbiological laboratories.

Models Available. Ecolotec presently offers two models: the smaller SE-80 machine and the larger MW-130 machine. However, Ecolotec is concentrating on the MW-130 machine as standard. A dual-vessel MW-130 and machines with larger or smaller vessels can be built when justified by market conditions.

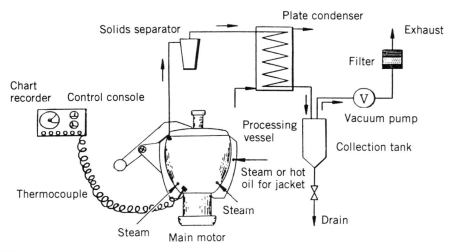

Figure 10.4   Ecolotec medical waste processor. Courtesy of Ecolotec, Inc. Ecolotec Processor, copyright 1995 by Ecolotec, Inc., Cherry Hill, New Jersey. Reprinted by permission.

Specifications

Model MW-130

Use:	On-site use by small or medium-sized biomedical waste generators
Size (overall):	85 in. × 45 in. footprint × 90 in. height; recommended minimum space: 15 ft × 20 ft area
Electrical requirements:	240 V, 60 Hz, three-phase, 60 kW; 115 V, 60 Hz, single-phase, 1 kW
Water requirements:	Cooling water for condenser, approximately 30 to 78 gpm flow for less than 5 percent of cycle time (preferably, treated boiler makeup water which can be returned)
Steam requirements:	60 to 80 lb/hr at 60 psi minimum
Load capacity:	Approximately 5 ft³ per charge
Total cycle time:	15 to 20 minutes per batch, or 3 to 3.5 batches/hour
Sterilizing time/temperature:	2 to 3 minutes at 270°F will provide a reliable minimum 1×10^6 kill of *Bacillus stearothermophilus* spores [the exceptionally rapid thermal kill time is attributed to the very efficient exposure of the disintegrating waste particles to a saturated steam atmosphere

	caused by the violent agitation and cutting action of the tools rotating at 3500 revolutions per minute (rpm)]
Venting requirements:	Vacuum pump exhaust to outside of 15 cfm
Filter system:	Mechanical mesh prefilter; activated carbon filter, HEPA filter [0.2 to 0.3 micron (μm)], activated carbon filter

Capital Cost. Depending on features and circumstances, the capital cost including installation, startup, and operator training is approximately $375,000.

Waste Stream Compatibility, Challenge, Incompatibility

Compatible Waste Types

- Cultures and stocks of infectious agents and associated biologicals
- Human and animal waste, including blood and blood products and limited blood fluids
- Pathological waste (depending upon local restrictions)
- Sharps
- Other biomedical waste (accidental inclusion of steel or hard-alloy parts, such as bolts and pins, which can not be reduced in size will help the reduction process of softer materials and should not harm the equipment)

Challenging Waste Types

- Densely folded or compacted plastic, glass fiber, or cellulosic textiles

Incompatible Waste Types

- Toxic chemical (e.g., antineoplastic drug waste, RCRA waste) or radiological waste (e.g., low-level radioactive waste)

Installation. The system is installed by the vendor. Costs for installation are built into the capital cost of the equipment.

Operation. After the waste has been loaded manually into the treatment vessel, the process is initiated by the operator from the control panel. The cover closes hydraulically and the rotary tools immediately begin the destruction process, rotating initially at 1750 rpm, then at 3500 rpm as the vessel is first evacuated before steam is injected, raising the temperature rapidly to sterilizing temperature. The heating process is aided by hot oil circulating through the insulated vessel jacket. The heating phase is maintained automatically until the sterilizing temperature has been reached and maintained for the predetermined period of time (usually 2 to 3 minutes at 270°F). The violent agitation and highly effective disintegration action of the tools expose the waste particles in an extremely thorough manner to the saturated steam atmosphere, causing very short lethality times for microorganisms. The tool action furthermore virtually eliminates air pockets, which can cause "cold spots." Once sterilization parameters have been achieved by the system, it shuts off the steam supply automatically and the chamber is vented through the solids separator, condenser, vacuum pump, and triple filter (prefilter, HEPA filter, activated carbon filter). Following pressure equilibrium, the vacuum pump evacuates moisture vapors that are generated as the chamber flash cools, which are condensed in a water-cooled heat exchanger (condenser). Additional drying can be achieved by hot oil circulation through the vessel's jacket. Liquids are collected in a condensate receiver for evaluation before discharge to the sewer system. (It is anticipated by the vendor that most regulatory authorities will allow direct discharge into the sewer, bypassing the receiver tank.) The system can be opened only after the predetermined amount of moisture has been removed and the system temperature has dropped to less the 160°F. Once achieved, the process vessel is opened, tilted, and its contents discharged into a suitable waste receptacle or conveyor and may be disposed of as municipal waste.

Operation of the system is controlled automatically by a programmable logic controller (PLC); preprogrammed process parameters cannot be altered by the operator. The system is capable of sensing if the rotary tools have become inoperative during the process. Should this condition occur, the PLC would switch automatically to the ARMS (automatic repair and maintenance safety program) mode, which extends the sterilization cycle to one that would be required for autoclaving without agitation. According to the vendor, the ARMS mode capability will protect operating and maintenance personnel against potential contact with bloodborne disease organisms, in accordance with the latest OSHA guidelines.

System Monitoring. The ideal process sterilization temperature is 270°F. However, specific sterilization time and temperature requirements are determined on a site-by-site basis, based on spore test kills following system installation, identification of wastes requiring treatment, and local regulations. The system is equipped with a recorder that provides hard-copy records of the temperature, time, and pressure parameters of each process cycle, providing proof of having achieved the preset sterilization conditions. The system is also equipped with three spore strip retainers inside the process vessel, allowing close monitoring of sterilization with bioindicators.

Training. On-site training of one to two days will be provided at the time of system installation.

Service. While requirements of routine service are part of the initial training, emergency service and parts can be supplied by the manufacturer from the Columbus, Ohio facility.

Treated Waste

Characteristics of Treated Material. Consistency of treated waste ranges from sloppy or putty-like to dry and crumbly or flaky, depending on the rate of desired moisture removal (energy balance). Maximum particle size for hard plastics and rubber is approximately ⅜ inch. Softer materials will disintegrate completely.

Volume Reduction. Depending on waste being treated, volume reduction is between 30 and 85 percent.

Disposal of Treated Material. Depending on local restrictions, the treated waste may be disposed of as municipal solid waste.

Environmental Discharges

By-Products and Discharges. Distilled water (possibly with traces of pharmaceuticals), filtered air, and unrecognizable sterile solid waste are discharged.

By-Product Controls. A triple filtration system (mechanical mesh prefilter, HEPA filter, activated carbon filter), and condensate collection tank are employed.

Efficacy Testing. Tests have been performed with spore strips of *Bacillus stearothermophilus* as bioindicators, placed inside the vessel in three strategic locations during the full process.

Corporate Profile. Ecolotec is a small company, incorporated in the state of Delaware and organized to commercialize the medical waste disposal technology patented by W. A. von Lersner, Ecolotec's president and principal shareholder. The technology is patented in the United States and applications are pending in various countries in Asia, Europe, and North America.

San-I-Pak Pacific, Inc. (20,21)

Source

San-I-Pak Pacific, Inc.
23535 South Bird Road
P.O. Box 1183
Tracy, CA 95378-1183
Tel: (209) 836-2310
Fax: (209) 836-2336

Product Overview. San-I-Pak Pacific, Inc. produces several models of a high-vacuum steam sterilizer/compactor system that sterilizes and compacts infectious

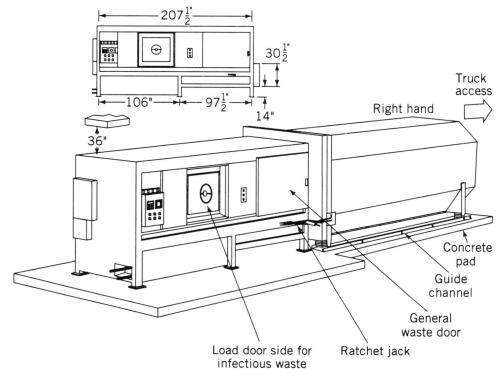

Figure 10.5 San-I-Pak Mark X sterilizer. Courtesy of San-I-Pak Pacific, Inc. San-I-Pak Mark X Sterilizer—copyright 1995 by San-I-Pak Pacific, Inc., Tracy, California. Reprinted by permission.

waste. A system diagram for the Mark X model is presented in Figure 10.5. To maximize waste contact with steam, the system first pulls a vacuum within the sterilization chamber to remove all air. The air removed is mixed with steam at 307°F to kill any airborne microorganisms that may be present. In the chamber, the steam is brought to a pressure of 38 psi. Once a temperature of 270°F has been reached, a 30-minute timer is activated automatically. During the sterilization period, the chamber will reach a maximum temperature of 281 to 284°F. After the 30-minute cycle has run its course, the system vents down and the steam is run through a diffuser, where it is condensed to water. The waste in the chamber is conveyed automatically to a compaction chamber, where it is compacted prior to disposal. Depending on a purchaser's needs, a San-I-Pak system can be retrofitted with waste destruction systems for unrecognizability. These systems to render waste unrecognizable include a single-stage shredder, a two-stage shredder, and/or a pulverizer.

Models Available. San-I-Pak offers several models in the following configurations:

- Sterilizers
- Sterilizer/compactors
- Sterilizer/compactors with single-stage shredder
- Sterilizer/compactors with two-stage shredder
- Sterilizer/compactors with shredder mill

The Mark system capacities range from 20 pounds per cycle to 600 pounds per cycle. Model numbers from this series include:

- Mark I Sterilizer (20 to 30 pounds/cycle)
- Mark II Sterilizer (100 pounds/cycle)
- Mark VI Sterilizer (200 pounds/cycle)
- Mark VII Sterilizer (320 pounds/cycle)
- Mark X Sterilizer (600 pounds/cycle)

San-I-Pak has recently begun marketing a 200/300 Series with options for design configuration. These models include:

- 200 series:
  - Mark 241L (160 pounds per hour) to Mark 248L (1,280 pounds per hour)
  - Mark 251L (200 pounds per hour) to Mark 258L (1,600 pounds per hour)
- 300 series:
  - Mark 341L (230 pounds per hour) to Mark 348L (1,840 pounds per hour)
  - Mark 351L (280 pounds per hour) to Mark 358L (2,240 pounds per hour)

Specifications. Specifications are provided for the small and large units from the Mark series of sterilizers: Mark I and Mark X.

Mark I Sterilizer

Use:	Medical waste sterilizer for small medical facilities
Size (overall):	$79\frac{1}{2}$ in. high $\times$ $33\frac{1}{2}$ in. wide $\times$ 50 in. deep
Weight:	980 lb
Electrical requirements:	120 V (15-A service)
Steam requirements:	$\frac{1}{4}$-in. steam line at 40 psi minimum
Water requirements:	$\frac{3}{8}$-in. water line, normal city water pressure at 3 gpm
Drain requirements:	Sanitary drain
Load capacity:	20 to 30 lb/cycle

Decontamination temperature: 281 to 284°F
Decontamination time: 30 minutes
Total cycle time: 39 minutes
Venting requirements: None
Filter system: None required

Mark X Sterilizer/Compactor

Use: Medical waste sterilizer and general waste com-
                                pactor
Size (overall): $94\frac{5}{8}$ in. high $\times$ $71\frac{1}{2}$ in. wide $\times$ $207\frac{1}{2}$ in. long
Weight: 16,500 lb
Electrical requirements: 120 V, single-phase (10- to 20-A service with
                                cold weather package) plus one three-phase
                                voltages: 208 V (70-A service) or 240 V (50-
                                A service) or 480 V (30-A service)
Water requirements: $\frac{1}{2}$-in. water line, normal city water pressure at
                                3 gpm
Drain requirements: Floor mount, typically connected to sanitary
                                drain line
Load capacity: 600 lb/cycle
Decontamination temperature: 281 to 284°F
Decontamination time: 30 minutes
Total cycle time: 68 minutes (includes loading and automatic dis-
                                charge sequences)
Venting requirements: None
Filter system: None required

Capital Cost. Capital costs for the Mark series ranges from $16,575 to $472,959, depending on model and configuration. Capital costs for the 200/300 series range from $92,490 to $410,320 depending on model and configuration. System options include:

- Roll-off container: ranges from $6485 to $6845.
- Guide channels: $1075
- Cold weather package (if applicable): $575

Installation, delivery, and freight (varies depending on location and configuration) ranges from $7000 to $24,900.

Waste Stream Compatibility, Incompatibility

Compatible Waste Types

- Cultures and stocks of infectious agents and associated biologicals

- Human and animal waste, including blood and blood products and limited blood fluids
- Pathological waste: unrecognizable (depending on local restrictions)
- Sharps
- Other (e.g., wastes from surgery or autopsy, dialysis wastes, discarded medical equipment, other biological waste)

Incompatible Waste Types

- Pathological waste: recognizable (e.g., amputated body parts)
- Contaminated animal carcasses, body parts, and bedding of animals exposed to infectious agents during research, production of biologicals, or testing of pharmaceuticals
- Toxic chemical (e.g., antineoplastic drug waste, RCRA waste) or radiological waste (e.g., low-level radioactive waste)

Installation. Installations are conducted by San-I-Pak Pacific. Required is a concrete pad—the preferred dimensions are 10 ft wider and 5 feet longer than the combined length of the unit and receiving container (e.g., 42 feet long). The pad should be a minimum 3000-psi concrete, steel reinforced, 6 inches thick. The concrete pad must be level, with no more than $\frac{1}{2}$ inch slope side to side or end to end.

The systems also require a steam supply with a minimum pressure of 60 psig, capable of handling a maximum of 125 psig. A dedicated phone line is required, positioned at the site. Other system requirements are identified under "Specifications."

San-I-Pak suggests a five-day schedule for system installation, which includes securing the unit to the concrete pad (day 1), hookup to all utilities (day 2), safety checks and calibration (day 3), and in-service training for operators (days 4 and 5).

Operation. Infectious waste is placed into autoclave bags and loaded by the operator into the sterilization chamber. The load door is closed, the personal access code is entered, and the enter button depressed. The system automatically continues to completion of the cycle. Once the cycle is completed, the operator will either open the load door and initiate the rotate and dump cycle, or the waste will be discharged automatically, depending on the model. The waste within the sterilization chamber is then automatically discharged and compacted into the roll-off container, or shredded, depending on the unit's configuration.

System Monitoring. The San-I-Pak system is programmed to discharge waste only after complete vacuum, pressure, time duration, and temperature parameters have been met. These parameters can not be changed by the operator and can be changed by San-I-Pak only if directed to do so. A strip printer prints a record of the date, time, temperature, pressure, and vacuum level of every sterilization cycle. An optional digital display is available to inform operators of system status during operation. (The digital display is standard equipment for the Mark II, VI, VII, and X sterilizer/compactors.)

The vendor recommends periodic monitoring with heat-resistant biological indicators such as *Bacillus stearothhermophilus* spores to demonstrate proper system operation. Other auxiliary monitors include autoclave tape and time/temperature-sensitive chemical integrators.

Training. San-I-Pak provides a two-day on-site in-service training session to both system operators and engineers. Items covered include safety, system/equipment overview, loading procedures, equipment operations, general maintenance, trouble shooting, repairs, and housekeeping.

San-I-Pak offers a one- or two-week complimentary factory technical training program in which an employee may tour the manufacturing plant and travel with a service technician on preventive maintenance or service calls. The purchaser is required to provide for expenses associated with transportation or other travel expenses.

Service. San-I-Pak offers a full-service contract to purchasers of the equipment that includes all necessary parts, labor, equipment, tools, and materials to perform all scheduled and unscheduled preventive maintenance services. A 24-hour response time is provided for emergency service. Systems are covered under a limited warranty for one year for defects in materials and workmanship, and extended warranties and preventive maintenance contracts are available.

Treated Waste

Characteristics of Treated Waste. The sterilized waste is either compacted or shredded so that it is no longer recognizable, depending on the system configuration.

Volume Reduction. Volume reductions of up to 6:1 are achieved for systems equipped with a compactor.

Disposal of Treated Material. Depending on local restrictions, the sterilized waste may be disposed of as municipal solid waste.

Efficacy Testing. San-I-Pak systems have received U.S. Department of Agriculture efficacy approval for processing international wastes generated by Delta Airlines. San-I-Pak was the first nonincineration technology accepted in New York. Efficacy testing was conducted successfully by placing a spore test in the center of a bag of wood chips for processing with a load of waste.

Approvals and Installations. More than 250 systems are operating in over 38 U.S. states and five countries. Systems are accepted by regulatory agencies in all 50 states and the Virgin Islands, Canada, Colombia, Italy, England, and New Zealand.

Corporate Profile. San-I-Pak is a privately held corporation established in 1978. Systems are patented.

SAS Systems, Inc. (22,23)

Source

SAS Systems, Inc.
8552 Katy Freeway, Suite 200
Houston, Texas 77024
Tel: (713) 973-1467
Fax: (713) 973-8557

Product Overview. SAS Systems, Inc. manufactures a continuous-shredding, steam sterilization biomedical waste treatment technology, the SAS-1 shredding autoclave sterilizer. (SAS is an acronym for "shredding, autoclaving, steriliza-tion.") A system schematic is presented in Figure 10.6.

Containerized biomedical waste is lifted hydraulically and dumped by the auto-feed/weighing unit into an autoclave shredder chamber. The waste is then simul-taneously shredded and steam sterilized in an enclosed shredder/sterilization cham-ber. As treatment proceeds, a vacuum is drawn on the electrically heated integrated SAS-1 unit. Steam is then introduced into the chamber at the required operating parameters of "pressure-temperature-time" for autoclaves. As this occurs, the rotary disk vane shredder begins shredding the waste. The shredding process ensures steam

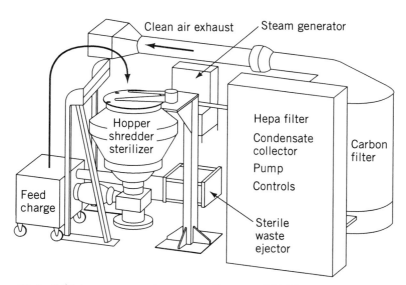

Figure 10.6 SAS-1 waste processing system. Courtesy of SAS Systems, Inc. SAS-1 Waste Processing System, copyright 1995 by SAS Systems, Inc., Houston, Texas. Reprinted by permission.

penetration of the entire waste load. Once the sterilization/shredding process has been completed, a final vacuum phase is applied to the chamber which further dehydrates the waste, producing a sterile, nonrecognizable product that has achieved an 80 percent volume reduction.

Models Available

• SAS-1

Specifications

MODEL SAS-1

Use:	Stationary plant for small, medium-sized, and large hospitals or medical waste generators, or as a mobile unit mounted on a truck
Size (overall):	11.5 ft × 7.5 ft
Weight:	7.5 metric tons
Electrical requirements:	220/480 V ac, 60 Hz, three-phase, 63 A, 50 kWh
Power consumption:	50 kWh
Water requirements:	Potable water source, 0.75-in.-diameter supply line
Drain requirements:	None, except for washdown of work area and steam flush
Load capacity:	Approximately 300 lb/hr, the actual rate depending on the characteristics of the infectious waste and local regulatory requirements
Total cycle time:	Minimum cycle time: 15 minutes
Decontamination time:	5 minutes
Decontamination temperature:	250°F/15 psig
Venting requirements:	Venting is required
Filter system:	HEPA filter and a series of activated carbon filters

Capital Cost

• Basis cost: $400,000

Waste Stream Compatibility, Incompatibility

Compatible Waste Types

• Cultures and stocks of infectious agents and associated biologicals
• Liquid human and animal waste, including blood and blood products and blood fluids

- Pathological waste
- Sharps
- Other biomedical waste (e.g., contaminated waste from international activities, contaminated food products)

Incompatible Waste Types

- Toxic chemical (e.g., antineoplastic drug waste, RCRA waste) or radiological waste (e.g., low-level radioactive waste)

Installation. Installation can be accomplished within a minimum of time and at no additional cost. The SAS-1 comes skidmounted. Installation requirements are a 220/480-volt ac electrical outlet, an outlet for potable water, and a vent to the outside for steam and air emissions. Installation requires approximately 4 hours and is conducted by the purchaser.

Operation. The SAS-1 is operated continuously through a programmed process control computer which ensures that proper operating conditions are attained prior to waste being fed into the system. Because of this, no designated labor is required. The duties of the waste handler are to position the waste containing cart on the automatic lifting device and push the "on" button. The processor also maintains proper processing conditions, which are based on the actual weight of the infectious waste. The system computer safeguards the unit from operating until proper operating conditions have been met and will make the necessary adjustments to maintain proper conditions. The SAS-1 self-sterilizes after each process schedule and prior to opening the enclosed system for maintenance. After each batch of waste processed and self-sterilization cycle, the system produces a hard-copy printout of key cycle parameters, which include the weight of waste, operating temperatures, and residence time.

System Monitoring. The system is controlled by a microprocessor, which assures exposure to saturated steam.

Training. SAS Systems, Inc. offers one week of training to equipment purchasers.

Service. Service is conducted via a standard maintenance agreement. Twenty-four-hour telephone technical assistance is available. A repair service is offered.

Treated Waste

Characteristics of Treated Material. The SAS-1 unit produces a sterile shredded solid waste that is dry and unrecognizable as medical waste.

Volume Reduction. Volume is reduced by approximately 70 to 80 percent.

Disposal of Treated Material. Depending on local and state restrictions, treated waste may be disposed of as municipal solid waste.

Waste Recycling. Waste plastics may be recycled through a proprietary UTAP process (mixed postconsumer thermoplastics recycling process), designed specifically for waste processed by the SAS-1 System. The process first granulates the material to a smaller size, then removes contaminants such as glass, metals, and fine fibers through a gravity process. The remaining material is ready to be used as raw material in the recycling process in the system's molding unit. For cost containment, the UTAP process does not require separation of different plastic resins or any special washing. As a result, the process generates a less expensive, lower-grade plastic raw material that may be used in certain product lines.

Environmental Discharges

By-Products and Discharges. Only air and excess steam are generated.

By-Product Controls. Air and excess steam pass through a HEPA filter and a series of activated carbon filters. Volatile organic hydrocarbons and odorous agents are removed by an activated carbon filter. Particulate matter is removed by a fine-mesh fabric filter.

Efficacy Testing

Model: SAS-1. Testing was conducted during 1992 by an independent certified laboratory, North American Laboratory Group, with the exception of the duck hepatitis B Virus test protocol, which was conducted by Patricia L. Marion of the Stanford University School of Medicine. Test results concluded that under the time and temperature conditions achieved during the test, sterilization of pure test culture isolates had taken place. Test microorganisms included *Candida albicans, Giardia lamblia, Mycobacterium bovis*, *M. fortuitum,* and duck hepatitis.

Spore efficacy testing was conducted at the Manfred Jacobs Microbiology Laboratories in Germany during July 1994 using Spore-O-Chex bioindicator strips with 2.0×10^6 spores of *Bacillus subtilis* var. *niger*. Under the test parameters, the SAS-1 achieved complete elimination of all spores.

Approvals and Installations. The SAS-1 system has received state authority to operate in Arkansas, Delaware, Florida, Georgia, Kentucky, Massachusetts, Mississippi, Montana, Nebraska, New York, North Carolina, New Mexico, North Dakota, Oregon, Pennsylvania, Tennessee, Texas, Vermont, West Virginia, and Wyoming.

Roland steam sterilization units have been performing all across Europe since 1986, with a total of approximately 30 systems currently in operation. While most European units are mobile, two stationary units are currently operating at the Mayo Clinic, Rochester, Minnesota.

Corporate Profile. Following his introduction of two Roland steam sterilization units to the United States in 1992, Otmar Kolber formed SAS Systems, Inc. as the exclusive distributor for Roland steam sterilization technologies in North America. The SAS-1 was designed for the purpose of meeting the specific demands of the American market.

Tempico Rotoclave Medical Products, Inc. (12,24,25)

Source

Tempico, Inc.
Tempico Rotoclave Medical Products
251 Highway 21 North
P.O. Box 428
Madisonville, LA 70447-0428
Tel: (800) 728-9006
Tel: (504) 845-0800
Fax: (504) 845-4411

Product Overview. Tempico's Remedy-One Rotoclave sterilizes biomedical waste through a process of steam sterilization. Waste is loaded into a pressure vessel with a rotating internal drum. High-pressure steam is then introduced into the chamber, which causes the bags and containers to become soft. As the internal drum rotates during processing, the bags and containers of waste rupture and spill their contents, allowing complete contact of the waste with the pressurized steam, which is maintained at a temperature between 275 and 300°F and a pressure between 45 and 50 psi for at least 30 minutes. After the sterilization parameters of time, temperature, and pressure have been met, the waste is dried through a vacuum/condensing system. The sterile, dry waste is then discharged through a size-reduction system to render all waste unrecognizable. The final sterilized waste product has been reduced by up to 85 percent in volume and may be disposed as municipal solid waste. There is no increase in weight.

Models Available

- Remedy-One Rotoclave Model 1500-D2
- Remedy-One Rotoclave Model 2500-D2

The Model 2500-D2 includes all the features of the Model 1500-D2, plus two rotoclave vessels, providing a total capacity of up to 2000 pounds per 350 cubic feet of volume. Both models may be ordered without the second-stage of size-reduction by specifying Model 1500-D1 or Model 2500-D1. Optional accessories include (1) an integrated scale and automatic loading system, and (2) cooling water, steam generation, and compressed-air systems for freestanding operation.

Specifications

Model 1500-D2

Use:	On-site, at a central location in the hospital, or off-site as a commercial regional treatment system	
Electrical requirements:	Vessel drive:	480 V ac, three-phase
	Four conveyors:	110 V ac, one-phase
	Shredder:	480 V ac, three-phase
	Grinder:	480 V ac, three-phase
Power consumption:	Vessel drive:	3.73 kWh
	Four conveyors:	0.56 kWh
	Shredder:	5.61 kWh
	Grinder:	7.01 kWh
Load capacity:	175 ft^3 of waste per 65 to 80-minute cycle; the actual weight of waste can vary from 500 to 900 lb per load, depending on the waste density	
Total cycle time:	Vessel drive:	80 minutes
	Four conveyors:	15 minutes
	Shredder:	15 minutes
	Grinder:	15 minutes

Capital Cost. Depending on the model, requirements, and circumstances, prices range from $395,000 up to $1,485,000.

Waste Stream Compatibility, Incompatibility

Compatible Waste Types

- Cultures and stocks of infectious agents and associated biologicals
- Liquid human and animal waste, including blood and blood products and blood fluids
- Pathological waste (human or animal)
- Sharps
- Other (general waste)

Incompatible Waste Types

- Waste that would shorten the life of or damage the shredding device, such as large or high-grade stainless steel items
- Toxic chemical (e.g., antineoplastic drug waste, RCRA waste) or radiological waste (e.g., low-level radioactive waste)

Installation. Rotoclave systems are fully assembled and tested prior to shipment. Once on-site, installation can be performed by hospital engineering personnel or local contractors arranged by Tempico. Tempico personnel are present on-site during installation to ensure proper installation, supervise startup, and conduct operator training.

Operation. During system installation and startup, Tempico personnel are on-site to supervise startup and conduct operator training.

System Monitoring. The system is equipped with electronic monitors that record and print out operational parameters automatically. This includes the time that the process is maintained at the required temperature and pressure. The manufacturer recommends that a microbiological indicator (e.g., *Bacillus stearothermophilus*) be used to confirm system effectiveness at a frequency of once every 30 days, although this frequency is subject to local regulations. A fail-safe mechanism will not allow the waste to be discharged until time, temperature, and pressure sterilization parameters have been met. Should a system malfunction occur, all repairs can be done without opening the treatment chamber.

Training. Tempico trains all maintenance, engineering, and operating personnel in all aspects pertaining to the operation of its systems.

Treated Waste

Characteristics of Treated Material. Processed waste is an unrecognizable mass of dry, shredded material.

Volume Reduction. Waste volume is reduced by approximately 85 percent.

Disposal of Treated Material. Depending on local restrictions, processed waste is disposed of as municipal solid waste.

Environmental Discharges

By-Products and Discharges. There is a release of 20 gallons of sterile condensate after each cycle, which is discharged to the sanitary sewer.

Efficacy Testing. Efficacy test results were reported from three laboratories as follows:

1. Testing was conducted at the University of New Orleans, Department of Biological Sciences, during January 1991. A perforated stainless steel container containing a cotton wad impregnated with 5 mL of *Bacillus megaterium* at a concentration of 5×10^5 colony-forming units per milliliter was introduced into the system with batches of municipal waste. Batches were processed at 300°F and 50

psi for 1 hour, and samples of effluent and the cotton wad were collected for testing. Posttreatment growth of *B. megaterium* was not detected.

2. Testing was conducted at the University of New Orleans, Department of Biological Sciences, during February 1992. Sterility was achieved using *Bacillus stearothermophilus,* available in Kilit vials containing the organism in nutrient broth. The tester concluded that the Remedy-One processor was capable of achieving sterilization of waste.

3. Testing was conducted at the Forrest General Hospital, Hattiesburg, Mississippi, during August 1993. The Remedy-One Rotoclave System falls under the same state regulations as those for autoclaves that require monthly sterility testing. Since testing began in October 1992, monthly tests were run and 100 percent kills were observed.

Approvals and Installations. The distributor has technology approval in Alabama, Arizona, Florida, Louisiana, New York, and other states as well as many foreign countries.

STEAM REFORMING TECHNOLOGY

The steam detoxifier developed by Synthetica Technologies incorporates steam reforming technology. Superheated steam is used at temperatures of up to 1200°F in a process referred to as steam reforming. The vapors formed during the process are then subjected to temperatures of up to 2100°F in a detoxification reactor.

System Review. More about this technology is described in the Synthetica Technologies, Inc. system overview.

Synthetica Technologies, Inc. (26–29)

Source

> Synthetica Technologies, Inc.
> 5327 Jacuzzi Street
> Richmond, CA 94804
> Tel: (800) 336-1141
> Tel: (510) 525-3000
> Fax: (510) 526-2277

Product Overview. The Synthetica detoxifier (STD) heated shredder/screw conveyor (HSE) system is for the destruction of medical waste by steam reforming and no incineration. The system is designed to convert all of the organics from the waste streams to a synthetic vent gas (syngas), leaving a small volume of inert residue for final disposal. A system diagram is presented in Figure 10.7.

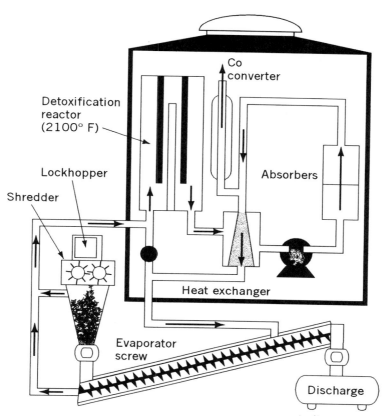

Figure 10.7 Synthetica steam reformer. Courtesy of Synthetica Technologies, Inc. Synthetica Steam Reformer, copyright 1995 by Synthetica Technologies, Inc., Richmond, California. Reprinted by permission.

The system is designed to destroy the organics from the waste materials, leaving a small volume of residue for final disposal. The waste fraction is assumed to be 30 to 40 percent solids. The total waste stream is assumed not to contain halogens (chlorides, bromides, etc.) in excess of 2 percent. The Synthetica detoxifier system includes one Synthetica steam reformer detoxifier (SSR), one heated screw conveyor evaporator (HSE), and one solids shredder feeder (SSF).

HSE. The HSE is designed to receive shredded solid waste materials and/or slurries. The HSE consists of a custom-designed 316 stainless steel screw conveyor and housing, electrical heaters, feed hopper and level controls, lock valves (two sets: one set located at the waste entry to the conveyor and one set at the residue discharge of the conveyor), all syngas piping and valving to/from the STD, all controls, and insulation.

SSF. The solid shredder is designed as a two-stage shredder. The first stage shreds the solid materials with cutter disks, and the second stage consists of a granulator that forces the first-stage shredding through selectively sized screens to assure that the material is sized to uniform and proper sizing compatible with the HSE. The shredder, constructed of 316 stainless steel and other compatible alloys, comes complete with all required controls, waste feed elevator, and transfer auger to load shredded material into the HSE's feed hopper.

STD-HSE. The medical wastes are loaded into the shredder-hopper, shredded into small pieces (approximately 0.25 inches), and fed into the heated screw conveyor. The hot gas from the SSR enters the SSR and vaporizes the organics and moisture. The hot gas enters through interconnecting piping with the SSR. The inert and sterilized residue discharges the screw conveyor into a residue collection container. The residue can then be disposed of in an approved municipal landfill site. The mass and volume reduction of the organic waste are 90 to 100 percent.

Models Available.

- Synthetica steam detoxification system (STD)

A mobile Synthetica van is also available.

Specifications

Use:	Suitable for 150- to 500-bed hospitals
Size (overall):	Approximately 511 ft²; allow a 6-ft aisle all around; minimum ceiling height, 16 ft
Weight:	8800 lb
Electrical requirements:	100 kilovolt-amperes (kVA), 480 V, three-phase
Steam requirements:	400 to 3000 lb/day
Power consumption:	Approximately 80 kWh for the entire process; energy consumption is estimated at 0.7 to 1.6 kWh per pound of waste treated
Water requirements:	5 gpm at 60 psig
Load capacity:	400 to 2400 lb/day
Decontamination time:	Up to 2 hours
Total cycle time:	Shredded waste is exposed to superheated steam (up to 1200°F) in the evaporator; vapors formed in the evaporator are heated to about 2100°F in the detoxification reactor
Exhaust air requirements:	25 standard cubic foot/minute (scfm) at 100 psig
Filter system:	Vapors from the detoxification reactor are passed through adsorber beds to remove acid gases, trace metals, and organics

Capital Cost. Cost ranges from $500,000 to $900,000, depending on waste input and overall application, with no special aspects to installation.

Waste Stream Compatibility, Incompatibility

Compatible Waste Types

- Cultures and stocks of infectious agents and associated biologicals
- Liquid human and animal waste, including blood and blood products and body fluids
- Pathological waste
- Sharps
- Other: in addition to biomedical waste, system has been tested successfully with a variety of industrial wastes, including spent activated carbon, toxic solvents, low-level radioactive wastes, and contaminated soils

Incompatible Waste Types

- Toxic chemical (e.g., antineoplastic drug waste, RCRA waste) and radiological wastes (e.g., low-level radioactive waste) are compatible with the system but may be restricted by regulation.

Installation. One hundred hours of startup at the customer site is included in the detoxifier system purchase price. Travel and living expenses are additional. Installation and startup labor beyond the 100 hours is available at field service rates.

Operation. The Synthetica detoxification process consists of two steps. The hydrocarbons of the waste are initially evaporated in a separate first stage waste feed evaporator (WFE), which can have a variety of forms, depending on the type and form of the waste. Within the WFE, steam reforming chemistry begins at temperatures of 315 to 1100°F. The steam reforming of the hydrocarbon forms hydrogen, carbon monoxide, carbon dioxide, water, and a small amount of methane. The gases, generated in the WFE, are heated to higher temperatures and mixed with excess superheated steam as they are pulled into the Synthetica steam reformer (1800 to 2300°F), where the final destruction reactions proceed to completion to achieve high destruction removal efficiencies.

System Monitoring. The operation is highly automated through application of the latest computer technology: distributed process controllers, fail-safe circuitry, four-level redundancy, on-line process modeling, on-line chemistry sensors, error checking, and so on.

Training. A three-day training session held at Synthetica's Richmond, California facility or at the customer's site is included in the purchase price of the STD system. Up to eight people may attend the training session free of charge. If the training sessions are conducted at the customer's facility, travel and living expenses for Synthetica's training staff will be charged at cost. Additional training, either in Richmond or at the customer's location, is available at the company's standard rates in effect at the time.

Service. Technical product support is an option that Synthetica recommends to assure that system operations and maintenance are optimized at all times. This service includes continuous remote monitoring of the system, with monthly reports provided. These reports include system operating data and graphs and projected maintenance scheduling requirements. The monitoring program assures that if any unusual operating parameters occur, customer personnel will be alerted and advised of the corrective measures required. If required, Synthetica technical personnel will arrange for an immediate visit to the customer site to assure that the proper corrective measures are implemented. This service option is available at the rate of $2500 per month or $30,000 per year.

If the technical product support option is not selected, technical support visits by Synthetica personnel are available at the standard rates in effect at the time service is required, plus travel and living expenses. A listing of recommended spare parts with pricing is provided upon completion of the bill of materials. Replacement parts are charged to the customer.

Treated Waste

Characteristics of Treated Material. Remaining solids are unrecognizable as medical waste and consist of graphitic carbon and varying amounts of metal, glass, and other inorganic solids.

Volume Reduction. The mass/volume reduction of the organic waste is 90 to 100 percent.

Disposal of Treated Material. Depending on local restrictions, the solids may be disposed of as municipal solid waste.

Environmental Discharges

By-Products and Discharges. There are no visible emissions. Steam reforming may produce organics, including aromatic compounds, but according to the manufacturer, these are in barely detectable quantities. Because the system does not use air, flame, or combustion, it avoids NO_x emissions, and because the hydroxyl-radical-catalyzed destruction avoids the formation of products of incomplete combustion, the equipment does not need special downstream emission controls.

By-Product Controls. Hydrogen, carbon monoxide, and trace amounts of acid gases and hydrocarbons formed during the process are removed in the catalytic converter and carbon beds.

Efficacy Testing. Efficacy testing following protocols developed by the state of California for alternative medical waste treatment technology approval was conducted in 1993–1994. A demonstration project was conducted at the Liberty Memorial Hospital facility. Microbiological tests were conducted by the Northview Pacific Laboratories, Inc. (final report December 7, 1993) and BioVir Laboratories (final report March 2, 1994). Challenge loads consisted of operating room waste and additional solid waste simulating laboratory or infectious hospital wastes, containing plastics, intravenous bottles, PVC gloves, paper, bandages, alcohol, solvents, disinfectants, syringes, a full aerosol can, and gowns. For microbiological efficacy tests, the operating room waste was spiked with the following nine microorganisms at the inoculum level indicated:

- *Staphylococcus aureus* (8.3×10^8/mL)
- *Pseudomonas aeruginosa* (1.6×10^9/mL)
- *Candida albicans* (3.8×10^7/mL)
- *Mycobacterium laticola* (2.3×10^8/mL)
- *Bacillus subtilis* spores (1.8×10^8/mL)
- *Bacillus stearothermophilus* spores (1.7×10^9/mL)
- *Enterococcus faecalis* (1.6×10^7/mL)
- *Aspergillus fumigatus* (8.8×10^6/mL)
- Polio virus type 2 (oral vaccine strain) (1.0×10^{10}/mL)

The process was observed to achieve complete destruction of each of nine biological test organisms (10^7).

During the tests, temperature profiles within the heated screw were taken to identify temperature transients at various locations along the screw. Inlet gas temperatures were observed to range from 700°F at the beginning of the test to 900°F at the end. Steam reforming of solid waste was observed beginning at about 600 up to 830°F.

Approvals and Installations. System approval has been obtained from California and Georgia, with pending approval from Arizona, Hawaii, and Texas. The company has sold two systems for research and development studies in "protected environments" to two large industrial manufacturing companies, both outside the United States. Contracts are pending with the U.S. Air Force and the Department of Energy (Sandia National Laboratories). Letters of intent have been obtained from three hospitals to place STD systems as demonstration installations through a cosponsorship agreement between Synthetica and the Electric Power Research Institute (EPRI). The company has focused its initial marketing efforts on biohazardous, RCRA, and low-level radioactive wastes to establish a presence within these three important applications.

Corporate Profile. Synthetica Technologies is an environmental company focused on the research and development, engineering, marketing, and manufacturing of its patented technology of high-temperature steam reforming that converts hazardous wastes into clean environmentally compatible products. The company was founded in April 1990 by the STD inventor, Terry Galloway, with the goal to provide on-site detoxification of hazardous wastes that satisfy customer and regulatory requirements efficiently and economically. Now that the STD technology has been demonstrated successfully, Synthetica is looking to business opportunities throughout the world. The STD process is patented in the United States and in principal foreign countries, with patent applications for additional system features on file.

MICROWAVE DISINFECTION

Microwaves are electromagnetic waves situated in frequency between radio waves and infrared waves on the electromagnetic scale and overlap both. When applied to the treatment of medical waste, the mechanism of microbial inactivation is due to a thermal effect. During the process a typical industry standard radio-frequency energy of 2450 MHz is absorbed by the waste to produce friction in water molecules. Heat generated by friction of water molecules denatures proteins within the microbial cell, which results in the inactivation effect. The microwave technology has been used for the treatment of medical waste on both a large and small scale. With its origin for this purpose coming from Europe in the 1980s, microwave-based medical waste disinfection systems have come into more common use in North America during the 1990s.

As stated, medical waste disinfection by microwaves is a type of thermal process. An early comparative study by Goldblith and Wang of the microbiocidal effects of thermal and radio-frequency energy observed that inactivation was due to thermal effects alone, with no apparent affect from the microwaves (30). In this study, *Escherichia coli* and *Bacillus subtilis* were subjected to a microwave energy of 2450 MHz. Varying amounts of ice were added to the *E. coli* suspension to obtain longer exposure time in the microwave field while maintaining temperatures ranging between 68 and 125°F, in which thermal inactivation would not come into play. As indicated in Table 10.2, time in the radarange microwave field of 50, 70, and 100 seconds had no effect on microbial inactivation, indicating that the bacterial inactivation is solely a thermal effect.

A subsequent study published by Vela and Wu in 1979 observed that the mechanism of microbial inactivation of microorganisms exposed to 1400-watt 2450-MHz microwave radiation was a thermal effect in the presence of water (31). When bacteria were subjected to direct microwave energy in both wet and dry soil conditions, it was determined that both the temperature increase and microbial inactivation were dependent on the presence of water. Lethal effects of the microwaves were reduced significantly as the water content in the soil approached zero. A variety of bacteria and two bacteriophages exposed to microwave radiation in a lyophilized (freeze-dried) sample were observed not to absorb enough energy in the dry state to bring about any significant level of cell inactivation (Table 10.3).

Jeng et al. evaluated nonthermal lethal effects of microwaves for developing a

TABLE 10.2   Inactivation of *Escherichia coli* in the Radarange with Ice Added to the Bacterial Suspension

Time in Radarange (sec)	Volume of *E. coli* Suspension (mL)	Weight of Ice Added (g)	Final Temperature of Suspension (°C)	Bacteria Count (no./mL) Before Exposure	Bacteria Count (no./mL) After Exposure
50	9.1	20	20	3.9×10^8	4.1×10^8
70	9.1	35	29	2.6×10^8	2.0×10^8
100	9.1	40	51.5	2.2×10^8	2.5×10^8

Source: From Goldblith SA and Wang DJC. Effect of microwaves on *Escherichia coli* and *Bacillus subtilis*. *Applied Microbiology*, 15(6):1374, copyright 1967 by Williams & Wilkins, Baltimore, Maryland. Reprinted by permission.

short sterilization cycle at a lower temperature than would be required for sterilization by thermal processes (32). Tests were conducted using dried spores of *Bacillus subtilis* because of their extreme resistance to heat. Pairs of vials containing spores were tested in parallel, one in a microwave oven generating a maximum output power of 4 kW at 2450 MHz and one in a convection oven. During the experiment

TABLE 10.3   Effect of Microwave Radiation (2450 MHz) on Microorganisms

Organism	LD$_{99.9\%}$[a] (joules $\times$ 10^3) Lyophilized State (Dry)	LD$_{99.9\%}$[a] (joules $\times$ 10^3) After Moistening (Wet)
Escherichia coli	>240	16
Pseudomonas aeruginosa	>240	8
Salmonella typhimurium	>240	11
Serratia marcescens	>240	10
Staphyloccocus aureus	>240	11
Bacillus cereus	>240	12
Azotabacter nielandii	>240	12
Azotobacter chroococcum	>240	10
Bdellovibrio spp.	>240	22
Bdellovibrio spp.	>240	20
Bacteriophage (*E. coli* K-12)	>240	18
Baceriophage (*E. coli*)	>240	18

Source: Vela GR and Wu JF. Mechanism of Lethal Action of 2,450-MHz radiation on microorganisms. *Applied and Environmental Microbiology*, 37(3):552, copyright 1979 by the American Society for Microbiology, Washington, D.C. Reprinted by permission.

[a]The 99.9% lethal dose, LD99.9%, was obtained from graphs depicting the surviving fraction (i.e., irradiated/nonirradiated control).

the temperature of the microwave oven tracked that of the convection oven. The computer temperature monitoring and measuring systems were capable of minimizing the temperature differences between the two ovens to less than 2°C. Exposure time was defined as the total time the dry-heat vial was situated in the convection dry-heat oven, regardless of temperature. Actual temperatures in the two ovens were 107, 117, 130, and 137°C. Total exposure time to inactivate 10^5 spores was 75 minutes at 130°C and 48 minutes at 137°C. No significant difference was observed between convection and microwave heat to inactivate spores in the temperature ranges studied.

The conclusions of Vela and Wu were supported in a study published by Najdovski et al. in 1991 (33). Aqueous suspensions in concentrations of 10^6 spores/mL of *B. subtilis* and *B. stearothermophilus* were completely inactivated (6-log reduction) when exposed to a microwave energy of 1400 watts, 2450 MHz for periods of 10 and 20 minutes. Complete inactivation was not observed when dry suspensions of spores were exposed to the same conditions. The authors noted that microwave systems of much higher power output than those studied could be used for decontaminating infectious waste if appropriate amounts of water are added during processing.

System Review. Two microwave medical waste treatment systems are reviewed in this section, which are offered by:

1. Sanitec, Inc.
2. Roatan Medical Services Corporation

Sanitec, Inc. (34,35)

Source

Sanitec, Inc.
26 Fairfield Place
West Caldwell, NJ 07006
Tel: (201) 227-8855
Fax: (201) 227-9048

Product Overview. The Sanitec microwave disinfection unit is designed to shred and disinfect biomedical waste so that it is no longer recognizable as medical waste. Two models are offered; a smaller model that is rated to process 450 pounds per hour and a larger model that is rated to process 900 pounds per hour. The systems process waste in an automatic mode in which all critical components are computer controlled. The operator control panel contains only two switches for normal operator use: one to lift the container and one to lower the waste container. A general system diagram is presented in Figure 10.8.

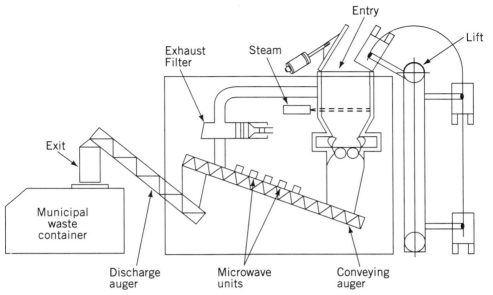

Figure 10.8 Sanitec microwave disinfection system. Courtesy of Sanitec, Inc. Sanitec Microwave Disinfection System, copyright 1995 by Sanitec, Inc., West Caldwell, New Jersey. Reprinted by permission.

During the treatment process, the waste is first shredded and then disinfected. Containers of waste are hoisted by a hydraulic lift mechanism and the contents dumped into a hopper located on top of the unit. During the loading process, air is evacuated from the chamber through a three-stage filter system that includes a HEPA filter. The hopper is purged with high-temperature steam (302°F) to kill any airborne pathogens. A rotating feeder arm then moves the waste toward and into the shredding device. The shredded waste is injected with high-temperature steam (302°F) to bring the chamber to the desired operating process temperature, and to add sufficient moisture to enhance the microwave treatment. The shredded material then passes through a screen and onto a screw auger which moves the waste through the microwave chamber. A series of microwave generators facilitates the treatment parameters: a minimum 203°F for 30 minutes. The action of the auger conveyor also provides additional mixing to ensure that the waste is heated uniformly. After disinfection, treated waste may be shredded through a secondary shredder to further reduce its particle size. The final cooled waste is then deposited via an exit screw conveyor into a dumpster or other waste transport container for disposal. Waste volume is reduced by approximately 80 percent. Throughout the process, key operating parameters (time, temperature, screw speed) are monitored continuously by the system's microprocessor.

Models Available. The Sanitec microwave disinfection system is offered in two models:

- HGA-250S
- HGA-100S

Both models are available in mobile versions.

Specifications

Model HGA-100

Use:	Central location in a medium to large medical facility or as a regional treatment center
Size (overall):	24 ft long × 10 ft wide × 17 ft high (flap open) or 11 ft high (flap closed)
Weight:	18,000 lb; particlizer adds 5000 lb
Microwave generators:	Four generators with a high-frequency output (each) of 1.2 kW and an output frequency of 2450 MHz
Electrical requirements:	460/480 V ac, 150 A, 60 Hz, three-phase
Power consumption:	60 kWh
Water requirements:	$\frac{3}{4}$-in. NPT (national pipe thread) connect; uses from 8 to 9 gal of water per hour for steam generation
Drain requirements:	No sewer connection is necessary; all water generated from the steam mechanism is recycled and consumed within the unit without generating a liquid discharge
Load capacity:	220 to 450 lb/hr
	Note: Maximum amount treated per hour depends on waste contents (e.g., density, moisture)
Total cycle time:	Continuous treatment process; however, all waste is subjected to 203°F for a minimum of 30 minutes
Decontamination temperature:	203°F for a minimum of 30 minutes
Exhaust air requirements:	6-in. exhaust louver for air, which is discharged from a HEPA filter
HEPA filter:	12 in. wide × 24 in. long × 11.75 in. deep; 99.9995 percent efficiency at 0.12 micron

Model HGA-250

Use:	Central location in a large medical facility or as a regional treatment center
Size (overall):	9 ft 4 in. wide × 23 ft 6 in. long × 17 ft high (flap open) or 10 ft 11 in. high (flap closed)
Weight:	22,000 lb; particlizer adds 5000 lb
Microwave generators:	Six generators with an HF output (each) of 1.2 kW and an output frequency of 2450 MHz
Electrical requirements:	460/480 V ac, 200 A, 60 Hz, three-phase
Power consumption:	75 kWh nominal, 60 kWh average
Water requirements:	$\frac{3}{4}$-in. NPT connect. The system uses from 8 to 9 gal of water per hour for steam generation
Drain requirements:	No sewer connection is necessary; all water generated from the steam mechanism is recycled and consumed within the unit, without generating a liquid discharge
Load capacity:	550 to 900 lb/hr
	Note: Maximum amount treated per hour depends on waste contents (e.g., density, moisture)
Total cycle time:	Continuous treatment process; however, all waste is subjected to 203°F for a minimum of 30 minutes
Decontamination temperature:	203°F for a minimum of 30 minutes
Venting requirement:	6-in. exhaust louver for air, which is discharged from a HEPA filter
HEPA filter:	12 in. wide × 24 in. long × 11.75 in. high; 99.9995 percent efficiency at 0.12 μm

Capital Cost

- Model HGA-A-100: $375,000
- Model HGA-250: $650,000

Waste Stream Compatibility, Challenge, Incompatibility

Compatible Waste Types

- Cultures and stocks of infectious agents and associated biologicals.
- Liquid human and animal waste, including blood and blood products and blood fluids; large volumes of blood will decrease throughput somewhat.

- Pathological waste. For aesthetic reasons the manufacturer recommends that recognizable anatomical waste (e.g., limbs, organs) be incinerated.
- Sharps.

Challenging Waste Types

- From a shredding standpoint only, any waste that requires excess shredding (e.g., dense or compacted material) will slow throughput.
- It is recommended that liquid content be minimized because large amounts of liquid may decrease throughput.

Incompatible Waste Types

- From a shredding standpoint only, any waste that would shorten the life of the shredder, such as:
  - Large or high-grade stainless steel items
  - Implants such as pins, rods, or joints
  - Tools and broken pieces of hospital equipment
- Toxic chemical (e.g., antineoplastic drug waste, RCRA waste) or radiological waste (e.g., low-level radioactive waste)

Installation. Delivery time of a unit depends on the availability of a unit with the facility's specifications; can be as short as two weeks. Complete setup takes less than two working days. Operation can begin immediately after setup. Sanitec technicians are on-site for a two-week training period, after which the owner is equipped to operate and maintain the unit. The unit may be installed indoors or outdoors. There are no special site specifications. Installation requires a flat level surface, a water source, and electrical hookups.

Operation. System startup requires approximately 30 minutes. The first startup phase (switching on the main power, starting the microwave generators, recording hours of operation, etc.) requires about 15 minutes. Waste is then added. The second startup phase requires conducting a series of 24 system inspection tasks prior to initiating the treatment process.

System Monitoring. Quality assurance is guaranteed through initial verification and periodic testing using spores of *Bacillus subtilis*. The unit contains a chart recorder that records the temperature at the inlet and temperature holding sections both automatically and continually. After the system has been demonstrated to perform adequately under certain parameters of time and temperature, the system will continue to operate effectively provided that the parameters are met. The manufacturer suggests that efficacy verification using microbiological spore test indicators should be conducted as required by state or local regulation.

Training. Sanitec, Inc. offers a two-week training session to purchasers of its system. The operator does not require any experience beyond the training provided by Sanitec upon installation of the unit.

Service. The system comes with a 12-month warranty on most materials and workmanship.

Treated Waste

Characteristics of Treated Material. Shredded, nonrecognizable, confetti-like particles with a 14 percent average moisture content and a 0.3 gram per milliliter (20 pounds per cubic foot) average density. The treated waste has a high Btu value (8000 to 12,000 Btu per pound) and can be used as an alternative fuel.

Volume Reduction. Waste volume is reduced by up to 80 percent.

Disposal of Treated Material. Depending upon state and local restrictions, treated waste is handled as municipal solid waste and may be disposed of in a landfill or municipal incinerator. Due to its high Btu value, the treated waste may also be used as an alternative fuel.

Environmental Discharges

By-Products and Discharges. No liquid, chemical, or aerosol discharges. Slight odors may be generated.

By-Product Controls. The units contain a prefilter, HEPA filter (99.9995 percent efficient to 0.12 micron) to capture any potential airborne particles that may be generated when air is released from the unit's infeed area as it accepts waste, and a carbon filter to control odors.

Efficacy Testing. Numerous tests have been conducted on the microwave unit. Baseline testing was conducted during 1991 by an independent certified laboratory, North American Laboratory Group, with the exception of the duck hepatitis B virus test protocol, which was conducted by the Stanford University School of Medicine. Testing concluded that the high level of disinfection achieved by the Sanitec system is capable of meeting the parameters established by the state of New York's Department of Health for alternative treatment technologies.

The following microorganisms were tested during that phase of efficacy testing:

- Bacterial
  - *Bacillus subtilis* (spores) (ATCC 6633)
  - *Enterococcus faecalis* (ATCC 19433)

- *Pseudomonas aeruginosa* (ATCC 27317)
- *Staphylococcus aureus* (ATCC 25923)
- *Nocardia species* (ATCC 31531)
- Mycobacterial species
  - *Mycobacterium bovis* (ATCC 35737)
  - *M. fortuitum* (ATCC 35755)
- Fungi
  - *Candida albicans* (ATCC 14053)
  - *Aspergillus fumigatus* (ATCC 1022)
- Protozoa
  - *Giardia intestinalis* (ATCC 50114)
- Virus
  - Duck hepatitis B virus (provided by Stanford University)

Subsequent testing to meet alternative treatment technology efficacy testing parameters for the Illinois Environmental Protection Agency was conducted in 1993. That testing demonstrated a 6-log kill of spore strips of *B. subtilis* ATCC strain 19659 enclosed in glassine envelopes, and *B. subtilis* ATCC strain 9372 enclosed in polypropylene vials.

Approvals and Installations. The Sanitec technology has received system technology approvals in over 40 states. More than 40 systems have been installed, with several others pending.

Corporate Profile. Sanitec, Inc., (formerly ABB Sanitec, Inc.), with headquarters in West Caldwell, New Jersey, has been involved in the design and manufacture of microwave disinfection systems since 1990. Sanitec is a unit of H.S. Holdings of Roseland, New Jersey, whose principal, Henry F. Henderson, Jr., owns and operates Henderson Industries; an electrical and mechanical engineering firm that supplies turn-key manufacturing services to the high technology marketplace. Henderson Industries, also of West Caldwell, employs 150 people and has annual revenues of over $25,000,000.

Roatan Medical Services Corporation (36,37)

Source

Roatan Medical Services Corporation
1022 Santerre Drive
Grand Prairie, TX 75050
Tel: (214) 647-4033
Fax: (214) 647-4454

Product Overview. Roatan Medical Services Corporation has developed the Redloc 2 Sterilizer System which uses microwave energy to create steam under pressure to sterilize biohazardous waste. The system is comprised of (1) one or a series of modular sterilization units; (2) reusable containers; (3) an optional granulator; (4) a container washing system; and (5) a container cart. Each individual sterilizer is capable of processing 50 pounds of waste per hour at a rate of 2 to 3 cycles per hour. Several sterilizers may be used in tandem to increase treatment capacity. A system diagram is presented in Figure 10.9.

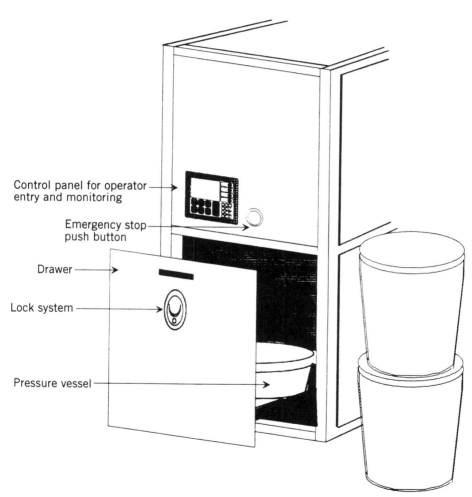

Figure 10.9 *Redloc 2 sterilizer system.* Courtesy of Roatan Medical Services Corporation. Redloc 2 Sterilizer System, copyright 1995 by Roatan Medical Services Corporation, Grand Prairie, Texas. Reprinted by permission.

The sterilization units are designed for on-site use and are self-contained. Biohazardous waste is collected at the point of generation into reusable, puncture-proof, leak-resistant 25 gallon containers and placed into the system's pressure vessels. Energy is released by two microwave generators and water is injected at periodic intervals creating steam under pressure which is heated until sterilization parameters are achieved. After the pre-set treatment parameters of pressure and temperature have been met, the steam is discharged through a three-filter evacuation system consisting of a coalescing filter, a HEPA filter and a carbon filter. Following venting, the operator opens the front door. The sterilized waste is either automatically lifted to an optional granulator where it is rendered unrecognizable, or removed from the system for disposal. Saturated, superheated steam is the physical, microbial killing agent. Efficacy testing indicates that the system is capable of achieving sterility based on observations of greater than 6-$\log_{10}$ *Bacillus stearothermophilus* spore kills.

Models Available. Redloc 2 Sterilizer System

Specifications

Use:	On-site for individual health care providers and facilities
Size (overall):	Sterilizer: 40 in × 40 in × 85 in
	Granulator: 112 in × 55 in × 190 in
Weight:	Sterilizer: 700 lb
	Granulator: 9800 lb
Electrical requirements:	Sterilizer: 220V, 30A, three-phase
	Granulator: 440V, 150A, three-phase
Power consumption:	20 to 40 kWh average
Water requirements:	Sterilizer requires 1 gal/hr through $\frac{1}{2}$ in hose supply at 40 psig
Drain requirements:	None
Load capacity:	50 lbs/hr for each individual unit
Total cycle time:	Approximately 20 to 30 minutes, depending upon moisture content, waste composition, etc.
Treatment temperature:	260°F
Exhaust air requirement:	None
Filter system:	Coalescing filter, HEPA filter, carbon filter

Capital Cost. The price range is based on the size of the system which includes sterilizers, optional granulator, containers, transportation cart, and container washer/dryer.

Purchase price:	$75,600–$462,000 without granulator
	$138,000–$522,000 with granulator

Lease per month: $1,800–$11,00 without granulator
$3,250–$12,500 with granulator

Waste Stream Compatibility, Challenge, Incompatibility

Compatible Waste Types

- Cultures and stocks of infectious agents and associated biologicals.
- Liquid human and animal waste, including blood and blood products and blood fluids.
- Pathological waste
- Sharps
- Contaminated waste from animals
- Other biomedical waste (e.g., isolation wastes)

Challenging Waste Types

- The rate of treatment (lbs/hr) will vary depending upon certain characteristics (e.g., bulk liquids, densely packed containers, etc.), though sterilization efficacy is not affected.
- Certain metal objects may interfere with the granulator, which is an optional component.

Incompatible Waste Types

- Toxic chemical (e.g., antineoplastic drug waste, RCRA waste) or radiological waste (e.g., low-level radioactive waste)

Installation. Each sterilizer should be located in a treatment area capable of a supporting floor load of approximately 65 pounds per square foot in an area that generally allows 18 inches on the back side for access. There are no major side renovations necessary.

Operation. The system is designed for ease of operation with a minimum of instruction. The operator is required to understand and follow prompts displayed on the operator screen.

System Monitoring. Critical parameters (e.g., date/time of run, total run time, maximum temperature achieved, maximum pressure achieved, operator code, container code) are monitored on an LCD display screen and are recorded as a permanent record on disk. Time, pressure, and temperature are continuously monitored throughout each treatment cycle. Biological efficacy testing may be periodically conducted.

Training. No special training to operate the system is required beyond that provided by the service representative. The manufacturer will provide on-site training for 5 to 10 working days.

Service. System pricing includes maintenance for a sixty month period based on normal use. Regular maintenance, filter changes, etc., are provided by the manufacturer.

Treated Waste

Characteristics of Treated Material. Prior to granulation, the physical characteristics of the waste are changes due to heat (e.g., certain plastics may meld) but the waste retains recognizability. Following granulation the waste is generally unrecognizable.

Volume Reduction. Approximately 80 percent volume reduction with granulator.

Disposal of Treated Material. Depending upon local restrictions, the sterilized waste may be disposed of into the general waste stream.

Environmental Discharges

By-Products and Discharges. There is no water discharge and the system does not require connection to a drain. Air is passed through a HEPA and carbon filter and released to the atmosphere. According to the manufacturer, during hundreds of hours of testing and usage, no microwave leakage has been detected. The microwaves are contained within an inner and outer frame and skin. The skin is a special laminate that functions as an emissions shield.

By-Product Controls. Steam is condensed on a coalescing filter with excess water re-injected into the unit; remaining air is passed through a HEPA filter to trap particulates and a carbon filter to control odors before venting to the atmosphere.

Efficacy Testing. Efficacy testing was conducted between January and March 1995 by the University of Oklahoma Health Sciences Center, College of Public Health, Occupational and Environmental Health Department, and in September 1995 by the Cleveland State University Biology Department. Tests were conducted on a challenge waste load designed to be consistent to a bio-burden challenge of typical medical waste, which included non-infectious unshredded linens, sharps, glass, gauze, laboratory material, sharps, plastic and rubber items and potable water both in sealed containers and added directly into the test vessel. Test organisms included:

- Bacteria
  - *Pseudomonas aeruginosa* (ATCC 10145-U)
  - *Mycobacterium phlei* (ATCC 11758)

- Fungi
  - *Penicillium chrysogenum* (ATCC 10002)
- Virus
  - Polio 2 (ATCC VR-301)
- Protozoa forming cysts
  - *Colpoda steinii* (ATCC 30920)
  - *Colpoda cucullus* (ATCC 30916)
  - *Giardia muris*
- Spore Forming Bacillus
  - *Bacillus stearothermophilus*

The efficacy testing observed a greater than 6-$\log_{10}$ reduction in vegetative microorganisms and a 6-$\log_{10}$ reduction in *Bacillus stearothermophilus* spores.

Approvals and Installations. According to the manufacturer, the system is approved as an alternative technology in California, Texas and Florida and is approved as a steam sterilizer in most other states. Based on efficacy testing results, Roatan anticipates a broad approval.

Corporate Profile. Roatan Medical Services Corporation is a privately-financed corporation based in Texas. Sales offices are located in California, Wisconsin, and Florida.

DRY HEAT TECHNOLOGY

Several commercially available biohazardous waste treatment systems are currently available in which the mechanism for microbial destruction is based on a dry heat technology. Each of the sterilizers reviewed in this section is heated electrically and relies on the presence of hot air, which creates the inactivating conditions. Because diffusion and penetration of dry heat is slow, dry heat sterilization relies on higher temperatures held for longer periods of time than the temperature required for steam sterilization (38).

System Reviews. Four biohazardous waste treatment systems are reviewed in this section. Three are designed for on-site use for the smaller generator, and one for on-site use by larger generators or for use as a regional treatment facitlites. These systems are offered by the following companies:

1. Disposal Sciences, Inc.
2. D.O.C.C., Inc.
3. MedAway International Inc.
4. Spintech, Inc.

Disposal Sciences, Inc. (39,40)

Source

Disposal Sciences, Inc.
Cannon Plaza South
6352 320 Street Way
P.O. Box 487
Cannon Falls, MN 55009
Tel: (800) 537-7324
Tel: (507) 263-4721
Fax: (507) 263-0333

Product Overview. The DSI System 2000 disinfects plastic medical waste using a dry heat disinfection technology that converts the waste into an unrecognizable plastic brick. The system is designed for large-scale use either on an off-site regional basis or on-site use by larger medical waste generators. It has been designed as a low-cost, easily operated disposal system for specific classes of medical waste (e.g., plastic medical waste such as disposable sharps, sharps containers, bedpans, urinals, and other rigid plastic items). A system photograph is presented in Figure 10.10.

Models Available

• DSI System 2000

Specifications

Use:	Large scale off-site regional use or on-site use for larger generators
Size (overall):	66 in. high × 37 in. wide × 32 in. deep
Weight:	Approximately 380 lb
Electrical requirements:	*United States:* 208 to 220 V ac, three-phase, 50/60 Hz, 25 A; power cord: length 8 ft with NEMA L21–30 plug three-conductor with neutral and ground
	International: 380 to 415 V ac, three-phase, 50/60 Hz, 16 A; power cord: length 8 ft, three-conductor with neutral and ground harmonize cord (no plug is supplied)
Power consumption:	9 kWh per cycle
Water requirements:	None
Drain requirements:	None
Load capacity:	Up to 16 gal/cycle
Decontamination time:	40 minutes
Decontamination temperature:	535°F

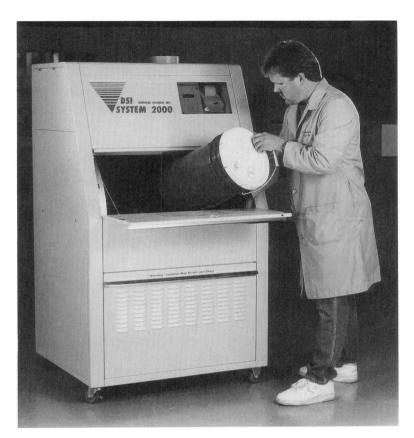

Figure 10.10 DSI System 2000. Photograph from Disposal Sciences, Inc. DSI System 2000, copyright 1995 by Disposal Sciences, Inc., Cannon Falls, Minnesota. Reprinted by permission.

Total cycle time: Approximately 105 minutes

Venting requirements: Unit is to be vented out of the building; a 6-in. vent connection is supplied with the unit; up to 13 ft of 6-in. rigid and 2 ft of flex duct, for 15 ft total, may be used without an auxiliary blower

Filter system: 18 in. × 24 in × 6 in. combination 99.99 percent HEPA and activated carbon filter; 18 in. × 24 in × 0.25 in. foam prefilter

Capital Cost

• $25,000

Waste Stream Compatibility, Challenge, Incompatibility

Compatible Waste Types

- Plastic medical waste such as disposable sharps, sharps containers, bedpans, urinals, and other rigid plastic items

Challenging Waste Types

- High liquid content (may cause the system not to reach the required temperature within the system time limits which would cause a system error condition)

Incompatible Waste Types

- Cultures and stocks of infectious agents and associated biologicals
- Liquid human and animal waste, including blood and blood products and blood fluids
- Pathological waste
- PVC plastic

Installation. The system should be installed in a clean temperature-controlled location with an ambient temperature range of 60 to 80°F, with access to a power source, and central exhaust air system or an exterior exhaust air location with a 350-cfm exhaust air connection.

Operation. Operating the system involves the following steps:

1. When "load" appears on the LED display, the upper door on the DSI system 2000 is opened and the crucible is loaded with the plastic medical waste to be processed.
2. The door is closed. The start button is depressed to begin the cycle. The system's heat/cool cycle operates automatically to completion.
3. The drive motor rotates and holds the crucible to the upright "heat" position until the target temperature has been reached (535°F) and maintained for 40 minutes.
4. The cooling system is initiated at the end of the 40-minute heating period. The motor drives the crucible to the "cool" position and the waste is cooled by three fans, two that supply ambient outside air to the crucible and one that draws hot air from the cabinet through a HEPA filter bank. The fans create a negative air pressure inside the unit, which contains odors and gases.
5. After the waste has cooled to prescribed temperatures, the motor rotates the crucible 180 degrees to the "dump" position and the now-hardened plastic brick falls into the collection bin. If the plastic brick fails to eject, the system

is equipped to heat the bottom of the crucible to up to 350°F to help loosen the brick.

System Monitoring. Routine biological monitoring is not a requirement in the system operator's manual. Quality assurance is monitored via system software and sensors that monitor time, temperature, crucible location, and airflow. The system displays its status at each step in the cycle (e.g., movement, heating, cooling). The parameters of a treatment cycle are stored in memory for printout or download. Treatment monitors have been correlated with biological efficacy testing.

Training. In-service training is provided to the purchaser by Disposal Sciences prior to operation of the system.

Service. Disposal Sciences provides a 12-month limited warranty after delivery against defects in material and workmanship. During this period, Disposal Sciences, Inc. will repair or replace defective parts at no cost to the purchaser.

Treated Waste

Characteristics of Treated Material. The plastic medical waste is melted into an unrecognizable plastic brick.

Volume reduction. There is up to 80 percent reduction of rigid plastics.

Disposal of Treated Material. The plastic brick may be disposed of directly into the general waste stream or as directed by local regulations.

Environmental Discharges

By-Products and Discharges. Slight odors and low volatile organic emissions are produced.

By-Product Controls. Odor volatile organic emission testing performed by MMT Environmental Services, Inc. of St. Paul, Minnesota observed both odor emissions below those established by the Minnesota Pollution Control Agency and volatile organic emissions below those established under current OSHA occupational exposure limits. The system is equipped with a prefilter and a HEPA filter. Following use, both filters should be disposed of as hazardous waste.

Efficacy Testing. Efficacy testing was performed by ViroMed Laboratories, Inc., Minneapolis, Minnesota, upon the following microorganisms:

- *Bacillus stearothermophilus* spores
- *B. subtilis* spores

- *Staphylococcus aureus*
- *Mycobacterium bovis*
- *M. fortuitum*
- *Candida albicans*
- Poliovirus type 3
- Bacteriophage MS2

The laboratory report indicated "no growth" for any organism following treatment. A 6-log reduction was observed for tests of *B. stearothermophilus* and *B. subtilis* spores.

Approvals and Installations. As of August 1995, the distributor claims system approvals in 35 states with 2 applications pending. No information on installations was provided.

D.O.C.C., Inc. (41,42)

Source

D.O.C.C., Inc.
240 East 76th Street
New York, NY 10021
Tel: (212) 988-4420
Fax: (212) 734-4801
Contact: Jonathan Bricken

Product Overview. The Demolizer System is a desktop treatment system that plugs into a standard 110-volt line designed for small biomedical waste generators. Waste to be processed is collected in special disposable containers designed specifically for the unit. After the filled containers are placed in the unit, the waste is subjected to a dry heat disinfection cycle of 350°F for 90 minutes. Following the cycle, the disposable container holding the disinfected waste is removed from the unit and disposed in its entirety as municipal solid waste. The Demolizer System is shown in Figure 10.11.

Models Available

- Demolizer System, Model 47

Specifications

Use: On-site for small-scale use by small or large
                                            generator facilities.

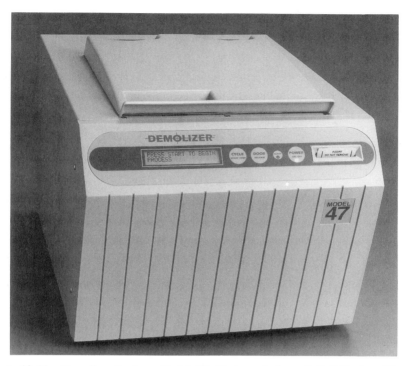

Figure 10.11   Demolizer System, Model 47. Photograph from D.O.C.C., Inc. The Demolizer System, Model 47, copyright 1995 by D.O.C.C., Inc., New York, New York. Reprinted by permission.

Size (overall):	19 in. deep × 13 in. wide × 10.5 in. high
Weight:	Approximately 30 lb
Electrical requirements:	110 V, 15 A
Power consumption:	1.5 kW/cycle
Water requirements:	Does not use water
Drain requirements:	None
Load capacity:	1 gal/cycle
Total cycle time:	2 hours 35 minutes minimum
Decontamination temperature:	350°F
Decontamination time:	90 minutes
Venting requirements:	None
Filter system:	Carbon filtration system

Capital Cost

• $2,500

Waste Stream Compatibility, Incompatibility

Compatible Waste Types

- Cultures and stocks of infectious agents and associated biologicals
- Liquid human and animal waste, including blood and blood products and blood fluids
- Sharps
- Other biomedical waste (e.g., red bag waste, such as gauze, gloves, cotton balls, etc.)

Incompatible Waste Types

- Pathological waste (human and animal body parts)
- Toxic chemical (e.g., antineoplastic drug waste, RCRA waste) or radiological waste (e.g., low-level radioactive waste), depending on regulatory restrictions

Installation. The system is a desktop unit that plugs into a standard 110-volt line. No other special installation requirements are necessary.

Operation. The system employs two 1-gallon collection containers, one for sharps waste and one for red bag waste, that are placed in treatment areas for waste collection. When full, the containers are closed to prevent spillage during transfer to the unit. Full containers are placed in the unit and the start button is depressed, which automatically engages the lid interlock system to begin processing. Following the 2-hour 35-minute treatment cycle, and after the temperature has dropped below 95°F, the canister containing the decontaminated waste is removed from the unit and disposed of as municipal solid waste.

System Monitoring. Routine biological testing with spore strips is not recommended by the manufacturer. The system is electronically monitored for temperature and time. The system is designed so that any interruption in the heating cycle will cause the unit to be reset and begin the full 90-minute cycle at 350°F. After the cycle has completed, the unit prints out a hard copy of cycle parameters verifying that the treatment has occurred within the time and temperature requirements. Each container has a heat-sensitive color change strip to provide visual separation between treated and untreated canisters.

The system is equipped with a fail-safe mechanism that engages upon malfunction: No printout is provided; the lid interlock system does not disengage, preventing removal of the waste containing canister; and the power supply shuts down, rendering the system inoperable. In this event, the system will have to be replaced by D.O.C.C.

Training. The distributor provides training to at least two persons at each facility that purchases the Demolizer system.

Service. If the system fails, the unit will be replaced under warrantee or service will be performed off-site. An hourly rate of $95 per hour will be charged if the warrantee expires.

Treated Waste

Characteristics of Treated Material. The waste is dry and retains its physical appearance inside the sealed metal processing container. Sharps waste melts down in a puck in which reside all nonplastic portions of the sharps waste.

Volume Reduction. All sharps waste is reduced by 75 percent.

Disposal of Treated Material. Depending upon local restrictions, the sealed metal container and its treated contents are disposed of as municipal solid waste.

Environmental Discharges

By-Products and Discharges. Residual heat is discharged.

By-Product Controls. All emissions from the heat cylinder are channeled to an activated charcoal filter surrounded by a bacteriological filter material.

Efficacy Testing. Efficacy testing was conducted by Patricia L. Marion of the Stanford University School of Medicine for duck hepatitis and by Leberco Laboratories in New Jersey for the other microorganisms tested. The purpose of the testing was to meet the state of New York's Department of Health efficacy testing criteria, Public Health Law, Section 1389-dd(i) and to 10 NYCRR Subpart 70–2, which the system met successfully. Efficacy testing was performed using the following test organisms:

- Vegetative bacteria
  - *Pseudomonas aeruginosa* (ATCC 9027)
- Fungi
  - *Candida albicans* (ATCC 10231)
- Parasites
  - *Giardia* spp. cysts (obtained from USEPA)
- Bacterial spores
  - *Bacillus stearothermophilus* (ATCC 6538)
  - *B. subtilis* (ATCC 6633 and 9772)
- Virus
  - Duck hepatitis

Approvals and Installations. D.O.C.C. reports that the technology has been approved in all states with approval processes except New Jersey and Rhode Island in which approvals are pending.

Corporate Profile. For 10 years, D.O.C.C. has been servicing the medical industry. In 1986, the company identified the need to help healthcare providers solve the serious problems of infectious medical waste destruction and disposal. Responding to the demands of increased governmental regulation, D.O.C.C. introduced a patent pending disposal system, the Demolizer. D.O.C.C. products provide safe and efficient solutions for the on-site conversion of infectious medical waste into sterile garbage. D.O.C.C. concentrates exclusively on the infectious medical waste business and has demonstrated a commitment to solve this problem at the point of generation.

MedAway International Inc. (43)

Source

MedAway International Inc.
4020 Galt Ocean Drive, Suite 1401
Ft. Lauderdale, FL 33308
Tel: (305) 565-4990
Fax: (305) 565-1174

Product Overview. The MedAway-1 system, developed by Fuji Medical Systems of Tokyo, Japan, and IMSEC of Kyoto, Japan, is distributed by MedAway International, which has worldwide distribution rights. The system decontaminates waste using a dry heat sterilization process that employs both resistance and quartz infrared heaters. A system diagram is presented in Figure 10.12. The procedure is labeled by the distributor as target-specific dry heat sterilization, in which the waste is heated to temperatures of up to 300°C. The resistance heaters first heat the air surrounding the waste, which then heats the waste itself. The quartz infrared heaters generate far-infrared waves, defined as that part of the electromagnetic spectrum with wavelength greater than 3 microns, which are absorbed directly by the targeted waste. Waste volume is reduced by up to 90 percent and fluids are evaporated. Efficacy testing data supplied by the distributor shows the system capable of achieving sterilization of test material through a 6-log reduction of *Bacillus subtilis* spores.

Models Available

• MedAway-1 waste processor

Specifications

Use:	Designed for on-site use by a small generator
Size (overall):	33.5 in. long × 41 in. wide × 51 in. high
Weight:	660 lb

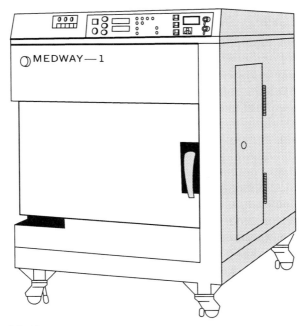

Figure 10.12 MedAway-1 waste processor. Courtesy of MedAway International, Inc. MedAway Waste Processor, copyright 1995 by MedAway International, Inc., Crestwood, Missouri. Reprinted by permission.

Electrical requirements:	220 V ac (wall socket), 30 A, one-phase, 60 Hz
Power consumption:	5.8 kWh
Water requirements:	21 to 42 psi
Drain requirements:	Floor drain with 1.5-in.-diameter minimum
Load capacity:	4.5 ft^3, up to 35 lb
Decontamination time:	Up to 30 minutes
Decontamination temperature:	Up to 572°F
Total cycle time:	90 minutes to 2 hours
Venting requirements:	None required (system uses a proprietary internal scrubber/filter)
Filter system:	Employs a water air scrubbing system, which recirculates the air through four charcoal/HEPA filters; filter life is approximately 30 days or 50 cycles

Capital Cost

• $49,400

Waste Stream Compatibility, Challenge, Incompatibility

Compatible Waste Types

- Cultures and stocks of infectious agents and associated biologicals
- Liquid human and animal waste, including blood and blood products and blood fluids, although not in bulk form
- Sharps

Challenging Waste Types

- The system is not designed to handle bulk liquid blood or blood products, which should be poured into a sanitary sewer.
- Pathological waste. The system would sterilize but not render human or animal body parts unrecognizable
- Woven gauze in heavy concentrations creates a smoke problem upon opening door of chamber.

Incompatible Waste Types

- Toxic chemical (e.g., antineoplastic drug waste, RCRA waste) or radiological waste (e.g., low-level radioactive waste)

Installation. The MedAway-1, a mobile unit on wheels weighing 660 pounds, can be installed in less than 1 hour. The system requires access to a 220-volt 30-ampere, single-phase line. A garden hose is needed for water intake and a drain hose for wastewater to a floor drain. At the time of installation, a flow regulator is installed by the distributor to keep the water pressure at 15 to 21 psi. A spore test is also conducted at that time.

Operation. The system is operated through the following steps:

1. Turn the water on.
2. Turn the key to "on" and open the door (the unit will go through a self-diagnostic check).
3. Load waste; close the door.
4. Choose the mode of operation (H, M, or L) (sharps, micro, or mixed).
5. If all operating lights are green, push "start."
6. At the conclusion of the cycle, open the door and remove the waste.

According to the vendor, operating the system does not require special skills and any employee can operate the unit. The operator does not have to wear masks for protection from chemical spray or debris from grinding or shredding, although good laboratory procedures must be followed and protective gloves worn.

System Monitoring. The system self-monitors time and temperature levels. Two identical printout graphs of the time/temperature plot are generated for each cycle to demonstrate that the unit reached decontamination temperature and stayed at that temperature for the programmed time, then cooled to 40°C. The thermograph printout also provides the date, process start time, a process number, the highest temperature attained, and the serial number of the machine. One printout can be attached to the load and one can be kept for quality control inspection.

For safety, the system is self-diagnostic and will not allow the operator to start the process if any of the warning lights are red. The door interlock will not open if the cycle has been aborted or a power failure has caused the unit to lose power.

Training. On-site training is provided by a field service engineer following system installation.

Service. The unit has a 12-month parts and labor warranty and includes four preventive maintenance visits. MedAway also offers a flexible service contract for following years. For immediate questions, MedAway provides a 24-hour telephone troubleshooting service.

Treated Waste

Characteristics of Treated Material. Treated plastic material is melted into a flat plastic, dry solid. Treated mixed waste is transformed into a mound of ash. The product is unrecognizable as medical waste.

Volume Reduction. The waste undergoes a volume reduction of up to 90 percent.

Disposal of Treated Material. Depending on local restrictions, the treated waste can be disposed of as general rubbish.

Environmental Discharges

By-Products and Discharges. Ash, dust, and very slight odors are generated during the process.

By-Product Controls. The MedAway-1 unit employs a water air scrubbing system, which recirculates the air through four charcoal/HEPA filters. The filters require monthly replacement. Any remaining decontaminated liquids are discharged through a floor drain.

Odor detection tests have been completed with the Cosmos Electric Company's Stench Monitor XP-329. When the monitor is set at air standard 0000, the readings from the MedAway-1 during operation should be 200 or less.

Efficacy Testing. Efficacy data supplied by the distributor identified inactivation of 6 logs or better of *Mycobacterium* var. *bovis* and *B. subtilis* spores and 5 logs or better of poliovirus 1 following inoculation into glass Vacutainer tubes containing a 100% serum protein load using a soil substrate following 5-, 10-, and 15-minute cycles at 480°F.

Approvals and Installations. The vendor reports technology approvals in Alabama, Arizona, Arkansas, California, Colorado, Connecticut, Florida, Hawaii, Indiana, Iowa, Kansas, Louisiana, Massachusetts, Mississippi, Missouri, Nebraska, Nevada, New Hampshire, New Mexico, North Carolina, North Dakota, Oklahoma, Rhode Island, South Carolina, South Dakota, Tennessee, Washington, and Wyoming.

Spintech, Inc. (45,46)

Source

Spintech, Inc.
100 Leader Heights Road
York, PA 17403
Tel: (717) 741-5900
Fax: (717) 741-5670

Product Overview. TAPS (thermal activated plastic sanitizer), developed by Spintech, Inc., is a medical waste treatment process designed for on-site use by the small generator at the point of waste generation. The treatment process sterilizes the waste and renders it unrecognizable as medical waste. The final product can be disposed of in the general waste stream. A system diagram is presented in Figure 10.13.

The TAPS medical waste processor is designed to sterilize and destroy regulated medical waste on-site. Medical waste is deposited into the TAPS transporter which is designed to replace the sharps container used in each treatment room. When filled, one or more TAPS dispos-a-disks are placed on top of the waste and the transporter is inserted into the TAPS processor for a four hour processing cycle. The unit temperature rises to approximately 410°F for about 1 hour. As the internal temperature reaches 400°F to 415°F, the plastic dispos-a-disk(s) melts and encapsulates the medical waste into an amorphous, opaque plastic block. The combination of dry heat thermal inactivation and contact with molten plastic effectively sterilizes the waste.

Models Available

- TAPS
- TAPS-II

Figure 10.13 TAPS thermal-activated plastic sanitizer. Courtesy of Spintech, Inc. TAPS Thermal Activated Plastic Sanitizer, copyright 1995 by Spintech, Inc., York, Pennsylvania. Reprinted by permission.

Specifications

TAPS

Use:	On-site for the small generator
Size (overall):	8.5 in. in diameter × 15 in. high
Weight:	10 lb
Electrical requirements:	115 V, 500 W, 60 Hz
Water requirements:	None
Drain requirements:	None
Load capacity:	Up to 1.5 quarts
Decontamination temperature:	415°F maximum
Total cycle time:	Approximately 4 hours
Final slug dimension:	4 in. in diameter × 3 to 4 in. high
Venting requirements:	None

TAPS-II

Use:	On-site for the small generator
Size (overall):	9.0 in. in diameter × 35.25 in. high

Weight: 26 lb
Electrical requirements: 115 V, 500 W, 60 Hz
Water requirements: None
Drain requirements: None
Load capacity: Up to 6 quarts
Decontamination temperature: 415°F maximum
Total cycle time: Approximately 6 hours
Final slug dimension: 4 in. in diameter × 5 ft 8 in. high
Venting requirements: None

Capital Cost

- TAPS: $995.00
- TAPS-II: $2195.00

Waste Stream Compatibility, Challenges, Incompatibility

Compatible Waste Types

- Sharps and similar regulated waste (e.g., some bloody gauze pads, extracted teeth)

Challenging Waste Types

- Cultures and stocks of infectious agents and associated biologicals
- Liquid human and animal waste, including blood and blood products and blood fluids
- Other medical waste with high organic content

Incompatible Waste Types

- Pathological waste
- Hazardous or radioactive waste

Installation. The TAPS unit is a desktop model and the TAPS II a floor model each only requiring a plug into a wall receptacle.

Operation. The TAPS systems are operated as follows:

1. Remove top cap assembly of transporter, place one dispos-a-disk in the bottom of the metal canister (label facing down), and replace top cap assembly by locking it in place.
2. Place waste material (e.g., syringes, needles, scalpel blades, gauze pads, etc.) through limited access opening at the top of the safety chute.

3. Lift chute to deposit waste into canister.

4. Place transporter into TAPS oven. Remove top cap assembly carefully so that all waste is deposited into canister.

5. Place required number of dispos-a-disks on top of waste (label facing up).

6. Lock lid into place using the simple latching mechanism.

7. Once activated, the primary heat cycle raises the unit temperature to 410°F and holds the temperature for 10 minutes.

8. Following the secondary heating cycle, a cool-down process is initiated. The heater turns off and the side vent opens, allowing cool air to enter the unit. During cool-down, the plastic block retracts in size, allowing easy removal from the metal canister.

9. When the unit cools to 130°F, the TAPS unit is opened, the metal canister removed, and the resulting plastic disk disposed of as common waste.

System Monitoring. In the TAPS system, the waste is placed in a metal canister with one dispos-a-disk at the bottom of the canister and two disks placed on top of the waste. The disks have a certifiable melting point significantly above time and temperature coefficients known to sterilize the waste. The dispos-a-disks are designed to melt at temperatures of between 330°F and 360°F at processing times of one hour. The change in shape of the waste and plastic disks to a solid block provides observational proof that the melting point temperatures of the plastic were met, and therefore provides a verifiable biophysical indication that temperatures were reached to cause destruction of the organisms contained within it.

Training. Training is provided by the sales representative following purchase of the system. No special training is required due to the simplicity of operation.

Service. Spintech provides a limited warranty for a period of one year from date of purchase against manufacturing defects in materials and workmanship.

Treated Waste

Characteristics of Treated Material. The waste is transformed into an amorphous, solid plastic cylinder measuring $4\frac{1}{2}$ in. in diameter by 3 to 4 in. high for the TAPS model, and $4\frac{1}{2}$ in. in diameter by 5 ft 8 in. high for the TAPS-II model. It is not recognizable as medical waste.

Volume Reduction. The waste is reduced in volume by a ratio of 4:1 in a typical load of plastic disposable syringes.

Disposal of Treated Material. Depending on local restrictions, the treated waste can be disposed as general waste in a common waste receptacle.

Environmental Discharges

By-Products and Discharges. Only heat is discharged—described by the manufacturer as being similar to that of a crock pot or coffeemaker. Based on test results, gas emissions generated during venting did not contain measurable hydrocarbons.

Efficacy Testing. Efficacy testing on Spintech's TAPS Medical Waste Processor was conducted by the National Environmental Technology Applications Corporation (NETAC), Bioremediation Products Evaluation Center. NETAC, a nonprofit research organization, is a subsidiary of the University of Pittsburgh Trust in Pittsburgh, Pennsylvania. NETAC was established jointly by the U.S. Environmental Protection agency and the University of Pittsburgh Trust for the express purpose of assisting in the commercialization of new environmental technologies and alternative treatment methods.

The primary findings conducted during 1992 on the TAPS system are reported as follows (NETAC):

- The plastic shells melt at 350 to 375°F.
- The treated simulated medical waste product was considered unrecognizable compared to the original characteristics of the untreated test waste.
- The TAPS system consistently killed spores of *B. subtilis* var. *niger* and *B. stearothermophilus.*
- The aerosol test results obtained in accordance with the protocol provided by Spintech did not indicate the presence of resistant *B. subtilis* var. *niger* on the agar plates that surrounded the unit during the test.
- Gas emissions generated during the test did not contain measurable hydrocarbons.
- The plastic shell consistently melted at unit temperatures and heating times above documented death curve temperatures for *B. subtilis* var. *niger*. This indicates that the melted shell can be considered a biophysical indicator that sterilization has been accomplished.

Based on study findings, NETAC concluded that the TAPS unit tested does operate in accordance with the operating procedures and principals defined by Spintech prior to the evaluation and that the unit does effectively sterilize simulated medical wastes inoculated with *B. subtilis* var. *niger*. NETAC also noted that the resultant product is generally no longer reusable or recognizable as defined in U.S. EPA's "Standards for the Tracking and Management of Medical Waste," 40 CFR Part 259.

Approvals and Installations. According to the vendor, the technology has received verification for use by regulatory agencies in 34 U.S. states.

Corporate Profile. Spintech, Inc. was founded in 1991 by Ronald Spinello and others to develop new technology in the medical and dental fields. It owns patent rights to the TAPS and TAPS-II systems, and to eXit, a needle incineration system. Spintech has several other new technologies under development in both the medical and dental markets.

ELECTROTHERMAL DEACTIVATION

Electrothermal deactivation (ETD) is a technology developed by Stericycle to treat biohazardous waste (47). ETD is a form of dielectric heat that employs a high-strength, high-voltage electrical field at radio frequencies of 10 to 11 MHz and an electrical field strength of 50,000 V per meter frequencies to heat and disinfect medical waste. In this process, materials absorb high-frequency electromagnetic radiation which generates heat. Such heating can be achieved by both microwave (frequencies ranging from 300 to 30,000 MHz) and radio-frequency waves (ranging from 5 to 100 MHz) and is appropriate for electrically nonconducting materials containing polar molecules, such as water. During the dielectric heating process, the direction of the applied field rapidly changes millions of times per second. Polar molecules (e.g., molecules with an asymmetric electronic structure) in the field rotate rapidly and align with those changes. The rapid agitation of polar molecules in a rapidly changing electrical field produces heat similar to friction throughout the material being processed. This heating technology contrasts with convection heating, in which heat is applied to the surface of a material and must rely on conduction to transfer heat to the center.

System Review. The Stericycle ETD technology is reviewed in this section.

Stericycle, Inc. (7,12,47,48)

Source

Stericycle, Inc.
1419 Lake Cook Road
Suite 410
Deerfield, IL 60015
Tel: (708) 374-5112
Fax: (708) 945-6583

Product Overview. Stericycle has developed a treatment method that deactivates biomedical waste via a heating process based on electrothermal deactivation (ETD) technology, which is a form of dielectric heating. Waste is initially shredded to small fragments averaging about 4 to 8 in. in greatest dimension, sprayed with water to achieve a 10 to 15 percent moisture content, and compacted to an average density of about 25 pounds per cubic foot in uniformly sized polyethylene plastic process vessels. This uniformity of size, moisture content, and compaction improves the effectiveness of the ETD process. The shredding/compacting process is carried out in an enclosed area under negative pressure with a HEPA filter system to filter any exhaust air.

The process vessels are capped before being conveyed through the ETD ovens, where they are subjected to a high-strength, high-voltage electrical field and an electrical field strength of 50,000 volts per meter. In the Stericycle dielectric heating process, electrical energy is transferred directly to the waste, causing polar molecules to rotate rapidly in synchronization with the field. This rapid rotation of the molecules uniformly generates heat throughout the waste mass, which results in rapid, uniform heating of the waste to temperatures above 194°F. After treatment the heated containers of waste are stacked in a thermal containment area and held for a minimum of 1 hour. The heating process and storage period eliminate all vegetative microorganisms and causes a 5-log reduction in bacterial spores.

Following the heating and holding period, the containers are uncapped and the waste is baled into bales. A portion of the waste stream is diverted for recycling, and another portion for the development of refuse-derived fuel, which is called Sterifuel. The Sterifuel, which has a Btu value approximately equivalent to coal, is packaged into bales and burned as fuel in cement kilns. When end-users for recycling and Sterifuel are not available, the final waste material is landfilled.

On January 16, 1995, Stericycle became the first certified manufacturer in the medical products sector to receive certification from Scientific Certification Systems for the recycled content in its plastic resin. Stericycle's Steri-Plastic, an injection-grade polypropylene, has been certified as being made from 100% recycled postconsumer medical waste, such as syringes, sharps containers, and other medical plastics. The treated plastic material is collected from its processing plants and shipped to a Stericycle-owned state-of-the-art plastic recycling plant located in West Memphis, Arkansas. The manufactured Steri-Plastic is sold for use in the manufacture of new plastic hospital products, such as sharps containers.

Models Available. Stericycle custom designs its systems to meet any applicable state medical waste management requirements.

Specifications

Use:	Meant to operate as a commercial, regional treatment facility, separate from hospitals or other generators, for large-scale volume
Size (overall):	A warehouse of 10,000 to 20,000 ft^2 is required to house Stericycle equipment, which includes shredders, conveyors, and an ETD oven
Weight:	Proprietary
Electrical requirements:	Proprietary
Power consumption:	Proprietary
Water requirements:	Water is recycled; use is minimal
Drain requirements:	None

Load capacity: About 1000 to 6000 lb/hr or 20 to 40 million pounds/year

Total cycle time: Sealed containers are passed through the dielectric heaters for 8 to 15 minutes and maintained within the deactivation temperature range for an additional 1 hour

Decontamination time/temperature: Approximately 1 hour 15 minutes at 203 to 212°F

Venting requirements: Air is exhausted through air ducts equipped with HEPA filters to the atmosphere

Filter system: HEPA filter ducts control airflow out of the sealed processing room

Capital Cost. This is not applicable because the system is owned and operated by Stericycle.

Operating Cost. This is not applicable because the system is owned and operated by Stericycle. Costs for services charged by Stericycle would be based on an integrated waste processing system, which considers the costs to segregate, package, and store the waste according to Stericycle's specifications, train generator personnel, transport waste to the Stericycle facility via a Stericycle transporter, process waste at the facility, and dispose of the treated waste through recycling, refuse-derived fuel, or landfill disposal.

Waste Stream Compatibility, Incompatibility

Compatible Waste Types

- Cultures and stocks of infectious agents and associated biologicals
- Liquid human and animal waste, including blood and blood products and body fluids, if not contained in tightly closed containers
- Sharps
- Other biomedical waste (e.g., isolation wastes)
- Pathological waste: accepts small amounts of human and animal pathological waste when mixed with other categories of biomedical waste. Anatomical waste, animals, etc. are segregated and transhipped for incineration.

Incompatible Waste Types

- Anatomical wastes, including human and animal body parts, tissues, diagnostic specimens, and other animal wastes
- Hazardous wastes as defined under RCRA, radioactive wastes as defined by the NRC, and certain chemicals, solvents, and metals

Installation. A Stericycle facility is housed in a warehouse of 10,000 to 20,000 square feet, depending on design. Installation is conducted exclusively by Stericycle, the facility's owner and operator.

Operation. Operating parameters for the dielectric heater (e.g., field strength, conveyor speed, dwell time of the wastes in the unit) are preset and controlled by a programmable logic circuit. These parameters cannot be altered by the operator. The operator is responsible for ensuring that all parameters are met.

Waste, maintained at negative pressure, is brought in from the loading area to the processing room for shredding. All equipment in the processing room is controlled by a programmable logic controller. Airflow out of the processing room is controlled through HEPA filter ducts. Waste is shredded by a multistage size reduction system. The waste is sprayed with water to achieve a typical moisture content of 10 to 15 percent. Waste is discharged to a press, which fills approximately 500 pounds of uniformly shredded material into 23-cubic foot polyethylene process vessels. Once filled and capped, the polyethylene process vessels are conveyed to the dielectric heater for processing. Once waste vessels have been heated in the oven and stored at temperature for the required 2-hour holding period, the waste is conveyed to a baler, which bales the processed material into bales.

System Monitoring. Each container is probed for temperature and must register at least 194°F as the container exits the heater. If this temperature is not achieved, the container is reprocessed through the heater. In addition, biological indicators (spore vials of *B. subtilis* var. *niger*) are placed inside a representative number of containers. Evidence of temperature levels and effective spore kills indicates successful deactivation.

Training. All Stericycle employees must undergo a training program that emphasizes the hazards associated with processing infectious waste and safety procedures. Operators of the dielectric heaters must undergo additional specialized training in equipment operation. Stericycle also provides training to hospital and other generator staff to ensure that waste is segregated and packaged according to Stericycle specifications.

Service. This is not applicable because the system is owned and operated by Stericycle.

Treated Waste

Characteristics of Treated Material. The composition of the treated waste is identical to that of the untreated waste except that it has been treated, shredded, and is unrecognizable as medical waste.

Volume Reduction. A total volume reduction of approximately 80 to 85 percent is achieved.

Disposal of Treated Material. Treated waste may be disposed of as municipal solid waste or burned in cement kilns as Sterifuel. Stericycle has also developed a plastics recovery and recycling process.

Recycling. The hospital segregates the plastics at the hospital. The containers are marked to indicate "medical plastics or sharps." These containers are processed separately and the treated plastics shipped to a dedicated Stericycle recycling plant in West Memphis, Arkansas. Certification by Scientific Certification Systems for the recycled content in its plastic resin is discussed in the "Product Overview" section.

Environmental Discharges

By-Products and Discharges. Water use is limited to Steri-tub washing, and water is recycled into the process. According to the manufacturer, monitoring of volatile organic compounds has indicated no issues. Microbiological air monitoring indicates no air issues.

By-Product Controls. A HEPA filter system is used.

Efficacy Testing. According to the manufacturer, efficacy testing protocols recommended by the State and Territorial Association on Alternate Treatment Technologies have been completed, with total required kill, defined as inactivation of vegetative bacteria, fungi, lipid/nonlipid viruses, parasites, and mycobacteria at a 6 $\log_{10}$ reduction or greater, and *B. stearothermophilus* or *B. subtilis* spores at a 4 $\log_{10}$ reduction or greater. All tests were conducted by the Illinois Institute of Technology Research Institute.

Approvals and Installations. Stericycle ETD facilities are operating in Woonsocket, Rhode Island; Morton, Washington; Yorkville, Wisconsin; and Loma Linda, California. Stericycle also has the only dedicated medical waste plastics recycling plant in the United States, West Memphis, Arkansas.

Corporate Profile. Stericycle, Inc. was founded in March 1989 as a privately owned medical waste treatment and recycling company headquartered in Deerfield, Illinois. Its concept was developed by James Sharp. Stericycle's treatment technology, first used in West Memphis, Arkansas in February 1990, was based on a cobalt-60 sterilization technology, then converted to the ETD technology in August 1990. The ETD technology is used at all of its facilities today. Stericycle employs approximately 200 employees throughout the country.

PLASMA TECHNOLOGY

In the plasma–based waste treatment process, electricity is used to ionize gas to a plasmalike state by running electric current through a gas (49–51). The ionized gas causes an electric current to arc between a positive and a negative pole. The superheated gas, at temperatures up to 12,000°F is then directed into a waste treatment chamber to pyrolyze the waste material. Medical waste is brought to temperatures of 2500 to 3200°F, which sterilizes the waste and reduces it to a glassy rock or slag, a ferrous metal, and gases.

System Review. The Plasma Energy Applied Technology waste treatment system is reviewed in this section.

Plasma Energy Applied Technology (7,49–51)

Source

> Plasma Energy Applied Technology
> 4914 Moores Mill Road
> Huntsville, AL 35811
> Tel: (205) 859-3006
> Fax: (205) 859-9588

Product Overview. Plasma Energy Applied Technology (PEAT) has developed a waste treatment system based on plasma technology. A plasma arc torch is an electric heater that uses the resistance of ionized gas to convert electrical energy to heat. The term *plasma* has been defined as an electrically generated discharge at atmospheric pressure that produces temperatures ranging from 7000 to 21,000°F (1). Although new as applied to the treatment of medical waste, plasma heating systems have been used during the past 30 years in several other industrial applications, including metal ore smelting, metal refining, cutting and welding, and metal and ceramic production. PEAT's Thermal Destruction and Recovery (TDR) System is composed of six major components: (1) a waste material handling feed system, (2) a plasma heating system, (3) a processing chamber (furnace), (4) a slag collection system, (5) a gas emissions treatment system, and (6) a data acquisition and process control system. A system diagram is presented in Figure 10.14. In the process, electricity is used to ionize gas to a plasmalike state by running electric current through it. The ionized gas causes an electric current to arc between a positive and a negative pole. The superheated gas is then directed into a waste treatment chamber to pyrolyze/vitrify the waste material. Waste is fed into the chamber by an auger extruder system. The gas being fed into the treatment chamber has an initial temperature of 12,000°F. The temperature of the chamber during treatment of medical waste is approximately 2500 to 3200°F. Waste is rendered sterile by the process and is reduced to a glassy rock or slag, a ferrous metal, and gases. Organic constituents are broken down to a gas composed primarily of diatomic hydrogen and carbon monoxide. Off-gases exit the chamber at a temperature of approximately 1800°F, are rapidly quenched, and are then passed through a scrubber, producing a clean product gas composed primarily of hydrogen (40 to 50%) and carbon monoxide (30 to 35%). Off-gases may be burned in a flare or

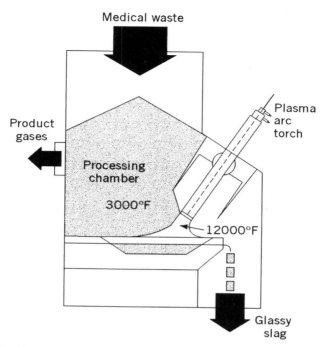

Figure 10.14 Plasma energy waste processing vessel. Courtesy of Plasma Energy Applied Technology. Plasma Energy Waste Processing Vessel, copyright 1995 by Plasma Energy Applied Technology, Huntsville, Alabama. Reprinted by permission.

steam boiler and exhausted through an emission stack to the atmosphere. Inorganic constituents are melted to a recyclable ferrous alloy and a silica-based glassy slag which bonds heavy metals that may be present in the waste. Heavy metals bonded in the slag are prevented from leaching to the environment when recycled or disposed in a landfill.

PEAT structures its facilities to meet the capacity needs of the purchaser. Plasma heating systems are commercially available in sizes ranging from 90 kilowatts (kW) to 5 megawatts (MW), making it possible to vary the size and waste processing capacity of the TDR System. For example, a 500-kW plasma arc torch system would be capable of processing approximately $\frac{1}{2}$ ton of waste per hour. A 2-MW plasma arc torch would be capable of treating up to 45 tons of waste per day, which would be sufficient for use as a regional treatment center. A mobile system could be designed based on a small 90-kW plasma arc torch.

Models Available. No specific model is available. Each TDR System is custom designed by PEAT to meet the waste treatment capacity needs of the purchaser.

Specifications. Specifications are presented for a TDR System designed to treat biohazardous waste at 1000 pounds per hour.

Use: On-site, off-site, large or small generators,
                                       modular or mobile
Size (overall):   90 in. long × 30 in. wide × 15 to 18 ft high
Weight: 110 tons
Electrical requirements: 480 to 4160 V
Power consumption: 1200 kVA
Water requirements: 35 gal/minute at 60 to 80 psi
Drain requirements: 25 gal/minute
Load capacity: 1000 lb/hr
Total cycle time: Operates on a continuous cycle
Sterilization time: Instantaneous
Venting requirements: 15,000 to 25,000 scfm
Filter system: Not applicable

Capital Cost

- $2 to $8 million, depending on size and requirements

Waste Stream Compatibility, Incompatibility

Compatible Waste Types

- Cultures and stocks of infectious agents and associated biologicals
- Human and animal waste, including blood and blood products and limited blood fluids
- Pathological waste (depending on local restrictions)
- Sharps
- Other medical waste, including toxic chemical and chemotherapy waste

Incompatible Waste Types

- Radiological waste (due to regulatory restrictions)

Installation. Installation and testing is performed by PEAT engineers in conjunction with the customer's facility personnel.

Operation. TDR Systems can be operated by technician level personnel employed by either the customer or PEAT.

System Monitoring. In many applications, PEAT will monitor system operation continuously by remote control.

Training. Training of operators is conducted by PEAT and included in the system pricing. Training is similar to that for an incinerator.

Service. PEAT offers a yearly service agreement.

Treated Waste

Characteristics of Treated Material. The process produces gas, a vitrified material that resembles a glassy rock or slag, and ferrous metal. The organics pyrolyze into hydrogen and carbon monoxide and make excellent fuel. The inorganics (glass and metal) are bound on the molecular level, so are non-leachable. The ferrous metal alloy is recyclable.

Volume Reduction There is 150:1 volume reduction.

Disposal of Treated Material. There is no significant secondary waste. The small amount of glassy slag and ferrous alloy that remains following treatment may be disposed of as municipal solid waste or recycled entirely.

Recycling Options. In California, the vitrified material may be used as construction aggregate.

Environmental Discharges

By-Products and Discharges. By-products are metal, glassy aggregate, and hydrogen gas. Scrubber water is discharged to the sewer.

Efficacy Testing. Glassy aggregate has consistently passed TCLP tests.

ELECTRON BEAM STERILIZATION

The process of electron beam sterilization involves exposing medical waste to high-energy electrons generated by an electron accelerator. As the electrons enter the material being processed, they create highly reactive molecules that attack the nucleic acids within the living cell and rupture the cell walls. In the process, small amounts of ozone and hydrogen peroxide are also created, which add further to the inactivating effect (52). At least one company, Nutek Corporation of Palo Alto, California, has proposed an electron beam sterilization system for biohazardous waste treatment application.

System Review. The Nutek Corporation electron beam technology is reviewed in this section.

Nutek Corporation (12,52)

Source

Nutek Corporation
1237 North San Antonio Road
Palo Alto, CA 94303
Tel: (800) 696-8835
Tel: (415) 969-8903
Fax: (415) 969-9087

Product Overview. Nutek Corporation has developed a process for treating medical waste using electron beam technology. Treatment may be conducted to achieve either disinfection or sterilization, depending on the outcome desired. The technology itself has been employed for the past two decades as a medical sterilizer for such items as implant devices, pharmaceuticals, medical equipment, and other medical supplies, and has been licensed by the U.S. Food and Drug Administration as a sterilization facility for medical devices. The Nutek Corporation has reapplied this technology to the treatment of medical waste. A system diagram is presented in Figure 10.15.

The process involves exposing medical waste to high-energy electrons generated by an electron accelerator. As the electrons enter the material being processed, they create highly reactive molecules that attack the nucleic acids within the living cell and rupture the cell walls. In the process, small amounts of ozone and hydrogen peroxide are also created, which add to the inactivating effect.

The electron beam generator is an electron accelerator that closely resembles accelerators used in cancer therapy at hospitals. Waste that has been collected in boxes or drums is first weighed and then loaded onto a conveyor that transports the waste into a shielded processing area for treatment by electron beam scanning. The speed of the conveyor is controlled automatically so that the waste is treated for the designated amount of time. Treatment usually requires between 1 and 3 minutes. During the process, the intensity of the beam is also controlled to ensure effective sterilization of the waste, which requires a dosage of approximately 1 megarad for effective processing. The sterilization process is continuously monitored by on-line electronic equipment (dosimeters) which monitors the dose rate of the beam, the distribution of electrons within the material, and the amount of electrons absorbed. The effectiveness of microbiological kill can also be monitored with biological spore strip indicators. Following sterilization, the waste may be loaded into an optionally purchased shredder/compactor, where it is rendered unrecognizable as medical waste. The final waste product may be handled and disposed of as municipal solid waste.

Models Available. Systems are custom designed by the manufacturer. Note: As of August 1995, the Nutek Corporation has ceased marketing its system for biohazardous waste treatment (53).

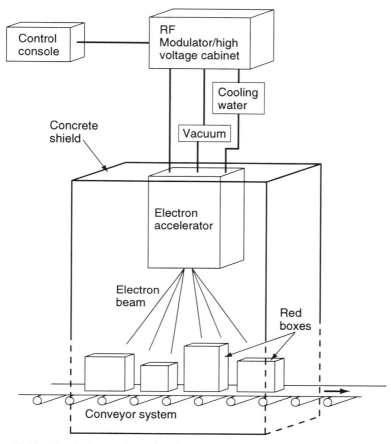

Figure 10.15 Nutek electron beam irradiator. Courtesy of Nutek Corporation. Nutek Electron Beam Irradiator, copyright 1995 by the Nutek Corporation, Palo Alto, California. Reprinted by permission.

Specifications

Use:	Designed for both generator in-house treatment and larger quantity, off-site treatment, as well as mobile units
Size (overall):	The electron beam equipment package is 20 ft long by 40 ft wide
Load capacity:	Approximately 800 lb/hr
Decontamination time:	The on-line sterilization time for a package is from 1 to 3 minutes

Capital Cost

• $250,000 to $500,000.

A shredder may be included at an additional cost of $150,000 to $200,000.

Waste Stream Compatibility, Incompatibility

Compatible Waste Types. No biohazardous wastes are identified as being unsuitable for the process.

Incompatible Waste Types

• Toxic chemical (e.g., antineoplastic drug waste, RCRA waste) or radiological waste (e.g., low-level radioactive waste)

System Monitoring. The sterilization process is monitored continuously by on-line electronic equipment (dosimeters) that monitors the dose rate of the beam, the distribution of electrons within the material, and the number of electrons absorbed. The effectiveness of microbiological kill can also be monitored with biological spore strip indicators.

Treated Waste

Characteristics of Treated Material. The treated waste is not altered following treatment. If the waste needs to be rendered unrecognizable, it may be shredded in an optionally purchased shredder to $\frac{1}{4}$ in.

Volume Reduction. There is no volume reduction unless the waste is shredded. The volume of waste is reduced by 70 percent through the process of shredding and compaction.

Disposal of Treated Material. Depending on local restrictions, the treated waste may be disposed of as municipal solid waste.

COBALT-60 IRRADIATION

Gamma irradiation systems have been in commercial use since the early 1960s for sterilization of various medical products, drugs, pharmaceuticals, and tissue for transplantation (54). Sterilization is accomplished by exposing such products to a radiation source (cobalt-60). Microorganisms are inactivated by the radiation which hydrolyze water molecules within the organisms, creating intermediate hydrolysis products that result in complete inactivation. The radiation dose can be calculated with great reliability, which makes the process highly predictable. The antimicrobial efficacy of gamma irradiation is well documented. No residual radiation is found in the treated product or waste.

This technology was applied to the treatment of biohazardous waste in the United States by one company during the early 1990s, although that company later abandoned gamma irradiation for a different treatment technology. Although no systems are currently in operation, one company, Nordion International of Kanata, Ontario, Canada, has proposed a system design that may be used for treating biohazardous waste in the future.

System Review. The Nordion International, Inc. cobalt-60 irradiation medical waste treatment system is reviewed in this section.

Nordion International, Inc. (54–57)

Source

Nordion International, Inc.
447 March Road
P.O. Box 13500
Kanata, Ontario
Canada K2K 1X8
Tel: (613) 592-2790
Fax: (613) 592-0440

Product Overview. Nordion International, Inc. is developing a biomedical waste treatment system based on sterilization by gamma irradiation using a controlled cobalt-60 radiation source. To date, no systems are in operation. The system is most economic for large-scale use by several hospitals (e.g., greater than 5 tons per day) and will require about 1.5 to 2 acres of land for a free-standing facility, although less land would be necessary if the facility were to be incorporated into an existing warehouse. A diagram of a typical system is presented in Figure 10.16.

Cobalt-60 radiation technology itself is not new; it has been used since the 1960s to sterilize medical supplies such as syringes, sutures, and surgeons' gloves as well as other pharmaceutical and consumer goods. Currently, there are approximately 160 industrial gamma radiation systems operating commercially in over 40 countries.

The process for treating biomedical waste consists of four major components: (1) an irradiation chamber, including the source storage pool; (2) a product handling mechanism, including control and safety subsystems; (3) a cobalt-60 gamma radiation source, including the source movement mechanism; and (4) a treated-product handling area, typically consisting of a waste shredder/compactor/baler unit. The purpose of the irradiation chamber is to provide safe shielding for the irradiation source and a chamber in which to irradiate the biomedical waste. The irradiation chamber is a concrete room that typically measures about 24 feet long by 24 feet wide by 10 to 20 feet in height, with a wall thickness of about 6 feet. Inside the

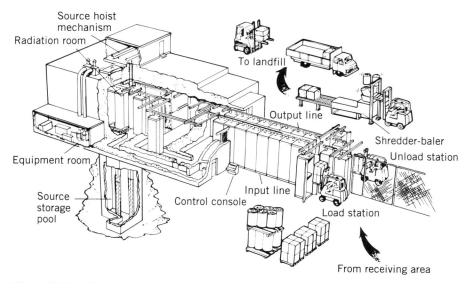

Figure 10.16   Nordion industrial irradiator treatment system. Courtesy of Nordion International, Inc. Nordion Industrial Irradiator Treatment System, copyright 1995 by Nordion International, Inc., Kanata, Ontario, Canada. Reprinted by permission.

irradiation chamber is a water-filled pool, typically measuring about 20 to 30 feet deep where the cobalt-60 is safely stored when not in use. Energy given off by the cobalt-60 during storage is converted to heat and the pool is cooled by a heat exchanger/chiller.

The cobalt-60 gamma radiation source is typically made up of a vertical stainless steel frame measuring about 7 to 10 feet wide by 7 to 16 feet high by 1 inch thick. It holds a number of stainless steel sealed sources that measure approximately 18 inches long by 0.4 inch in diameter, each containing a measured amount of radioactive cobalt-60. During waste processing, the cobalt-60 is raised from the pool into the chamber, and the waste is conveyed by the product handling mechanism around the source in a predetermined pattern. Following sterilization, the waste is shredded to render it unrecognizable, and then compacted and baled for landfill disposal.

The manufacturer describes cobalt-60 as a deliberately produced radioisotope of naturally occurring cobalt-59 that is produced to meet industry and medical requirements. Cobalt-60 has a half-life of five years and decays to stable nickel-60. It is not a by-product or waste from nuclear power or weapons programs.

Models Available.   Systems are custom designed by the manufacturer.

Specifications

Use: Designed for large-scale use by several hospitals

Size (overall):	Requires about 1.5 to 2 acres of land for a freestanding facility, although less land would be necessary if the facility was to be incorporated into an existing warehouse
Load capacity:	Becomes economically feasible when processing greater than 5 tons of waste per day

Capital Cost. The manufacturer describes its system as being capital intensive on a scale with larger incineration systems (approximately $2,000,000 to $5,000;000) and not economically suitable for small-volume scenarios such as municipalities of fewer than 50,000 people, single hospitals, and smaller international airports and seaports. The vendor generalizes a "best economic scenario" as follows:

- Incineration is not a feasible alternative.
- A dense population base of more than 1 million people must be serviced.
- The lowest cost will win the business.
- A group of cooperating hospitals exists.

Operating Cost. According to the manufacturer, the minimum economic volume is approximately 5 tons per day. In this range, irradiation costs are described as being competitive with other existing waste treatment methods.

Waste Stream Compatibility, Incompatibility

Compatible Waste Types.

- Cultures and stocks of infectious agents and associated biologicals.
- Liquid human and animal waste, including blood and blood products and blood fluids. Large volumes of blood will decrease throughput somewhat.
- Sharps.
- Other wastes such as plastics, paper, glass.

Incompatible Waste Types

- Pathological waste (based on regulatory restrictions only)
- Toxic chemical (e.g., antineoplastic drug waste, RCRA waste) or radiological waste (e.g., low-level radioactive waste)

Installation. Installation teams, led by Nordion engineers, provide project management, installation, and commissioning services.

Training. Nordion engineers conduct on-site training for the customer's operating personnel.

Service. Services by Nordion cover the complete life cycle of its irradiation systems and cobalt-60 sources. Once an irradiator is in operation, Nordion technicians are available 24 hours a day to provide ongoing technical support.

Treated Waste

Characteristics of Treated Material. The irradiated sterile waste looks identical to the waste prior to treatment. A secondary destructive process (e.g., shredder, grinder, compactor) can be added.

Volume Reduction. There is no reduction unless the waste is subjected to a secondary destructive process.

Disposal of Treated Material. The processed waste may be disposed of as municipal solid waste.

Environmental Discharges

By-Products and Discharges. The process is environmentally clean, no chemicals are used or generated, no stack gases are emitted, and no toxic changes are made in the waste.

By-Product Controls. Because the process generates no toxic products, no special pollution control devices are required.

Efficacy Testing. Although Nordion's cobalt-60 irradiation systems are designed and widely used to sterilize medical products, no efficacy testing information was provided by the manufacturer specific to biomedical waste sterilization.

Approvals and Installations. According to the vendor, irradiation is an approved method of biohazardous waste treatment prior to landfill disposal in Arkansas, Indiana, and Pennsylvania. In Canada, although not specifically approved as a sterilization method for biomedical waste, the technology is approved and widely accepted for sterilizing medical supplies. To date, no biomedical waste treatment systems have been constructed or are in operation.

Corporate Profile. Nordion International is a leading international producer, marketer, and supplier of radioisotope products and technology for use in healthcare and industrial applications. Nordion designs, supplies, and installs cobalt-60 irradiation systems and related equipment for the sterilization of disposable medical and consumer products, and supplies radioisotopes used in nuclear medicine procedures to diagnose and treat disease. A subsidiary of MDS Health Group Limited of Etobicoke, Ontario, Nordion employs over 700 people, most located at the company's main operations in Kanata, Ontario. Annual sales at Nordion are approximately $200 million.

REFERENCES

1. U.S. Environmental Protection Agency. EPA Guide for Infectious Waste Management. EPA/530-SW-86-014. May 1986.

2. Kostenbauder HB. Physical factors influencing the activity of antimicrobial agents. In: Block S., Ed., *Disinfection, Sterilization, and Preservation*, 3rd ed., pp. 811–828. Lea and Febiger, Philadelphia. 1983.

3. Personal communication with Dr. Zig Vaituzis, Antimicrobial Program Branch, Registration Division, U.S. Environmental Protection Agency. June 1995.

4. Jette LP and Lapierre S. Evaluation of a mechanical/chemical infectious waste disposal system. *Infect Control Hosp Epidemiol*, 13(7):387–393. 1992.

5. Medical Materials & Technology, Inc. product information. Houston, Texas. 1994.

6. Personal communication with Charles Miller, President, Medical Materials & Technology, Inc., Houston, Texas. September 1995.

7. State of New Jersey. Solid Waste Management Plan Update, 1993–2002. Section II: Comprehensive Regulated Medical Waste Management Plan—Draft. Department of Environmental Protection and Energy, Division of Solid Waste Management, in cooperation with the Department of Health, Trenton, New Jersey. April 1993.

8. DiDomenico A. Inactivation of pathogenic microorganisms in infectious medical waste: A literature review of current on-site treatment technologies. Non-thesis paper, Master of Science in Civil Engineering. University of Washington, Seattle, Washington. 1992.

9. Denys, GA. Microbiological evaluation of the Medical Safe TEC mechanical/chemical infectious waste disposal system. Presented at the 89th Annual Meeting of the American Society for Microbiology, New Orleans, Louisiana. 1989.

10. Denys GA and Street BA. Inactivation of coliphage in mechanical/chemical infectious waste disposal system. Presented at the 91st Annual Meeting of the American Society for Microbiology, Dallas, Texas. 1991.

11. French MLV and Eitzen HE. Report on the microbiological effectiveness of the Medical Safe-TEC waste management system. Department of Infection Control, Indiana University, Bloomington, Indiana. 1984.

12. Minnesota Healthcare Partners. Study of Non-burn Technologies for the Treatment of Infectious and Pathological Waste and Siting Considerations. St. Paul, Minnesota. 1992.

13. Premier Medical Technology, Inc. product information. Houston, Texas. 1995.

14. Steris Corporation product information. Mentor, Ohio. 1994.

15. Personal communication with Eric Schuldt, Director, Steris Corporation, Mentor, Ohio. September 1995.

16. Winfield Environmental Corporation product information. San Diego, California. 1994.

17. Personal communication with Harold Callicoat, Executive Vice President, Winfield Environmental Corporation. San Diego, California. September 1995.

18. Ecolotec, Inc. product information. Cherry Hill, New Jersey. 1994.

19. Personal communication with W. A. von Lersner, President, Ecolotec, Inc., Cherry Hill, New Jersey. September 1995.

20. San-I-Pak Pacific, Inc. product information. Tracy, California. 1994.

21. Personal communication with Shari Hood, Project Coordinator, San-I-Pak Pacific, Inc., Tracy, California. September 1995.

22. SAS Systems, Inc. product information. Houston, Texas. 1994.

23. Personal communication with Otmar Kolber, President, SAS Systems, Inc., Houston, Texas. September 1995.

24. Tempico, Inc. product information. Madisonville, Louisiana. 1994.

25. Personal communication with Murray F. Cleveland, Jr., Senior Vice President, Tempico, Inc. September 1995.

26. Galloway TR and Green TJ. On-Site Bio-Hazardous Waste Destruction with the Synthetica Steam Detoxifier. Synthetica Technologies, Richmond, California. June 1993.

27. Galloway TR and Depetris S. On-Site Bio-Hazardous Waste Destruction with the Synthetica Steam Detoxifier—Test Organism Kill and Chlorocarbon Destruction. Synthetica Technologies, Richmond, California. 1994

28. Synthetica Technologies, Inc. product information. Richmond, California. 1994.

29. Personal communication with Christopher Miles, Synthetica Technologies, Inc., Richmond, California. September 1995.

30. Goldblith SA and Wang DI. Effect of microwaves on *Escherichia coli* and *Bacillus subtilis*. *Appl Microbiol*, 16(6):1371–1375. 1967.

31. Vela GR and Wu JF. Mechanism of lethal action of 2,450 MHz radiation on microorganisms. *Appl Environ Microbiol*, 37(3):550–553. 1979.

32. Jeng DKH, Caczmarek KA, Woodworth AG, and Balasky G. Mechanism of microwave sterilization in the dry state. *Appl Environ Microbiol*, 53(9):2133–2137. 1987.

33. Najdovski L, Dragas AZ, and Kotnik V. The killing activity of microwaves on some non-sporogenic and sporogenic medically important bacterial strains. *J Hosp Infect*, 19:239–247. 1991.

34. ABB Sanitec, Inc. product information. Wayne, New Jersey. 1994.

35. Personal communication with Mark Taitz, Sanitec, Inc., West Caldwell, New Jersey. September 1995.

36. Roatan Medical Services Corporation product information. Grand Prairie, Texas. 1995.

37. Personal communication with Suzanne Helton Beck, Western Region Manager, Roatan Medical Services Corporation, San Bruno, California. September 1995.

38. Joslyn LJ. Sterilization by heat. In: Block, SS, Ed., *Disinfection, Sterilization, and Preservation*, 4th ed., pp. 513–515. Lea & Febiger, Philadelphia. 1991.

39. Disposal Sciences, Inc. product information. Cannon Falls, Minnesota. 1994.

40. Personal communication with Michael S. Simmons, Director of Regulatory Affairs and Quality Assurance, Disposal Sciences, Inc., Cannon Falls, Minnesota. September 1995.

41. D.O.C.C., Inc. product information. New York. 1994.

42. Personal communication with Jonathan Bricken, D.O.C.C., Inc., New York. September 1995.

43. MedAway International Inc. product information. Menlo Park, California. 1995.

44. Personal communication with Joe Militello, MedAway International Inc., Yorba Linda, California. September 1995.

45. Spintech, Inc. product information. York, Pennsylvania. 1995.

46. Personal communication with Thomas R. Bowman, Spintech, Inc., York, Pennsylvania. September 1995.

47. Stericycle, Inc. product information. Deerfield, Illinois. 1993.

48. Personal communication with Linda Lee, Vice President, Regulatory Affairs, Stericycle, Inc. Deerfield, Illinois. September 1995.

49. Personal communication with R. D. Dupree, Jr., Director of Marketing and Sales, Plasma Energy Applied Technology, Huntsville, Alabama. September 1995.

50. Springer MD. Destruction of medical waste using plasma energy. Proceedings of the Thermal Treatment of Radioactive, Hazardous Chemical, Mixed and Medical Wastes 1992 Incineration Conference, Albuquerque, New Mexico. May 11–15, 1992.

51. Malloy MG. High-tech waste treatments set to take off. *Waste Age*, pp. 79–84. March 1992.

52. Ormsby AA. Electron beam ionizing radiation used to treat medical waste. *Med Waste Anal*, pp. 13–14. January 1994.

53. Personal communication with Edward Joehnk, President, Nutek Corporation, Palo Alto, California. August 1995.

54. Swinwood JF and Wilson BK. Treating biomedical waste. Dimensions Health Serv, pp. 14–16. October 1991.

55. Wilson BK. The economics of biomedical waste irradiation: Key issues influencing total cost. Presented at the 8th International Meeting on Radiation Processing, Beijing (China). September 13–18, 1992.

56. Nordion International, Inc. corporate profile. September 1994.

57. Personal communication with Jean F. Swinwood, Project Leader, Waste Treatment Applications, Nordion International, Inc., Kanata, Ontario, Canada. September 1995.

11

ALTERNATIVE TECHNOLOGY EVALUATION GUIDELINES

In this chapter we address the work of the State and Territorial Association on Alternate Treatment Technologies (STAATT) to develop a guidance manual for evaluating and approving the many new and creative alternative medical waste treatment technologies that are being developed. The chapter begins with a discussion entitled "Creating the Model Guidelines for States" authored by Diann J. Miele and Roger Greene, which is reprinted from an article published in the *Medical Waste Analyst* (1). The guidance manual stands as a valuable resource to government officials responsible for conducting new technology approval reviews, and to the developers and prospective purchasers of these new systems. The document, which is entitled "Technical Assistance Manual: State Regulatory Oversight of Medical Waste Treatment Technologies," was finalized by the STAATT in April 1994 and resides in the public domain. It has been reprinted in this chapter in its entirety as a resource to those working in this field.

CREATING THE MODEL GUIDELINES

In April 1994, the State and Territorial Association on Alternate Treatment Technologies published its report "Technical Assistance Manual: State Regulatory Oversight of Medical Waste Treatment Technologies." To the extent that the wide array of different state laws allow, this document can help standardize the process used by state agencies when reviewing alternative medical waste treatment and/or destruction technologies. (For the purpose of this article, *alternate* refers to technologies other than incineration or steam sterilization.) Such standardization may not only facilitate state review but may also provide manufacturers with the information they need to develop state-approvable technologies. In the long term, this guidance document should result in the approval of safe and effective technologies for the treatment and destruction of medical waste.[1]

While the publication of this document is a significant achievement, the process through which it developed is also of interest. During the course of $1\frac{1}{2}$ years, 27 agencies from 21 states and U.S. territories worked closely with five federal agencies, two Canadian ministries, and a consultant to prepare this document. The success of this project resulted from the conviction of these participants that any differences could be resolved in order to accomplish such an important goal.

Background

As with so many recent initiatives related to medical waste, the steps leading to the development of the "Technical Assistance Manual: State Regulatory Oversight of Medical Waste Treatment Technologies" can be traced to the enactment of the Federal Medical Waste Tracking Act (MWTA) in the fall of 1988 and the two-year Medical Waste Demonstration Program, administered by the U.S. Environmental Protection Agency (EPA), that followed. Specifically, two events contributed to the emergence of this document.

One event was the washups of medical waste on the beaches of five east coast states during the summers prior to the enactment of the MWTA. Consequently, passage of the act spurred several states into developing their own regulations. As stated in "Model Guidelines for State Medical Waste Management" (Council of State Governments, 1992), "since that time [the passage of the MWTA], almost every state has enacted regulations addressing some aspect of the medical waste management process. There are many areas, however, that have not been adequately addressed by most states, such as public education and facility operator training." In addition, review of alternate technologies for the treatment and destruction of medical waste has not been addressed by regulation in most states.

A second driving event contributing to the emergence of the guidance document was the passage of the Clean Air Act Amendments of 1990. Frank L. Cross, in the March 1994 issue of *Medical Waste Analyst*, discusses the implications of the MWTA with regard to medical waste incinerators. He concludes in his article that "incineration is still a viable option for medical waste management. Air pollution control may make this option costly, but it remains the only practical method to treat *all* hospital waste." The editors of *Medical Waste Analyst* state in the same issue: "Incineration remains a viable, albeit relatively expensive and controversial, treatment alternative for medical waste. It also continues to face an uncertain regulatory future, particularly with regard to ash management." These more stringent regulations affecting the operation of incinerators have stimulated entrepreneurs to develop alternate technologies to medical waste treatment and/or destruction. In addition, healthcare facilities, especially large hospitals, have begun to look for ways other than incineration to treat and destroy their medical waste either on-site or off-site.

Many new methods have been developed for the treatment and/or destruction of medical waste. Treatment technologies generally use either a heat or chemical process to achieve disinfection/sterilization. Treatment by heat includes dry heat, moist heat, and wet heat, provided by sources such as electricity, radio-frequency waves, and steam. Treatment by chemical processes includes the use of chemicals originally developed for other disinfection or sanitation processes. Destruction of medical waste occurs by a variety of mechanical methods used to shred, grind, or pulverize the waste so that it is virtually unrecognizable as medical waste after the destruction is completed.

A 1989 report entitled "Perspectives on Medical Waste," prepared by the Nelson A. Rockefeller Institute of Government at the State University of New York, showed foresight with regard to the development of the technologies. The report states: "We do not currently have a panacea method for treatment of medical waste (infectious or not). All current technologies have significant limitations and disadvantages. . . . Continued improvements in technology (particularly for incineration) and innovative new approaches must be sought, investigated and brought to fruition." It would be hard to imagine a more tempting invitation to entrepreneurs.

Treatment and *destruction* were defined by the EPA during the MWTA Demonstration Program. Treated medical waste referred to waste that had been altered to reduce substantially or eliminate its potential for causing disease. Destroyed medical waste referred to waste that had been ruined, torn apart, or mutilated so that it was no longer generally recognizable as medical waste. If medical waste was both treated and destroyed, the residue was considered solid waste and could be disposed of as such. For instance, ash from incineration was no longer considered medical waste and could be disposed of as solid waste according to the MWTA.

Residues from treatment and destruction processes that were no longer recognizable as medical waste were exempt from the tracking requirements of the MWTA. These tracking requirements, which followed the example of the hazardous waste cradle-to-grave manifesting system, were both a liability and a cost to generators. Therefore, any system that could treat and destroy medical waste on-site, or even off-site, would be extremely beneficial to generators.

Evolution of the STAATT

Connecticut, New Jersey, New York, Rhode Island, and the Commonwealth of Puerto Rico participated in the MWTA's two-year demonstration program, which expired in June 1991. Under the authority of the MWTA, the EPA was allowed to give grants to other states to undertake specific projects relating to medical waste issues. Michigan, California, and Washington became grantee states in the program. Each state worked on a specific project that dealt with some issue of medical waste management. For instance, Washington undertook a project related to the disposal of home-generated sharps (needles and syringes). This project involved testing the durability of containers that could be used for the disposal of sharps at home, as well as investigating other home-generated syringe disposal options, such as a pharmacy-return program.

Throughout the two-year demonstration program, the EPA sponsored periodic meetings of the above-mentioned states and other states, such as Ohio, Louisiana, and South Carolina, and various federal agencies, to assess the progress of the program and to identify and solve any problems inherent in the program. These meetings were found to be an effective way to work with other states in addressing many medical waste issues. These issues included the management of home-generated waste, chemotherapy waste, and cultures and stocks. However, what was often discussed in these EPA meetings but never sufficiently brought to conclusion was the problem of how to assess the many new and innovative technologies that treated and/or destroyed medical waste. While every state representative at these meetings had been petitioned by numerous companies and entrepreneurs seeking approval to market their device in that state, very few representatives felt comfortable with the way their state dealt with this issue. New York had developed a very thorough and systematic approach to the approval of these technologies; however, New York's method would not always work in other states due to personnel constraints or lack of appropriate legislative authority to implement such a program. In particular, states needed help in three areas: how to assess the technology (the "process), what should be required of a technology (the "requirements"), and what is acceptable (the "approval").

During the MWTA Demonstration Program, the California Department of Health Services surveyed 26 states to determine the extent to which each state had authority to evaluate and approve medical waste treatment technologies, as well as the states' procedures for undertaking these evaluations. Of the states surveyed, 17 responded that they had the authority to evaluate such technologies, but only four states had established procedures for carrying out these evaluations. Following this survey, California prepared a "Model Protocol for the Evaluation and Approval of New Medical Waste Treatment Technologies," which provided the framework for the application section of the "Technical Assistance Manual: State Regulatory Oversight of Medical Waste Treatment Technologies."

In May 1986, the EPA published the "EPA Guide for Medical Waste Management." An entire chapter in this guide was devoted to treatment of infectious waste. It discussed steam sterilization, incineration, thermal inactivation, gas/vapor sterilization, chemical disinfection, and sterilization by irradiation. Since then, the EPA has evaluated pyrolysis, ethylene oxide, dry heat, ultraviolet irradiation, gamma radiation, radio-frequency radiation, microwave irradiation, chemical microbial inactivation, and steam autoclaving. This evaluation is to be part of the agency's final report to Congress, which was required by Congress at the conclusion of the MWTA. It does not appear, however, that the EPA is going to specify how states should evaluate treatment and/or destruction technologies, and it is unclear when the final report to Congress will be completed. It does seem, however, from discussions with the EPA staff, that there will probably be no need for the EPA or any other federal agency to continue to regulate the management of medical waste to the extent that was required by MWTA regulations. The EPA's research has shown that most states had already promulgated regulations on the management of medical waste by the time the two-year demonstration program had ended.

When the MWTA ended in June 1991, one last meeting was held with this group of "core" states. At the end of that meeting, it was decided that the matter of these medical waste treatment and/or destruction technologies was important enough for the states to try to meet on their own to continue the dialogue that had already been established. As a result of these frequent meetings, a close association had developed with this group of people, an association that was needed to attempt to produce this guidance document.

Technical Assistance Manual Development

The discussions among this group of states continued throughout the next year; the need to develop a manual of this sort still existed. In the fall of 1992, Rhode Island hired a college intern to survey all states to determine the level of interest in attending a workshop in the spring of 1993 on developing a national protocol for approval of alternative technologies for the treatment and destruction of medical waste. The results of the survey were very positive; almost every state indicated that a guidance document would help them tremendously as the state attempted to review these alternative technologies. A similar survey, conducted by California, also revealed that interest in developing such a document was widespread. All except one of the states responding to California's survey were interested in the development of a uniform evaluation procedure.

With some administrative support from the EPA, an initial meeting was scheduled involving nine states, Puerto Rico, two Canadian ministries, the EPA, and a facilitator. The facilitator helped participants reach consensus on key issues in the evaluation of medical waste treatment technologies and would later produce a summary of the meeting.

The purpose of the first meeting was to:

- Establish a definition of adequate microbial destruction.
- Identify and discuss existing state requirements to document the efficacy of the medical waste treatment process.
- Develop a framework for state review of medical waste treatment technologies.
- Adopt a strategy for developing the framework into a draft document.

One major issue that was resolved at this meeting was the level of decontamination required for medical waste to be considered "treated." However, many more issues remained before the document could be completed.

Three actions were taken at this meeting to help ensure that the project would achieve its intended goal: A steering committee was formed, two participants (the authors) were asked to be coordinators for the group, and Nelson Slavik of Environmental Health Management Systems was hired to prepare the necessary drafts as well as the final document.

The second meeting was held two months later to continue the development process. At this meeting the attendees grappled with the issue of establishing a list of surrogate pathogens to be used in the testing process. This involved a lengthy discussion regarding the suitability of a large number of possible surrogates. The group also worked on a draft of a universal application form, originally prepared by California, which could be adopted by any state undertaking an evaluation of medical waste treatment technologies. Additional topics of discussion included periodic user verification (as opposed to the initial efficacy testing for technology approval), processes for maintaining biological indicator integrity during testing, challenge loads for testing, operator training, research/development projects, small-quantity treatment destruction devices, and federal jurisdiction for approval of chemical treatments through the EPA's Office of Pesticide Programs.

Soon after the meeting, a report entitled "Recommendations for State Regulatory Oversight of Medical Waste Treatment Technologies" was prepared and distributed to meeting participants. It summarized the decisions made during the previous six months. To obtain still more input, a final meeting was planned and state officials from all states were invited to attend. All officials who planned to attend were sent a copy of the draft document, which would be discussed at the final meeting.

In June 1993, the final meeting was held in Washington, D.C. Government officials from 22 states, Puerto Rico, and five federal agencies were represented at this meeting, which involved a complete review of the "recommendations" document for content, organization, and form.

Although consensus was reached on all aspects of the guidance document, there was not always total agreement on every issue. For instance, the suggested level of "log kill" for the surrogate pathogens and spores received much debate. States that had existing regulatory or statutory authority were concerned that their state's regulations regarding the level of log kill might be lower than those adopted in the guidance document. Another area that received debate was an exemption for treatment and/or destruction technologies that were marketed for small-quantity generators (SQGs). In the end, no exemptions from the approval process were granted for SQG devices.

After this three-day meeting adjourned, Dr. Slavik prepared the final guidance document. This final draft was edited extensively by the meeting's attendees before it was published in April 1994. It is hoped that the "Technical Assistance Manual: State Regulatory Oversight of Medical Waste Treatment Technologies" will be useful to all states, territories, and counties charged with reviewing and approving alternative medical waste treatment and/or destruction technologies.

Future activities of the steering committee may involve the creation of a clearinghouse to facilitate the sharing of information among all states regarding technology review. This would benefit not only the states but also the manufacturers, as the technology review time may be shortened.

The publication of this guidance document is an important step in establishing a network of state, local, and federal agencies all working toward the same goal: approving for use in their jurisdiction medical waste treatment and/or destruction technologies that are effective, reliable, environmentally friendly, and safe for workers and the public. Following is the text of the STAATT guidance document, "Technical Assistance Manual: State Regulatory Oversight of Medical Waste Treatment Technologies," which is presented in its entirety.

TECHNICAL ASSISTANCE MANUAL: STATE REGULATORY OVERSIGHT OF MEDICAL WASTE TREATMENT TECHNOLOGIES

A Report of the State and Territorial Association on Alternate Treatment Technologies   [April 1994]

Roger Greene, Rhode Island Department of Environmental Management, Diann J. Miele, M.S., Rhode Island Department of Health, and Nelson S. Slavik, Ph.D., President, Environmental Health Management Systems, Inc., were primarily responsible for facilitating consensus among participants during each of the three meetings that were held to discuss state review of medical waste treatment technologies.

Nelson S. Slavik, Ph.D., prepared this final document which reflects the discussions and consensus reached at these meetings.

The following state officials served as a steering committee for these meetings:

Charles H. Anderson
Louisiana Department of Health and Hospitals

Lawrence Chadzinski, M.P.H.
Michigan Department of Public Health

Robert M. Confer
New Jersey Department of Environmental Protection & Energy

Carolyn Dinger
Louisiana Department of Environmental Quality

Roger Greene
Rhode Island Department of Environmental Management

Diann J. Miele, M.S.
Rhode Island Department of Health

Phillip R. Morris
South Carolina Department of Health and Environmental Control

Ira F. Salkin, Ph.D.
New York Department of Health

Wayne L. Turnberg, M.S.P.H.
Washington Department of Ecology

John Winn, R.E.H.S.
California Department of Health Services

A complete listing of all participants attending the New Orleans, Atlanta, and Washington, D.C., meetings may be found in Attachment D.

1.0 Introduction

The development of new or modified medical waste treatment methods utilizing heat chemicals, or irradiation has provided potential alternative solutions to the medical waste treatment/disposal problem. However, with the development of these medical waste treatment methods, the concern has arisen that these new technologies may also lead to potential environmental or occupational health and safety exposures. Only a limited number of states have attempted to quantitatively and qualitatively assess the efficacy and safety of these new treatment technologies. For those states that have adopted criteria, there is no universality of approach in the assessment of treatment technology efficacy and safety.

Establishing a uniform guideline or a standard set of efficacy criteria can result in potential benefits to the state approval process. A uniform approach may provide economic benefits through facilitating the state review process via similarity in approval requirements and the avoidance of state-by-state review duplication. Minimizing state liability in the review process is also a potential benefit of standardized, documented efficacy criteria and testing protocol. As another potential benefit, developing nationally recognized protocols and assessment criteria might also enhance facilitation and cooperation between federal and other state agencies integral to or peripherally involved in the review process.

In an attempt to standardize processes for medical waste technology review, several states that had actively participated in the programs authorized under the federal Medical Waste Tracking Act of 1988 organized and conducted a meeting in New Orleans, Louisiana on December 13 and 14, 1992. With the purpose of establishing a framework or guideline for a state approval process for medical waste treatment technologies, particularly those other than steam sterilization or incineration, this meeting initiated discussions on defining medical waste treatment technology efficacy criteria and delineating the components required to establish an effective state approval process. Although much was accomplished at this meeting, many issues remained unresolved.

With the objective of attaining committee consensus on the technical and administrative elements of treatment technology approval, a second meeting was held on February 25 and 26, 1993, in Atlanta, Georgia to continue the discussions initiated at the December 1992 meeting. At this meeting the committee recognized the need for establishing its identity to coordinate and support these activities. As such, the name "State and Territorial Association on Alternate Treatment Technologies" was adopted for the purpose of defining the Committee and its objectives. The term "alternate" was defined as "other than steam sterilization or incineration."

The Atlanta meeting's agenda was based on attaining the committee's consensus on the technical and administrative elements of treatment technology approval. Specific topics addressed and discussed were as follows:

- Definition of the level of recommended microbial inactivation (i.e., Level II or Level III spore inactivation levels)
- Establishment of defined pathogen surrogates for microbial inactivation evaluation including:
  - Vegetative pathogen surrogates spore formers
  - Bacterial spore formers
- Determination of the use of bacterial spore formers, as ultimate pathogen surrogates, including the determination of which spore formers should be used for which treatment process, and at what level of required inactivation
- Adoption of enumeration formulae for efficacy testing protocol quantification
- Development of a comprehensive process approval application form
- Development of specific process approval mechanisms for:
  - Commercial facilities
  - Healthcare facilities
  - Research and development projects
  - Small quantity treatment devices
  - Previously approved technologies
- Development of criteria specifications and requirements for:
  - Waste residue disposal
  - Operator training
  - Challenge loads
- Development of specific testing protocols for:
  - State permitting/licensing of the technology
  - Site permitting
  - User verification
  - Processes maintaining/not maintaining biological test indicator integrity
- The timing and extent of USEPA FIFRA involvement in establishing efficacy criteria and protocols

At the conclusion of the Atlanta meeting a report was prepared entitled "Recommendations for State Regulatory Oversight of Medical Waste Treatment Technologies" which summarized the issues and recommendations discussed during both the New Orleans and Atlanta meetings. This report was distributed for review and comment to all state and territorial regulatory agencies involved in medical waste regulatory activities.

To gain additional input into the development of a uniform guideline for the assessment of medical waste treatment technologies, a third meeting was conducted on June 14–16, 1993, in Washington, D.C. with invited participants from all state and territorial medical waste regulatory agencies. The report prepared from the Atlanta meeting served as a basis of discussion. With invited input from all state and territorial representatives, the primary objective of the meeting was to seek consensus on the key topic areas listed above.

This report details the discussions and recommendations of the participants from the three meetings. It should be emphasized that the recommendations made in this report are an attempt to find commonality on many of the issues and criteria required in the medical waste treatment technology review process. As such, consensus agreement was sought on key issues to demonstrate support for the recommendations made in this report. However, consensus support for a recommendation does not necessarily imply unanimity for the position taken. Recognizing that all states may not totally agree with these recommended criteria or protocols, the guidelines developed through this series of meetings should serve only to provide guidance to states in the development of a review and approval process for medical waste treatment technologies.

Logistical support for all three meetings was provided by the USEPA. Roger Greene, Rhode Island Department of Environmental Management, Diann J. Miele, Rhode Island Department of Health, and Dr. Nelson S. Slavik, President, Environmental Health Management Systems, Inc., co-facilitated each of the meetings. A listing of all participants attending the New Orleans, Atlanta, and Washington, D.C. meetings is found in Attachment D.

2.0 Medical Waste Treatment Technology Efficacy Assessment

The establishment of specific criteria that define medical waste treatment technology efficacy is required to consistently evaluate new or modified medical waste treatment technologies. A number of terms are used in the literature to denote the level of treatment that may be assigned to a medical waste treatment technology (e.g., decontaminate, sterilize, disinfect, render harmless, and kill). However, these terms are non-descriptive and do not provide any mechanism for measuring the degree of treatment efficiency. It is critical that terms and performance criteria be established that quantitatively and qualitatively define the level of microbial destruction required of any medical waste treatment process.

Currently, there are no federal or national efficacy standards for medical waste treatment technologies and only a limited number of states have attempted to establish treatment efficacy criteria. The need exists to develop nationally recognized standard treatment performance criteria and operating protocols which establish the qualitative and quantitative parameters that ensure effective treatment. This section provides recommended medical waste treatment technology efficacy assessment criteria and discusses the rationale for their recommendation.

2.1 Classification of Emerging Medical Waste Treatment Technologies. To develop approval protocols and performance criteria for medical waste treatment technologies, it is necessary to classify known or anticipated technologies based on their mode of microbial inactivation. Medical waste treatment categories can be represented through the following categories:

- Thermal (wet and dry heat, microwaving, infrared, laser, plasma pyrolysis)
- Chemical (chlorine, chlorine derivatives, ozone, enzymes)
- Irradiation (UV, cobalt 60)
- Other treatment mechanisms designed for specific medical waste categories generated in small volumes (thermal/electrical)

For certain technologies, there may be a combination of inactivation modes used to inactivate microorganisms (i.e., chemical/thermal or chemical/irradiation). In addition to the treatment mode, there may also be—mechanical grinding introduced prior to, during, and/or at the end of the treatment process (*Note:* Grinding, shredding, and compaction are not viewed as treatment methods, but are used to facilitate the effectiveness of the treatment method or to render the waste destroyed, unrecognizable and nonfunctional). The total process by which the medical waste is treated will influence the selection of biological and physical indicators used in the testing and validation processes and will influence the protocols in which they are used.

2.2 Definition of Microbial Inactivation. Underlying the development of assessment protocols for approving an emerging medical waste treatment technology, is the establishment of efficacy criteria that provide a quantitative and qualitative measure of required performance. There is no consensus among the states on the level of microbial inactivation required of a medical waste treatment process. To properly define microbial inactivation requires that definitions established include both qualitative and quantitative aspects. From this perspective, definitions need to be established which qualitatively define microbial inactivation (i.e., form and type of microorganisms affected) and which quantify the required level of inactivation.

The terms sterilization and disinfection have provided some measure of prescriptive criteria as used in denoting sterilization or degree of disinfection required of medical instruments and supplies. Sterilization is commonly defined as the complete elimination or destruction of all forms of microbial life, including highly resistant bacterial endospores. Since complete elimination or destruction is difficult to prove, sterilization is usually expressed as a probability function in terms of the number of microorganisms surviving a particular treatment process. This function is usually expressed as a 6 $\log_{10}$ reduction (defined as 6 decade reduction or a one millionth [0.000001] survival probability in a microbial population; i.e., a 99.9999% reduction) of the most resistant microorganisms to the sterilization process in question. Spore suspensions of resistant *Bacillus* species are often used as biological indicators for determining the efficacy of the sterilization process (i.e., *B. stearothermophilus*, thermal inactivation; *B. subtilis*, chemical inactivation; *B. pumilus*, irradiation inactivation).

Disinfection can be defined as a procedure that reduces the level of microbial contamination. How disinfection is defined is dependent on the process in which the disinfectant is used, what microorganisms are affected, and what level of microbial inactivation is achieved. In the definition proposed by Spaulding (see Selected Bibliography), disinfectants are labeled as low-, intermediate- or high-level, determined in part on the survivability of microbial groups (i.e., bacterial spores [most resistant], mycobacteria, non-lipid or small viruses, fungi, vegetative bacteria, and lipid or medium-sized viruses [least resistant]) after treatment. Low-level disinfectant processes cause the death of all bacteria except *Mycobacterium tuberculosis* and *M. bovis*, lipid-enveloped and medium-sized viruses (e.g., herpes simplex virus, cytomegalovirus, respiratory syncytial virus, hepatitis B virus, and human immunodeficiency virus), and fungi. Intermediate-level disinfectant processes do not necessarily kill bacterial spores but are effective against tubercle bacillus and fungi. However, intermediate-level disinfectant processes vary in their effectiveness against viruses with small non-lipid viruses (e.g., rhinoviruses) being significantly more resistant than medium-sized, lipid viruses. High-level disinfectant processes cause the death of all microbial life, except for high numbers of bacterial spores. Sporicidal capacity is an essential property of high-level disinfection, although the amount of sporicidal activity is not quantified in any definition.

It was agreed during the New Orleans meeting that there was a need to establish a separate classification system that would specifically denote levels of microbial inactivation required of medical waste treatment. This classification system should quantitatively and qualitatively define the measure of required performance. To aid in the establishment of a separate classification system, the following categories of microbial inactivation were offered and discussed.

Level I	Inactivation of vegetative bacteria, fungi, and lipophilic virus
Level II	Inactivation of vegetative bacteria, fungi, all viruses, and mycobacteria
Level III	Inactivation of vegetative bacteria, fungi, all viruses, mycobacteria, and *B. stearothermophilus* spores at 10^4 or greater; or *B. subtilis* spores at 10^4 or greater with chemical treatment
Level IV	Inactivation of vegetative bacteria, fungi, all viruses, and mycobacteria, and *B. stearothermophilus* spores at 10^6 or greater

At the New Orleans meeting most participants generally favored level III criteria for medical waste treatment technologies. Although there was considerable discussion at that meeting, no consensus had been reached on the qualitative and quantitative aspects of the level II and level III definitions and the conditions to be applied, if any, for relaxation of the level III requirement to level II.

A primary objective of the Atlanta meeting was to specifically define the qualitative and quantitative aspects of the microbial inactivation definitions and to assign their application. To meet this objective, discussions centered on:

- Defining microbial inactivation levels by representative microbial groups and by the amount of microbial inactivation required for each
- Assigning representative pathogen surrogates to be used in the efficacy evaluation processes
- Assigning inactivation levels required of a medical waste treatment technology

To assist the committee in further defining levels I to IV, a summary was provided at the Atlanta meeting of USEPA sponsored research of emerging medical waste treatment technologies. Summarized were the treatment technologies evaluated, the surrogate organisms selected for testing and rationale for their selection, and in general, the results obtained from this research project. It was stated that the research material presented was not yet available for review since this material will serve as an appendix to the USEPA's "Final Report to Congress" when finalized.

Of particular interest to the committee was the availability of documentation that would support the use of an ultimate pathogen surrogate (i.e., *B. stearothermophilus* spores) that could be used to avoid the testing of representative pathogen surrogates from each of the microbial groups listed in the definitions above. As part of the USEPA sponsored study, comparative tests with vegetative bacteria, bacterial spores, fungal spores, and mycobacteria demonstrated that *B. stearothermophilus* and *B. subtilis* spores could be used to represent vegetative bacteria, fungi, and mycobacteria in evaluating both chemical and thermal (wet and dry heat) treatment systems.

No comparative testing, however, had been conducted with viruses or parasites. Without this supporting documentation for viruses and parasites, the committee could not recommend that *B. stearothermophilus* or *B. subtilis* be designated as an ultimate pathogen surrogate for efficacy testing. As such, the committee took the position to recommend that pathogen surrogates representing vegetative bacteria, fungi, parasites, viruses, mycobacteria, and bacterial spores be used to demonstrate efficacy of the treatment process. To determine if *B. stearothermophilus* and *B. subtilis* spores could be used in the future as pathogen surrogates representing all microbial groups, the committee recommended that further research be conducted to evaluate their relative resistance to representative parasitic agents (i.e., *Giardia* and *Cryptosporidium*) and viral agents (i.e., polio 2, MS-2).

In defining microbial inactivation levels, each level will require characterization by (1) the microbial groups to be inactivated and (2) the level of microbial inactivation required for each group. In the categories depicted as levels I to IV above, each level represents a hierarchy of increasing treatment resistance where treatment resistance is defined by the type of microorganism requiring inactivation and/or the amount of inactivation required for that type of microorganism. The definition of these categories requires that all groups of pathogen surrogate microorganisms recommended for testing be included in the definition. To be consistent with the committee's recommendation that a representative microorganism be tested from each microbial group, the definitions of levels II to IV were modified to include "parasites." Additionally, it was suggested that "all viruses" was too inclusive and it was recommended that all viruses be modified to "lipophilic/hydrophilic viruses." These changes are reflected in the definition for the Levels of Microbial Inactivation presented in Table 11.1.

It should be noted that the inactivation levels defined in Table 11.1 are not to be construed as having any relationship with microbial inactivation requirements for microorganisms in biosafety levels I to IV as defined within guidelines set by the Centers for Disease Control in *Biosafety in Microbiological and Biomedical Laboratories* (1993).

TABLE 11.1   Levels of Microbial Inactivation

Level I	Inactivation of vegetative bacteria, fungi, and lipophilic viruses at a 6 $\log_{10}$ reduction or greater
Level II	Inactivation of vegetative bacteria, fungi, lipophilic/hydrophilic viruses, parasites, and mycobacteria at a 6 $\log_{10}$ reduction or greater
Level III	Inactivation of vegetative bacteria, fungi, lipophilic/hydrophilic viruses, parasites, and mycobacteria at a 6 $\log_{10}$ reduction or greater; and inactivation of *B. stearothermophilus* spores or *B. subtilis* spores at a 4 $\log_{10}$ reduction or greater
Level IV	Inactivation of vegetative bacteria, fungi, lipophilic/hydrophilic parasites, mycobacteria, and *B. stearothermophilus* spores at a 6 $\log_{10}$ reduction or greater

Inactivation of spores from both *B. stearothermophilus* and *B. subtilis* is also defined in levels III and IV (refer to Table 11.1). It was questioned whether these microorganisms were the most chemically or thermally resistant biological indicators. From information provided, the use of these microorganisms as the most resistant indicators to thermal and chemical agents is supported in the literature.

To avoid assigning a specific bacterial species for each specific treatment process, documentation was sought that would support the use of spores from just one bacterial species for both chemical and thermal treatment processes. In the USEPA sponsored studies comparing *B. stearothermophilus* and *B. subtilis* resistance to hypochlorite (1000 ppm available free chlorine) and glutaraldehyde (3000 ppm, 2% alkaline glutaraldehyde), the resistance of spores from both was comparable. Data also supported that *B. stearothermophilus* spores were slightly more resistant to dry heat than *B. subtilis* var. *niger* spores (the *B. subtilis* variety traditionally used to determine dry heat resistance). These data indicate that *B. stearothermophilus* can be used as the sole spore indicator for chemical treatment processes and as the sole spore indicator for both dry and wet heat thermal processes.

B. stearothermophilus spores, however, are more resistant to wet heat than spores from *B. subtilis*. Debate centered on whether spores from either species could be used interchangeably for wet or dry heat thermal processes even though *B. stearothermophilus* spores are more resistant to wet heat. It was argued that the use of spore inactivation in the definition serves two functions: (1) to demonstrate that bacterial spore formers (originating primarily from laboratory wastes) can be inactivated and (2) to provide a margin of safety beyond the inactivation of vegetative bacteria, fungi, viruses, parasites, and mycobacteria.

From the first perspective, both *B. stearothermophilus* and *B. subtilis* spores are used as indicators of medical product sterility because of their documented resistance to heat and chemicals. Inactivation of either of these highly resistant bacteria spores serves to demonstrate that any spores found in medical waste will also be inactivated. From the second perspective, *B. subtilis* and *B. stearothermophilus* spores both display significantly more heat resistance than the microorganisms in the aforementioned microbial groups. The demonstration that highly resistant spores from either of these Bacillus species can be effectively destroyed ensures a margin of safety from the variables in the treatment of medical waste (i.e., waste packaging, waste composition, waste density, and factors influencing the homogeneity of the treatment process).

On the basis of these arguments presented above, the committee recommended that either *B. stearothermophilus* or *B. subtilis* spores be used as biological indicators for chemical or thermal treatment processes. The question arose, however, to whether a higher level of inactivation would be required when using *B. subtilis* for wet heat treatment processes. It was argued that *B. stearothermophilus* and *B. subtilis* spores both have a documented high degree of thermal resistance. As such, higher inactivation levels required of *B. subtilis* spores for wet heat treatment processes were considered unnecessary to further demonstrate effective spore inactivation or an expanded margin of safety. In addition, it was argued that assigning different threshold inactivation levels for each defined biological indicator would set a bad precedent and lead to an overly and unnecessarily complex definition. The revision to allow the use of either *B. stearothermophilus* and *B. subtilis* spores as biological indicators for chemical or thermal treatment processes is reflected in the recommended definition for the Levels of Microbial Inactivation as presented in Table 11.1.

The use of *B. stearothermophilus* or *B. subtilis* spores for demonstrating microbial inactivation by irradiation processes was also recommended. *B. pumilus* spores are used as the standard biological indicator to demonstrate irradiation efficacy in the sterilization of medical products. *B. pumilus* spores are, however, not as resistant to irradiation as the enteroviruses or the vegetative bacterium *Deinococcus radiodurans*. The use of an enterovirus (e.g., polio 2 or polio 3) or *Deinococcus radiodurans* can provide a more stringent measure of microbial inactivation than *B. pumilus* spores, making any requirement for this specific *Bacillus* species unnecessary for the purpose of providing an additional "margin of safety." To demonstrate that bacterial spores can be effectively inactivated, *B. subtilis* or *B. stearothermophilus* spores can serve as equivalent biological indicators. Inactivation of *B. stearothermophilus* or *B. subtilis* spores, although less resistant to irradiation than *B. pumilus* spores, serves to adequately demonstrate that any spores found in medical waste will also be inactivated.

Specific levels of inactivation are required of any adopted definition to quantitatively define the measure of required performance of a medical waste treatment technology. The definitions proposed by the committee state that inactivation is required of "vegetative bacteria, fungi, lipophilic/hydrophilic viruses, parasites, and mycobacteria." Although implied but not specifically stated, this definition requires complete inactivation of the representative microorganisms tested in each of the microbial groups listed. Since complete inactivation is impossible to prove, it can be expressed as a probability function in terms of the number of microorganisms surviving a particular treatment process. In defining sterilization, this function is usually expressed as a 6 $\log_{10}$ reduction. A 6 $\log_{10}$ reduction is defined as a 6 decade reduction or a one millionth (0.000001) survival probability in a microbial population (i.e., a 99.9999% reduction). Using this definition as a basis for quantifying complete inactivation, the recommendation was made that 6 $\log_{10}$ reduction be required of the representative microorganisms tested in each of the microbial groups listed (with the exception of *B. stearothermophilus* or *B. subtilis* spores). Table 11.1, Levels of Microbial Inactivation, incorporates these revisions.

For inactivation on levels required of *B. stearothermophilus* or *B. subtilis* spores, the original stated that inactivation was required at "10^4 or greater" (i.e., 4 $\log_{10}$ reduction or greater). It was questioned whether this level should remain as stated in the definition or be modified to be less or more stringent. In the USEPA sponsored studies it was demonstrated that of the medical waste treatment technologies studied, all could meet at least a 4 $\log_{10}$ reduction of *B. stearothermophilus* or *B. subtilis* spores. The committee supported the level as defined in the original definition. Language however, was modified to replace "10^4 or greater" with "4 $\log_{10}$ reduction or greater" to be consistent with the use of the definition of $\log_{10}$ reduction. A 4 $\log_{10}$ reduction is defined as a 4 decade reduction or a 0.0001 survival probability in a microbial population (i.e., a 99.99% reduction). The committee also revised the Level IV to replace "10^4 or greater" with "4 $\log_{10}$ reduction or greater" to be consistent with the use of the definition of $\log_{10}$ reduction. No further revision was suggested. These revisions are reflected in Table 11.1.

Recommendations made by the committee for establishing a quantitative and qualitative definition for the levels of microbial inactivation are incorporated into categories I to IV of Table 11.1. Summarizing, the committee recommended that:

- Pathogen surrogates representing vegetative bacteria, fungi, parasites, lipophilic/hydrophilic viruses, mycobacteria, and bacterial spores be used to demonstrate microbial inactivation
- Either *B. stearothermophilus* or *B. subtilis* spores be used as biological indicators for chemical or thermal treatment or irradiation processes
- A 6 $\log_{10}$ reduction be required of the representative microorganisms tested in each of the microbial groups listed (with the exception of *B. stearothermophilus* or *B. subtilis* spores)
- A 4 $\log_{10}$ reduction level be required of *B. subtilis* or *B. stearothermophilus* spores

Having quantitatively and qualitatively established a definition for the levels of microbial inactivation, arguments were presented and discussed to determine the position of the committee on which category would serve as the benchmark criteria for medical waste treatment technology efficacy. Debate centered on the recommendation of level II or level III criteria. Arguments for recommending level II criteria were as follows:

- Medical waste does not contain significant differences in amount and type of pathogens as household waste.
- Level II criteria provides a sufficient degree of microbial inactivation.
- Level III criteria may conflict with lesser inactivation criteria already defined by the state.

- Level III or IV criteria can be applied, if necessary, to those medical waste streams requiring an additional margin of safety.

Arguments for recommending level III criteria were as follows:

- Level III criteria serve as a margin of safety from the variables inherent in the treatment of medical waste (i.e., waste packaging, waste composition, waste density, and factors influencing the homogeneity of the treatment process).
- Segregation of some medical waste categories (i.e. laboratory cultures) requiring level III treatment would be impractical if level II criteria were in effect.
- Medical waste treatment equipment industry already achieves level III criteria.
- Level II or level IV criteria may still be allowed dependent on the technology application or waste type processed.

It was the consensus (not unanimous) of the committee that level III be required of all emerging medical waste technologies. The committee took the position that level III criteria were to be established as a benchmark and as such, were applicable to all medical waste treatment devices. The committee realized that there might be circumstances under which a state may allow relaxation of the level III requirement.

The committee rejected the allowance for exception to level II standards for those technologies that could be termed "counter-top" devices designed for a specific medical waste category. Relaxation from level III to level II criteria was not considered warranted on the basis of the equipment's:

- Inability to inactivate spores
- Designation as a small quantity treatment device
- Designation for treating minimally contaminated medical waste categories
- Exhibiting difficulty to demonstrate microbial inactivation through designated protocols (i.e., a needle thermal-destruction device)

The committee realized that there might be circumstances under which a state may allow relaxation of the level III requirement. These exceptions would by necessity need to be made on a case-by-case basis, would require the equipment manufacturer to provide a rationale for relaxation, and would require adequate supporting documentation to substantiate that rationale.

The committee also debated if laboratory wastes (i.e., discarded cultures and stocks of pathogenic agents) should require sterilization (i.e., meet level IV criteria) on the basis that these wastes may contain high concentrations of known pathogens. The committee took the position that level III criteria remained the standard for all medical waste categories. The committee emphasized, however, that laboratories should be aware that cultures and stocks of disease-causing agents may require sterilization before disposal. In addition to guidelines set by the Centers for Disease Control in *Biosafety in Microbiological and Biomedical Laboratories* (1993) and standards of the College of American Pathologists (CAP), some states require laboratory cultures to be incinerated or autoclaved (i.e., steam sterilized) before leaving the laboratory or before being disposed of. Although no specific recommendations for medical waste disposal are made under the Clinical Laboratory Improvement Amendments (CLIA), medical waste disposal practices are receiving increased scrutiny during routine inspections.

2.3 Representative Biological Indicators. In the absence of an ultimate pathogen surrogate to represent all defined microbial groups, the selection of pathogen surrogates representing vegetative bacteria, fungi, parasites, viruses, mycobacteria, and bacterial spores was considered necessary to define and facilitate any state approval process. Criteria defining surrogate selection should include that any surrogate recommended:

- Not affect healthy individuals
- Be easily obtainable
- Be an ATCC registered strain, as available
- Be easily cultured and maintained
- Meet quality control requirements

Microorganisms obtained from the American Type Culture Collection (ATCC) and methods prescribed by the Association of Official Analytical Chemists (AOAC) assist in fulfilling these recommendations by (1) providing traceable and pure cultures of known characteristics and concentration and (2) providing recognized culturing protocols and detailed sampling and testing protocols.

Provided in Table 11.2 are the biological indicators recommended by the committee for testing microbial inactivation efficacy in medical waste treatment processes. The selection of these representatives was based on each microorganism:

TABLE 11.2   Recommended Biological Indicators

Vegetative bacteria	*Staphylococcus aureus* (ATCC 6538)
	Pseudomonas aeruginosa (ATCC 15442)
Fungi	*Candida albicans* (ATCC 18804)
	Penicillium chrysogenum (ATCC 24791)
	Aspergillus niger
Viruses	Polio 2, polio 3
	MS-2 bacteriophage (ATCC 15597-B1)
Parasites	*Cryptosporidium* spp. oocysts
	Giardia spp. cysts
Mycobacteria	*Mycobacterium terrae*
	Mycobacterium phlei
	Mycobacterium bovis (BCG) (ATCC 35743)
Bacterial spores	*B. stearothermophilus* (ATCC 7953)
	B. subtilis (ATCC 19659)

- Meeting, where possible, the criteria established above
- Representing, where possible, those organisms associated with medical waste
- Providing a biological challenge equivalent to or greater than that associated with microorganisms found in medical waste

Biological indicators selected to provide documentation of relative resistance to an inactivating agent should be chosen after evaluation of the treatment process as it relates to the conditions used during comparative resistance research studies described in the literature. Literature studies support the assertion that the degree of relative resistance of a microorganism to an inactivating agent can be dependent on various factors (i.e., pH, temperature). Conditions used in literature studies that demonstrate a relatively high degree of resistance of a particular microorganism may be significantly different to the conditions found within the treatment process. A comparison of the conditions used in the literature to those used in the treatment process should be made to determine if relative microbial resistance can be altered (i.e., lowered) as a result of treatment process conditions.

The committee emphasized that although the microorganisms selected represent pathogen surrogates, these selected surrogates may have the potential to be pathogenic under certain conditions. As such, the committee recommended that all testing be conducted using recognized microbial techniques. For those pathogen surrogates that still retain some higher degree of pathogenicity (e.g., *Cryptosporidium*, *Giardia*, and *Mycobacteria*), efficacy testing should be conducted only by qualified laboratory personnel.

The committee recommended that one or more of the representative microorganisms from each microbial group be used in efficacy evaluation. Specific criteria for the selection of these microorganisms are provided below in Table 11.3.

After discussion on the rationale for selection of the representative biological indicators presented above, consensus by the committee was attained on recommending the use of these biological indicator strains for treatment technology efficacy testing.

2.4 Quantification of Microbial Inactivation. Establishing the mechanisms to quantify the level of microbial inactivation is essential in developing the format and requirements of the guidance protocols. As presented and discussed, microbial inactivation ("kill") is equated to "$\log_{10}$kill" which is defined as the difference between the logarithms of number of viable test microorganisms before and after treatment. This definition is translated into the following formula:

$$\log_{10}\text{kill} = \log_{10} \text{(cfu/g introduced)} - \log_{10}\text{(cfu/g recovered)}$$

where:

$\log_{10}$kill is equivalent to the term $\log_{10}$ reduction;

"Introduced" is the number of viable test microorganisms introduced into the treatment unit;

TABLE 11.3   Biological Indicator Selection Criteria

Vegetative bacteria	*Staphylococcus aureus* and *Pseudomonas aeruginosa* were selected to represent both gram-positive and gram-negative bacteria, respectively. Both are currently required by the Association of Official Analytical Chemists (AOAC) use-dilution method and both have been shown to be resistant to chemical inactivation.
Fungi	The selection of *Candida albicans* and *Penicillium chrysogenum* was based on reported data indicating these organisms representing yeast and molds, respectively, are the most resistant to germicides. Although *Trichophyton mentagrophytes* is the AOAC test organism for molds, *Penicillium chrysogenum* is reported to be more resistant to germicides. The inclusion of *Aspergillus* as an indicator organism was based on its familiarity as a common mold.
Viruses	Lipophilic (enveloped) viruses are less resistant to both thermal and chemical inactivation than the hydrophilic (nonenveloped) viruses. As such, enveloped viruses such as HIV, herpes simplex virus and hepatitis B virus are less resistant than enveloped viruses such as poliovirus, adenovirus, and coxsackievirus. Polio 2 (attenuated vaccine strain) and polio 3 virus were selected based on their relative higher chemical and thermal resistance. Additionally, the use of an enterovirus (e.g., polio 2 or polio 3) can provide a stringent measure of efficacy for irradiation treatment processes. MS-2 bacteriophage was selected as a hepatitis virus surrogate in that this bacteriophage offers a comparable degree of chemical and thermal resistance, is safe to handle and easy to culture.
Parasites	Both *Cryptosporidium* spp. oocysts and *Giardia* spp. cysts are used as test organisms to demonstrate germicidal effectiveness. *Cryptosporidium* has been demonstrated to have a higher chemical resistance and *Cryptosporidium* spp. oocysts are more readily available than *Giardia* spp. cysts. Both are significantly pathogenic (both have an infectious dose of 10 cysts) and care is advised when using these microorganisms as parasitic biological indicators.
Mycobacteria	*Mycobacterium* has a demonstrated measure of disinfectant resistance, is a rapid grower and is pigmented for easy identification. *M. bovis* (BCG) is used in the AOAC Tuberculocidal Method and is analogous to *M. tuberculosis* in that it is in the same group or complex. Individuals exposed to *M. bovis* (BCG, ATCC strain) may skin test convert although no actual infectivity or disease occurs. Risk of exposure would come from those mechanisms that grind the waste. *Mycobacterium terrae* is equivalent to *M. tuberculosis* in resistance to chemical inactivation. In Europe it is recommended for disinfectant testing. *M. terrae* does not grow as rapidly as *M. bovis* or *M. tuberculosis*.

(*continued*)

TABLE 11.3   *(Continued)*

Bacterial spores	Both *B. stearothermophilus* and *B. subtilis* spores are commonly used as biological indicators for both thermal and chemical resistance. *B. stearothermophilus* spores exhibit more thermal and chemical resistance than spores from *B. subtilis*.

"Recovered" is the number of viable test microorganisms recovered after treatment; and

"cfu/g" are colony forming units per gram of waste solids.

A $\log_{10}$kill of 6 or greater is equivalent or less than a one millionth [0.000001] survival probability in a microbial population or a 99.9999% reduction or greater of that population.

Using the level III definition recommended by the committee as shown in Table 11.1, a $\log_{10}$kill of 6 (e.g., 6 $\log_{10}$ reduction) is required of vegetative bacteria, fungi, lipophilic/hydrophilic viruses, parasites, and mycobacteria and a $\log_{10}$kill of 4 (e.g., 4 $\log_{10}$ reduction) is required of *B. stearothermophilus* or *B. subtilis* spores. Employing the above equation to quantify microbial inactivation will require the consideration of the methods of biological indicator introduction and recovery. For those treatment processes that can maintain the integrity of the carrier (i.e., ampules, plastic strips) of the desired microbiological test strain, commercially available biological indicators of the required strain and concentration can be easily placed, recovered, and cultured to demonstrate efficacy. Quantification is evaluated by growth or no growth of the cultured biological indicator. For example, if an ampule that contained 1×10^4 *B. stearothermophilus* spores were treated, retrieved, and cultured, no growth would demonstrate a 4 $\log_{10}$ reduction.

For those treatment mechanisms that cannot ensure or provide integrity of the biological indicator carrier, quantitative measurement of efficacy requires a two-step approach. The purpose of the first step is to account for the reduction of microorganisms due to equipment design (i.e., dilution of indicator organisms or physical entrapment).

This first step, the "control," is typically performed using microbial cultures (i.e., liquid suspensions) of predetermined concentrations necessary to ensure a sufficient microbial recovery at the end of this step. The microbial suspension is added to a standardized surrogate medical waste load that is processed under normal operating conditions without the addition of the microbial inactivation agent (i.e., heat, chemicals). Standard loads may vary depending on the various treatment challenges (i.e., high moisture content, high organic load, high density) required of the equipment. After processing, waste samples are collected and washed to recover the biological indicator organisms in the sample. Recovered microorganism suspensions are plated to quantify microbial recovery. The number of viable microorganisms recovered serves as a baseline quantity for comparison to the number of recovered microorganisms from wastes processed with the microbial inactivation agent. The required number of recovered viable indicator microorganisms from the "control" must be equal to or greater than the number of microorganisms required to demonstrate the prescribed Log reduction as defined in Level III (i.e., a 6 $\log_{10}$ reduction for vegetative microorganisms and a 4 $\log_{10}$ reduction for spores). See Attachment A (Section C3) and Attachment C for a detailed process description.

This step can be defined by the following equation:

$$\log_{10}RC = \log_{10}IC - \log_{10}NR$$

where

$\text{Log}_{10}RC > 6$ for vegetative microorganisms and > 4 for bacterial spores;

$\text{Log}_{10}RC$ is the number of viable "control" microorganisms (in colony forming units per gram of waste solids) recovered in the non-treated processed waste residue;

$\text{Log}_{10}IC$ is the number of viable "control" microorganisms (in colony forming units per gram of waste solids) introduced into the treatment unit; and

$\text{Log}_{10}NR$ is the number of "control" microorganisms (in colony forming units per gram of waste solids) not recovered in the non-treated processed waste residue.

Rearranging the equation above enables the calculation of microbial loss due to dilution, physical manipulation, or residue adhesion during the treatment process. $\text{Log}_{10}NR$ represents an accountability factor for microbial loss and is defined by the following equation:

$$\log_{10}NR = \log_{10}IC - \log_{10}RC$$

The second step ("test") is to operate the treatment unit as in the "control" run with the selected biological indicators, but with the addition of the microbial inactivation agent. After processing, waste samples are collected and washed as in the "control" to recover any viable biological indicator organisms in the sample. From data collected from the "test" and "control", the level of microbial inactivation (i.e., "$\log_{10}$kill") can be calculated by employing the following equation:

$$\log_{10}\text{kill} = \log_{10}IT - \log_{10}NR - \log_{10}RT$$

where:

$\text{Log}_{10}\text{kill}$ is equivalent to the term $\log_{10}$ reduction.

$\text{Log}_{10}IT$ is the number of viable "test" microorganisms (in colony forming units per gram of waste solids) introduced into the treatment unit. $\text{Log}_{10}IT = \log_{10}IC$.

$\text{Log}_{10}NR$ is the number of "control" microorganisms (in colony forming units per gram of waste solids) not in the non-treated processed waste residue.

$\text{Log}_{10}RT$ is the number of viable "test" microorganisms (in colony forming units per gram of waste solids) recovered in waste residue.

Attachment C (Section III) serves to illustrate the application of the equations presented above.

Formulas used in the discussion above for the quantification of microbial inactivation were modified from those used by Illinois EPA in their final regulations (June 1993) entitled "Potentially Infectious Medical Wastes" (see Selected Bibliography).

After discussion on the use and application of the formulas and calculations presented above, consensus by the committee was unanimous on recommending the use of the formulas and methods of calculation in the enumeration of medical waste treatment technology efficacy.

3.0 Process for Approving Medical Waste Treatment Technologies

State approval of an emerging medical waste treatment technology is necessary to ensure that the technology can effectively and safely treat medical waste. From discussions, the completed approval process can be viewed as fulfilling, where applicable, three components:

- Approval of the technology by the state to ensure the technology is effective in safely inactivating microorganisms to specified criteria
- Granting site approval to verify the sited equipment meets approved specifications and efficacy requirements under actual operating conditions
- USEPA FIFRA pesticide registration requirements, as applicable, for those medical waste treatment technologies that use chemicals as the microbial inactivator

Each of these components requires information be supplied to states demonstrating that the treatment technology is effectively treating medical waste by established criteria and that the process is environmentally sound and occupationally safe. Information necessary for proper review of medical waste treatment technologies is provided for each component described below.

3.1 Biological Inactivation Efficacy: Establishing Protocols. Methodology employed to determine efficacy of the technology will, by necessity, need to be developed by the equipment manufacturer to assure the protocols are congruent with the treatment method. Protocols developed for efficacy testing should incorporate recognized standard procedures such as those found in *Test Methods for Evaluating Solid Waste, Physical/Chemical Methods* and *Standard Methods for the Examination of Water and Wastewater* (see Selected Bibliography).

In establishing testing criteria to evaluate efficacy, the composition of the waste load(s) tested is critically important. Depending on the treatment mechanism, efficacy may vary with waste load composition (i.e., organic content, density, moisture or liquid content). Although the committee recognized that waste composition may affect efficacy results considerably, establishing specific requirements for challenge loads for all existing, pending, and future treatment technologies is not practical or necessarily all inclusive. The committee recommended that the equipment manufacturer prescribe those types of medical wastes that present the greatest challenge to efficacy of the equipment and present protocols that adequately evaluate efficacy under normal operating conditions. On submittal for evaluation by the state, the manufacturer's prescribed waste types and testing protocols could be accepted or modified at the discretion of the reviewing agency.

Dependent on the treatment process and efficacy protocols used, other factors may also influence the evaluation results. As such, the committee could not define specific protocols, but recommended that protocols evaluating medical waste treatment systems specifically delineate or incorporate:

- Waste compositions that typify actual waste to be processed
- Waste types that provide a challenge to the treatment process
- Comparable conditions to actual use (i.e., process time, temperature, chemical concentration, pH, humidity, load density, load volume)
- Assurance that biological indicators (i.e., ampules, strips) are not artificially affected by the treatment process
- Assurance of inoculum traceability, purity, viability and concentration
- Dilution and neutralization methods that do not affect microorganism viability
- Microorganism recovery methodologies that are statistically correct (i.e., sample collection, number of samples/test, number of colony forming units/plate
- Appropriate microbial culturing methods (i.e., avoidance of microbial competition, the selection of proper growth media and incubation times)

Based on the results obtained from challenge load testing, the medical waste treatment technology may be limited in its application to not treating all categories or types of medical wastes. Physical or aesthetic characteristics may also predicate the limitations applied or the conditions of the equipment's use. If certain medical waste categories are excluded from the treatment process, the state should specify for the manufacturer (vendor) and the user of the equipment the waste segregation parameters that will be employed to prohibit the waste from treatment and the mechanisms of treatment/disposal to be utilized for these excluded wastes.

Consideration should also be given to the equipment's use in a particular setting when applying challenge load testing. The composition of the challenge load would be conceivably different and more challenging if a particular application treats a medical waste stream containing a higher proportion of a waste type or composition that is difficult to treat by that process. Conversely, challenge loads for technologies whose primary application is hospital medical waste, might be relaxed if that technology was applied only to waste generated by physician offices. Efficacy testing protocols may also require modification dependent on the size or throughput of the equipment. Multiple testing points might be required due to the waste volume processed or the treatment process.

The committee recommended that efficacy testing protocols and all results of any evaluations conducted, including original data, be included for evaluation by the state agency reviewing the application for treatment technology approval. The methodologies and protocols developed are especially critical for state evaluation of medical waste treatment processes that pulverize, grind, or shred the waste during the treatment process and do not allow intact retrieval of the biological test indicator. The complexity of these protocols is illustrated in Attachment C, "Microbial Inactivation Testing Protocol for a Grinder/Chemical Medical Waste Inactivation Process".

To establish protocols that incorporate the recommended criteria above and meet any applicable recognized testing standards will, in most likelihood, require the equipment manufacturer to seek assistance from an independent laboratory. To ensure the required quality control and facilitate state review of the treatment process, the committee recommended that the qualified laboratory selected should:

- Be experienced in microbiological testing techniques and be familiar with required sampling and testing protocols
- Be an accredited laboratory or have experience with product registration through the federal Food and Drug Administration (FDA) or the USEPA Office of Pesticide Programs
- Be equipped to meet FDA "Good Laboratory Practices" requirements

3.2 Approval of Medical Waste Treatment Technologies. As a first step in the review process, information is required of the manufacturer to provide the state with the information it needs to properly assess the treatment technology proposed for approval. The state's use of a comprehensive information request form is essential in obtaining relevant information and in acquainting the manufacturer with the requirements and the responsibilities inherent in the review process. To meet these objectives, the form should at a minimum:

- Delineate state responsibilities and permitting requirements.
- Delineate manufacturer responsibilities and registration requirements.
- Request a detailed description of the medical waste treatment equipment to be tested, including manufacturer's instructions and equipment specifications, operating procedures and conditions, including, as applicable, treatment times, temperatures, pressures, chemical concentrations, irradiation doses, feed rates, and waste load composition.
- Request documentation demonstrating that the treatment method meets microbial inactivation criteria and required testing protocols, including a detailed description of the test procedures and calculations used in fulfilling designated performance standards verifying efficacy, of user verification methodology, and of microbial culturing protocols that ensure traceability, purity and concentration.
- Provide documentation of applicable emission controls for suspected pathological and toxic emissions.

- Provide documentation for occupational safety and health assurance by describing the medical waste treatment equipment's safety systems such as warning signage, operating zone restrictions, lock-out procedures, and personal protection equipment requirements.

To assist the committee in developing a format for an information request form, information forms from the states of California, Michigan, and New Jersey were reviewed for their content. In addition to the information requested on these forms, the committee recommended that the following information also be requested:

- A more extensive discussion on available parametric controls (to verify efficacy and ensure operator non-interference in the treatment process)
- A discussion on efficiency and other potential benefits the treatment technology has to offer to the environment
- More detailed information relating to waste residues including their potential hazards/toxicities and their specific mode of disposal or recycling

From the forms reviewed and the additional information requested by the committee, a recommended informational request form, termed an "Application for Evaluation and Approval of Medical Waste Treatment Technologies," was developed (see Attachment B).

In addition to fulfilling environmental and occupational safety requirements, all treatment technologies must meet level III efficacy criteria. Demonstration that these criteria are met is the responsibility of the equipment manufacturer. In meeting these requirements the manufacturer must:

- Demonstrate that all required pathogen surrogates and resistant bacterial endospores (as recommended in Table 11.2) are inactivated to level III criteria under all required challenge waste load compositions.
- Develop and demonstrate that site approval and user verification testing protocols are workable and valid.
- Demonstrate where technically practical, the relationship biological indicator data and data procured from real-time parametric monitoring equipment.

To assist in presenting the recommendations for efficacy review, an approval process guideline is presented in Attachment A.

3.3 Parametric Monitoring and Controls. Parametric monitoring of a medical waste treatment process can provide real-time data acquisition for assessing efficacy. However, correlation of the data acquired from the parametric monitoring device(s) with that of biological indicator studies is essential if parametric monitoring is to supplement or replace biological indicator monitoring. This demonstration is the responsibility of the manufacturer (vendor). To verify that a proper correlation has been established between the parametric monitoring device and biological indicator inactivation, the manufacturer (vendor) must demonstrate that parametric monitoring is:

- Correlated with biological indicator inactivation through documented efficacy studies linking microbial inactivation with the parameter(s) being monitored
- Accurately monitoring the treatment agent and/or treatment conditions, as applicable (i.e., provide the limiting conditions that influence accurate monitoring)
- Appropriate for the conditions that exist under operational circumstances

Demonstration of the above components may allow the use of parametric monitoring for auditing treatment conditions or alerting the equipment operator of equipment malfunction or abnormal behavior. However, the use of parametric monitoring to substitute or replace biological indicator inactivation must require the device to additionally:

- Have tamper-proof controls or automatic factory-set controllers
- Be integrated with the treatment unit to automatically shut-down or no longer accept or expel waste if treatment conditions are not maintained at specified performance levels
- Be calibrated periodically as specified by the monitoring device's manufacturer
- Provide a tamper-proof recording of all critical operating parameters

The committee recommended that parametric monitoring could substitute or replace biological indicator monitoring provided that all of the above conditions were achieved.

3.4 Site Approval for Medical Waste Treatment Technologies. The purpose of the site-approval process is to ensure that the treatment equipment sited is the same equipment and process approved by the state. Site approval may also require obtaining other state permits (i.e., solid waste treatment/disposal permits; emissions and discharge permits) in addition to those required under state medical waste regulations. Technology efficacy must also be demonstrated under actual operating conditions. However, the rigor of the biological indicator testing would be less than the testing required for technology approval, although tests conducted would be required to reflect the waste load compositions of waste treated. Effectiveness and reliability of the real-time monitoring systems must also be demonstrated to receive site approval. Additionally, agency review is necessitated to verify proper and safe operations, verify disposal of waste residues, and verify operator training.

Specifically, to fulfill microbial inactivation and information requirements recommended for site approval, the equipment user must:

- Demonstrate that required resistant bacterial endospores (as recommended in Table 11.2) are inactivated to level III criteria under typical waste load and challenge compositions.

- Verify that user verification protocols adequately demonstrate effectiveness of the treatment process.
- Verify the relationship between biological indicator data and data procured from real-time parametric treatment monitoring equipment (i.e., correlation of biological indicator inactivation with time and temperature via thermocouple monitoring).
- Document in a written plan:
  - Names or positions of the equipment operators
  - Waste types or categories to be treated
  - Waste segregation procedures required
  - Wastes types prohibited from treatment
  - Equipment operation parameters
  - Efficacy monitoring procedures
  - Operating documentation and record-keeping requirements
  - Contingency waste disposal plans
  - Personal protective equipment requirements
  - Shut-down, clean-out and maintenance procedures
  - Emergency response plans
  - Operator training requirements
- Provide for state review:
  - Equipment model number and serial number
  - Equipment specification and operations manual
  - Certification that equipment is identical to state approved system
  - User's written plan
  - Certification documentation of operator training.

The state may want to visit the site of proposed operation to validate operations, or approve the site by reviewing the submitted information and documents. As a condition of site approval, the state should affirm its right to inspect the facility and affirm the right to revoke site approval if health and safety violations are discovered, if permit conditions are not being fulfilled, or if the facility is not adhering to its written plan.

Recommendations for the site approval process are presented in the approval process guideline in Attachment A.

3.5 USEPA Pesticide Use Registration. The use of a chemical agent in any treatment process may involve pesticide registration with the USEPA Pesticide Registration Office under the Federal Insecticide, Fungicide and Rodenticide Act (FIFRA). The USEPA Pesticide Registration Office's involvement in the regulatory process is dependent on advertising claims made by the medical waste treatment equipment's manufacturer (vendor). If claims are made that specify a level of microbial inactivation by term (i.e., kills pathogens, disinfects), registration with the USEPA Pesticide Registration Office is required.

Registration for a label claim will require the manufacturer (vendor) to submit efficacy studies of the process for review. Currently, the only label claim allowed for any medical waste treatment technology is the claim of "sanitizer," which is defined as "an antimicrobial agent that is intended for application to inanimate objects or surfaces for the purpose of reducing the microbial count to safe levels."

Several questions remain to be addressed concerning the involvement of the USEPA Pesticide Registration Office in the medical waste treatment technology review process. These questions are summarized as follows:

- For what advertising claims (and by which media, e.g., newspaper, product labels, etc.) should federal pesticide registration be required for chemical treatment processes?
- What are the specific guidelines and protocols required or what information is necessary for efficacy assessment review by USEPA Pesticide Registration Office?
- What are the quality assurance/quality control requirements required for pesticide registration?
- What potential conflicts may arise from the microbial inactivation guidelines recommended by the committee and those claims allowed by the USEPA Pesticide Registration Office?

It was recommended that the committee continue its dialogue with the USEPA Pesticide on Office to ensure consistency in the regulatory review process.

Further information on USEPA pesticide use registration is presented in Chapter 10 (Chemical Disinfection).

4.0 Permitting and State Authorization Issues

Although the review process for medical waste treatment technology approval is primarily concerned with ensuring safe and effective medical waste treatment, several permitting issues were identified and discussed by the committee. Recommendations are summarized below for each issued discussed.

4.1 User Verification: Biological Inactivation Efficacy Monitoring. User verification methodology is necessary to periodically verify to the equipment user and the state that the treatment unit is functioning properly, that proper operating procedures are used, and that performance standards are achieved. User verification protocols will employ biological indicators in addition to available verified parametric monitoring. Protocols used will have previously been approved by the state to assure the protocols are congruent with the treatment method/mechanism.

Specifically, to fulfill microbial inactivation and documentation requirements recommended for user verification, the state operating protocol will require the equipment user to:

- Demonstrate on a periodic basis that required resistant bacterial endospores (as recommended in Table 11.2) are inactivated to level III criteria under standard operating procedures.
- Document the frequency of biological and/or parametric monitoring.
- Document and record all biological indicator and critical parametric monitoring data.

Although no formal verification of compliance with these recommendations was prescribed, the committee noted that numerous regulatory agencies (i.e., the federal Occupational Safety and Health Administration, the state department of health, the state environmental agency) and accrediting associations (i.e., Joint Commission on Accreditation of Healthcare Organizations, College of American Pathologists) would serve to provide oversight. User verification requirements recommended are contained in the "State Guideline for Approval of Medical Waste Treatment Technologies" presented in Attachment A.

4.2 Commercial Versus On-Site Facilities. Commercial and on-site facilities (i.e., hospitals) can be typically distinguished by the increased volume of waste throughput from commercial facilities. As such, additional process controls, efficacy monitoring, and permitting might be necessitated to ensure that microbial inactivation is maintained and that environmental and occupational/public health and safety concerns are met.

As a facility applying for a commercial medical waste treatment permit, additional requirements may be imposed under other solid or special waste treatment/ disposal regulations. As such, cooperative efforts between permitting agencies or divisions are necessitated to ensure the facility is meeting its environmental health and safety responsibilities. To assist in identifying the potential commercial application of a medical waste treatment technology, the committee recommended that the potential use of the technology be indicated in technology review information supplied to the state by the equipment manufacturer.

4.3 Previously Approved Technologies. With rapid evolution of emerging medical waste treatment technologies and with the establishment of more restrictive efficacy criteria, previously granted approvals become an issue. Within the framework of the approval or permitting process, some mechanism should be established that requires previously approved technologies to meet current efficacy criteria. A number of options should be available to the state to allow previously approved mechanisms to continue with the realization that at some point, previously approved technologies will have to meet current standards. The committee discussed several options that would allow the state to periodically review all medical waste treatment technologies to determine if they were fulfilling current standards of performance.

Option One involved the granting of approval for a technology with the provision that any modification to the equipment would require reapplication for approval under current standards. As an example, the State of New York Department of Health in its approval letter includes the following statement:

This approval is granted for this specific system used in your efficacy studies and should not be construed as a general endorsement of the technology employed or any other unit or system. Any modifications of the system will require separate approval of the Department and may involve further efficacy testing.

Option Two limits the granted site or use permit to a specific time period (e.g., 3 or 5 years). At the time of renewal, the unit must demonstrate that it meets the efficacy criteria and other permit conditions at the levels prescribed in the new standards.

As a third option, the state could mandate that on the issuance of the new medical waste efficacy standards, pre-equipment subject to regulation would be required to comply with current efficacy standards within a set time period. Following compliance, the user would have the option to replace the existing equipment with approved technology, retrofit the equipment to meet current standards, or take the equipment out of service. Incorporation of additional provisions as stated in Option One or Option Two with those in Option Three would ensure that technology meeting current standards would remain in compliance with future, more restrictive regulations.

Steam sterilizers or autoclaves were discussed as to whether they should be included as an emerging treatment technology. It was noted that the steam sterilization process has been used for decades to sterilize medical products, biological products, and medical or biohazardous waste and is generally recognized as a traditional sterilization process. Accordingly, many states currently do not consider steam sterilization to be a new technology and do not require any additional approval as such. It was recommended by the committee that steam sterilization not be included as an "emerging treatment technology" and thus, not be subject to registration and technology approval requirements. Site and operation permits would still be necessitated, as required, under applicable state regulations.

The committee, however, did recognize that the steam sterilization process is subject to waste load variables and operator control which could lead to inadequate processing of the waste. To assist in documenting that the process is effective, the equipment operator should:

- Adopt standard written operating procedures which denote:
  - Sterilization cycle time, temperature, and pressure
  - Types of waste acceptable
  - Types of containers and closures acceptable
  - Loading patterns or quantity limitations
- Document times/temperatures for each complete sterilization cycle.
- Use time/temperature sensitive indicators to visually note the waste has been decontaminated.
- Use biological indicators placed in the waste load (or simulated load) periodically to verify that conditions are met to achieve decontamination.

• Maintain all records of procedure documentation, time-temperature profiles, and biological indicator results.

4.4 Medical Waste Treatment Devices. As stated previously, the committee took the position that level III criteria were applicable to all medical waste treatment devices, including small "counter-top" devices. It was recognized by the committee that registration of all small medical waste treatment devices by the authorized state regulatory agency would be a significant effort in states which do not already have generator and disposal facility registration requirements. To minimize the state's effort, it was suggested that the equipment's manufacturer (or vendor) take responsibility in fulfilling siting requirements as a condition of technology approval. As such, the manufacturer would provide during the technology approval process, all information required for site approval for a typical site for which the equipment is designed. Information required of the small treatment device manufacturer would be similar to the information required of all medical waste treatment equipment manufacturers, but would include all materials and documents required for the user to ensure proper equipment use, operational safety, and treatment technology efficacy. These materials and documents would include:

• An operations and maintenance manual
• Information on proper use, safety precautions and the implications of potential misuse
• Efficacy testing instructions
• A training/education manual
• Available service agreements/programs

On installation of the treatment device, the manufacturer would complete a record of the buyer, the location, and the results of on-site challenge testing at the time of purchase. This information would be submitted annually to the state by the manufacturer as the notification record of site registrations of equipment installed that previous year. The committee recommended that small medical waste treatment devices be specifically identified on initial application for technology approval.

4.5 Waste Residue Disposal. The disposition of waste residues was an environmental concern expressed by many on the committee. To ensure that waste residues are properly identified and disposed of, the committee recommended they be addressed at both the technology approval stage and equipment siting stage of the review process. During the technology approval process, information on the characteristic(s) of the waste residues, the mechanism(s), and the mode(s) of their disposal should be provided by the manufacturer. This information should include:

• A description of residues (i.e., liquid, solid, shredded, hazardous constituents)
• Waste designation (i.e., hazardous, special, general)

- Disposal mechanisms (i.e., landfilling, incineration, recycling)
- Recycling efforts, if anticipated (i.e., waste types, amounts, percentages, name and location of recycling effort)

During the siting stage of the review process, specific information on residue disposal should also be required. This information should include all of the above information, but also specifically state with attached documentation the actual mechanism and location of disposal. To avoid recycling being used as a mechanism to potentially avoid regulatory permitting requirements and to assure that recycling efforts are legitimate, the state should request the following information from the on-site or commercial facility:

- The type of waste residue to be recycled
- The amounts of waste residue to be recycled
- The percentage of the total waste and waste residue to be recycled
- The recycling mechanism used
- The location of the recycler

Previously untreated medical wastes used in the development and testing of prototypical equipment should continue to be considered as potentially infectious and as such, be disposed of as untreated medical waste. To minimize environmental and occupational exposures that may result from using untreated medical wastes, it was recommended that prototypical equipment be tested using non-infectious or previously treated medical waste (i.e., treated by an approved process such as steam sterilization) that has been inoculated with recommended pathogen surrogates. Waste residues generated could then be disposed of as general solid waste after verification of microbial inactivation.

4.6 Operator Training. Mandated operator training was recommended (as appropriate: small treatment devices may be excluded from this recommendation) as a requirement for process approval because of its potential affect on both efficacy and operator safety. To assure proper operation of the treatment process, the manufacturer would be required to provide an operator training program which would include:

- Training and education materials adequately describing the process, process monitors, and safety precautions and controls
- Contingency plans in the event of abnormal occurrences. (e.g., power failure, jamming, inadequate chemical concentrations) and emergencies (e.g., fire, explosion, release of chemical or biohazardous materials)
- Shut-down, clean-out and maintenance procedures
- Personal protective equipment requirements
- A listing of all potential occupational safety and health risks posed by the equipment and its use

The proposed "ASME Standard for the Qualification and Certification of Medical Waste Incinerator Operators" (September 1992) was reviewed for its potential applicability as a guideline for developing required elements for operator training. Although the committee agreed that the proposed standard was far too extensive for emerging medical waste treatment equipment operations, certain components might provide the basis for an operator training program for medical waste treatment technologies.

4.7 Equipment Operations Plan. The proposed "ASME Standard for the Qualification and Certification of Medical Waste Incinerator Operators" (September 1992) offers elements for inclusion into an equipment operations plan. Using this proposed standard as a guide, the following components are recommended for incorporation into an equipment operations plan:

- A description of all mechanical equipment, instrumentation, and power controls
- A description of systems' operations including: acceptable waste types, loading parameters, process monitors, treatment conditions, and disposal
- A description of all parametric controls and monitoring devices, their appropriate settings, established ranges and operating parameters as correlated with biological indicators, and calibration requirements
- A description of the methods required, both to ensure process monitoring instrumentation is operating properly and to prevent tampering with controls
- A description of methods and schedules for periodic calibration of process monitoring instrumentation
- A description of proper mechanical and equipment responses, including identification of system upsets (e.g., power failure, jamming, inadequate treatment conditions) and emergency conditions (e.g., fire, explosion, release of chemical or biohazardous materials)
- A description of personal protective equipment requirements for routine, abnormal and emergency operations
- A thorough description of all potential occupational safety and health risks posed by the equipment and its use
- Specific responsibility assignments for operators:
  - Collecting and organizing data for inclusion into the operating record
  - Evaluating any discrepancies or problems
  - Recommending actions to correct identified problems
  - Evaluating actions taken and documenting improvement

4.8 Emergency and Contingency Response Plan. The development of a separate plan to assist the operating facility in properly responding to an unplanned, emergency, or abnormal event was recommended by the committee. The development of the plan will by necessity, be a shared responsibility between the manufacturer (vendor) and the equipment's user. The primary objectives of this emergency and contingency response plan are:

- To prevent or minimize biological and/or chemical agent release to the environment
- To prevent or minimize biological and/or chemical agent exposure to the equipment operator or other support or maintenance personnel
- To develop contingency medical waste treatment or disposal alternatives for untreated or inadequately treated waste

The plan should take into consideration those events that result in:

- Failure in the treatment technology (e.g., inadequate chemical agent concentration, temperature)
- Mechanical failure (e.g., jammed shredder, inadequate steam pressure)
- Equipment shut-down in mid-cycle
- Spill or release of biological or chemical agents
- Accumulation of untreated or inadequately treated medical waste

As the equipment designer, the manufacturer (vendor) should provide evidence of a failure mode and effect analysis to prevent or minimize inadequate treatment and biological/chemical exposures caused by equipment, process design, process control, and process monitoring failures. This analysis should examine all possible and expected effects of failures, specifying in detail the nature of the effect and causes of action to be taken to prevent biological/chemical exposures. The analysis must examine the effects of failure related to:

- All process controls and process monitoring devices, their appropriate settings, and established ranges and operating parameters
- All parametric controls and associated monitoring devices, their appropriate and established ranges and operating parameters as correlated with biological indicators, and calibration requirements
- Proper mechanical and equipment responses, including identification of upsets or malfunction (e.g., power failure, jamming, inadequate treatment conditions) and emergency conditions (e.g., fire, explosion, release of chemical or biohazardous materials)
- The methods required, both to ensure process and parametric monitoring devices are operating properly and to detect tampering with the devices
- The methods and schedules for periodic calibration of process and parametric control and monitoring instrumentation
- Equipment/inadequately treated waste decontamination procedures required in the event of a mid-cycle shut-down

The equipment user has the responsibility of incorporating the manufacturer-supplied information into a descriptive written emergency and contingency response plan. Additional information to be provided in the plan should at a minimum include:

- A description of all potential occupational safety and health risks posed by the equipment and its use
- A description of proper responses for system upsets and emergency conditions
- A description of personal protective equipment requirements for routine, abnormal, and emergency operations
- A description of proper medical response if required
- A pre-designated disposal method and site for untreated or inadequately treated medical waste if an equipment failure precludes use of the treatment equipment

5.0 Research and Development

The issue of state responsibility and regulation in the research and developmental phase of medical waste technologies was raised. It was recognized that there was a need to develop new technologies, but time, staffing and funding of the permitting state agency might preclude the state's involvement in a research and development project. Concerns raised in state involvement with research and development projects included:

- The process of establishing research and development variances, including limitations and allowances
- The knowledge of and permitting of potential environmental emissions and safety considerations
- Treatment process residue disposal
- Agency funding and staffing

Because of the above concerns, it was the consensus of the committee that each state view as optional its participation in experimental medical waste treatment research and development projects. For those states opting to participate in medical waste treatment technology research and development projects, the concerns raised above were discussed.

To provide a framework for discussion, the committee reviewed language currently proposed by the State of Illinois Environmental Protection Agency (IEPA) for "Experimental Permits" for medical waste treatment technologies. Language as proposed states that the "Agency may issue Experimental Permits" provided that the "applicant can provide proof that the process or technique has a reasonable chance for success." Additionally the IEPA requires evidence that "environmental hazards are minimal" and requires a "description of the type of residuals anticipated and how they will be managed and disposed of." As proposed, the Experimental Permits are to be granted for two years with a one-time renewal based on submittal of application of renewal and a report summarizing equipment performance, efficacy results, and management of residual materials.

In the discussion that followed, the question was raised of how proof can be provided that the equipment has a "reasonable chance of success." It was suggested that proof may consist of data acquired from scaled-down prototypical models or from analogous technologies that have a proven track record. It was noted from the prior discussion that IEPA stated it may issue Experimental Permits allowing the IEPA discretion in granting an experimental permit. To minimize concerns that research and development of a medical waste treatment technology may pose environmental and occupation risks, an application form similar to that required of a technology seeking formal approval might be submitted. The form would request available environmental and occupational safety data in addition to equipment specifications, residue management and disposal, and any available preliminary efficacy data and protocols.

To further minimize environmental and occupational safety concerns that might arise during research and development, it was recommended that the prototypical equipment be tested using non-infectious or previously treated medical waste (i.e., treated by an approved process such as steam sterilization) that has been inoculated with recommended pathogen surrogates. Waste residues generated could then be disposed of as general solid wastes on verification of microbial inactivation. Non-treated medical wastes used during research and development would require agency approved treatment after testing.

Concern that the research and development permit might be used as a mechanism to operate a commercial waste treatment venture was also raised. It was suggested that to avoid this possibility the following statements be adapted into guidance document language:

- Research and Development permits are to be granted for a period of two years with a one-time renewal.
- Granting of a Research and Development permit does not assure future site approval at that site or state approval of the process.
- Research and Development permitted facilities cannot accept waste for monetary gain.
- Research and Development permitted facilities must have any experimentally treated medical waste treated by a state approved medical waste treatment process before disposal or recycling.

Funding of the additional costs incurred by the state as a result of the increased oversight activities associated with a research and development project was also a concern. It was emphasized that the additional requirements of time, staff, and expertise to monitor and review the experimental technology would require that some mechanism (e.g., set fee or time and materials) be established to reimburse the state for these activities.

6.0 Recommendations for Future Activities

It was the committee's hope that these discussions and resultant report would be useful in establishing a nationally recognized foundation for the review and approval of emerging medical waste treatment technologies. To provide future support for the development and implementation of a nationally recognized guideline, the committee recommended:

- The establishment of a research program to evaluate the thermal, chemical and irradiation resistance of *B. subtilis* and *B. stearothermophilus* spores relative to all representative microbial groups for the determination of their use as ultimate pathogen surrogates for medical waste treatment technology efficacy testing
- The establishment of criteria and procedures for emergency and contingency response to ensure adequate equipment decontamination and operator safety in the event of a mid-cycle shut-down or other abnormal occurrence
- The establishment of criteria and testing procedures to monitor the potential release of biological aerosols from alternative medical waste treatment equipment
- Establishment of a clearinghouse to create a network for:
  - Future regulatory activities
  - Integration of technology approvals/denials
  - Information on equipment failures
  - Development of emergency equipment decontamination protocols
  - Provision of access to technical expertise and documentation
  - Assistance to manufacturers in the approval process
  - Protocol review/assessment/development/continuity
- Continued committee discussion and interaction with the USEPA Office of Pesticide Programs as that office further develops its registration requirements and protocols for medical waste treatment technologies using chemical agents
- The expanded integration of health and safety oversight of medical waste treatment activities by state regulatory agencies and professional accrediting associations to include defined oversight responsibilities and inspector training programs

7.0 Selected Bibliography

Agency for Toxic Substances and Disease Registry. The Public Health Implications of Medical Waste: A Report to Congress. U.S. DHHS, Public Health Service, Atlanta, Georgia. 1990.

AOAC Guide to Method Format and Definition of Terms and Explanatory Notes. In: Helrich K., ed., *Official Methods of Analysis of the Association of Official Analytical Chemists,* 15th ed., AOAC, Inc. 1990.

American Public Health Association, et al. *Standard Methods for the Examination of Water and Wastewater,* 18th ed. 1992.

American Society of Mechanical Engineers (ASME). Proposed ASME QMO-1 Standard for the Qualification and Certification of Medical Waste Incinerator Operators. April 1992.

Beloain A. Disinfectants. In: Helrich K., ed., *Official Methods of Analysis of the Association of Official Analytical Chemists,* 15th ed., AOAC, Inc. 1990.

Centers for Disease Control. *Biosafety in Microbiological and Biomedical Laboratories,* 3rd ed. 1993.

Bond WW, Favero MS, Peterson NJ, and Ebert JW. Inactivation of hepatitis B virus by intermediate-to-high level disinfectant chemicals. *J Clin Microbiol,* 18:535–538. 1983.

Favero MS. Sterilization, disinfection, and antisepsis in the hospital. *Manual of Clinical Microbiology,* 4th ed., American Society of Microbiology. 1985.

Favero MS and WW Bond. Chemical disinfection of medical and surgical materials. In: Block SS, ed., *Disinfection, Sterilization and Preservation,* 4th ed. 1991.

Illinois EPA. Notice of Proposed Amendments, Title 35, Subtitle K, Chapter I, Subchapter b: Potentially Infectious Wastes, Part 1420–1440. March 1993.

Spaulding EH. Chemical disinfection and antisepsis in the hospital. *J Hosp Res,* 9:5–31. 1972.

USEPA, Office of Pesticide Programs. Pesticide Assessment Guidelines. Subdivision G— Product Performance. September 1982.

USEPA. Test Methods for Evaluating Solid Waste, Physical/Chemical Methods. EPA Publication SW-846. 3rd ed., 1986 as amended by Update I. November 1990.

40 CFR Part 259 Standards for the Tracking and Management of Medical Waste. *Federal Register* 54(56):12326. March 24, 1989.

Attachment A: State Guideline for Approval of Medical Waste Treatment Technologies

Preface. This guideline summarizes the discussions and results of the State and Territorial Association on Alternate Treatment Technologies. It should be emphasized that the recommendations provided by the Association and adopted by the participating states are an attempt to find commonality on many of the issues and criteria required in the medical waste treatment technology review process. Recognizing that all states may not totally agree with these recommended criteria or protocols, this guideline can serve as a foundation or model for the development of state guidelines or regulations. It is also recognized that definitions, terms, and regulatory methodologies used within the framework of this guideline may not be compatible with granted legislative authority or existing regulatory language. As such, this guideline may require revision to conform with specific state statutes and regulatory requirements.

State Guideline for Approval of Medical Waste Treatment Technologies

A. *Definition of Microbial Inactivation*

    A1. Inactivation is required to be demonstrated of vegetative bacteria, fungi, lipophilic/hydrophilic viruses, parasites, and mycobacteria at a 6 $\log_{10}$

reduction or greater, a 6 $\log_{10}$ reduction is defined as a 6 decade reduction or a one millionth (0.000001) survival probability in a microbial population (i.e., a 99.9999% reduction).

A2. Inactivation is required to be demonstrated of *B. stearothermophilus* spores or *B. subtlis* spores at a 4 $\log_{10}$ reduction or greater; a 4 $\log_{10}$ reduction is defined as a 4 decade reduction or a 0.0001 survival probability in a microbial population (i.e., a 99.99% reduction).

B. *Representative Biological Indicators*

B1. One or more of the following representative microorganisms from each microbial group shall be used to determine if microbial inactivation requirements are met:

(a) Vegetative bacteria

- *Staphylococcus aureus* (ATCC 6538)
- *Pseudomonas aeruginosa* (ATCC 15442)

(b) Fungi

- *Candida albicans* (ATCC 18804)
- *Penicillium chrysogenum* (ATCC 24791)
- *Aspergillus niger*

(c) Viruses

- Polio 2 or polio 3
- MS-2 bacteriophage (ATCC 15597-Bl)

(d) Parasites

- *Cryptosporidium* spp. oocysts
- *Giardia* spp. cysts

(e) Mycobacteria

- *Mycobacterium terrae*
- *Mycobacterium phlei*
- *Mycobacterium bovis* (BCG) (ATCC 35743)

B2. Spores from one of the following bacterial species shall be used for efficacy evaluation of chemical, thermal, and irradiation treatment systems:

(a) *B. stearothermophilus* (ATCC 7953)

(b) *B. subtilis* (ATCC 19659)

C. *Quantification of Microbial Inactivation*

C1. Microbial inactivation ("kill") efficacy is equated to "$\log_{10}$kill" which is defined as the difference between the logarithms of the number of viable test microorganisms before and after treatment. This definition is equated as:

$$\log_{10}\text{kill} = \log_{10}(\text{cfu/g "I"}) - \log_{10}(\text{cfu/g "R"})$$

where:

$\log_{10}$kill is equivalent to the term $\log_{10}$ reduction.

"I" is the number of viable test microorganisms introduced into the treatment unit.

"R" is the number of viable test microorganisms recovered after treatment.

"cfu/g" are colony forming units per gram of waste solids.

C2. For those treatment processes that can maintain the integrity of the biological indicator carrier (i.e., ampules, plastic strips) of the desired microbiological test strain, biological indicators of the required strain and concentration can be used to demonstrate microbial inactivation. Quantification is evaluated by growth or no growth of the cultured biological indicator.

C3. For those treatment mechanisms that cannot ensure or provide integrity of the biological indicator (i.e., chemical inactivation/grinding), quantitative measurement of microbial inactivation requires a two step approach: step 1, "control"; step 2, "test." The purpose of step 1 is to account for the reduction of test microorganisms due to loss by dilution or physical entrapment.

(a) Step 1:

(1) Use microbial cultures of a predetermined concentration necessary to ensure a sufficient microbial recovery at the end of this step.

(2) Add suspension to a standardized medical waste load that is to be processed under normal operating conditions without the addition of the treatment agent (i.e., heat, chemicals).

(3) Collect and wash waste samples after processing to recover the biological indicator organisms in the sample.

(4) Plate recovered microorganism suspensions to quantify microbial recovery. (The number of viable microorganisms recovered serves as a baseline quantity for comparison to the number of recovered microorganisms from wastes processed with the treatment agent).

(5) The required number of recovered viable indicator microorganisms from step 1 must be equal to or greater than the number of microorganisms required to demonstrate the prescribed log reduction as specified in Section A (i.e., a 6 $\log_{10}$ reduction for vegetative microorganisms or a 4 $\log_{10}$ reduction for bacterial spores). This can be defined by the following equations:

$$\log_{10}RC = \log_{10}IC - \log_{10}NR$$
or
$$\log_{10}NR = \log_{10}IC - \log_{10}RC$$

where:

$\log_{10}RC > 6$ for vegetative microorganisms and > 4 for bacterial spores.

$\log_{10}RC$ is the number of viable "control" microorganisms (in colony forming units per gram of waste solids) recovered in the non-treated processed waste residue.

$\log_{10}IC$ is the number of viable "control" microorganisms (in colony forming units per gram of waste solids) introduced into the treatment unit.

$\log_{10}NR$ is the number of "control" microorganisms (in colony forming units per gram of waste solids) which were not recovered in the non-treated processed waste residue. $\log_{10}NR$ represents an accountability factor for microbial loss.

(b) Step 2:
(1) Use microbial cultures of the same concentration as in step 1.
(2) Add suspension to the standardized medical waste load that is to be processed under normal operating conditions with the addition of the treatment agent.
(3) Collect and wash waste samples after processing to recover the biological indicator organisms in the sample.
(4) Plate recovered microorganism suspensions to quantify microbial recovery.
(5) From data collected from step 1 and step 2, the level of microbial inactivation (i.e., "$\log_{10}$kill") is calculated by employing the following equation:

$$\log_{10}kill = \log_{10}IT - \log_{10}NR - \log_{10}RT$$

where:

$\log_{10}$kill is equivalent to the term $\log_{10}$ reduction.

$\log_{10}IT$ is the number of viable "test" microorganisms (in colony forming units per gram of waste solids) introduced into the treatment unit; $\log_{10}IT = \log_{10}IC$.

$\log_{10}NR$ is the number of "control" microorganisms (in colony forming units per gram of waste solids) which were not recovered in the non-treated processed waste residue.

$Log_{10}RT$ is the number of viable "test" microorganisms (in colony forming units per gram of waste solids) recovered in treated processed waste residue.

D. *Efficacy Testing Protocols*

D1. Methodology employed to determine treatment efficacy of the technology will need to assure required microbial inactivation and assure the protocols are congruent with the treatment method. Protocols developed for efficacy testing shall incorporate, as applicable, recognized standard procedures such as those found in USEPA "Test Methods for Evaluating Solid Waste, Physical/Chemical Methods" and APHA et al., *Standard Methods for the Examination of Water and Waste Water*.

D2. The state agency reviewing medical waste treatment technologies (the "Agency") shall prescribe those types and compositions of medical wastes that present the most challenge to treatment effectiveness under normal operating conditions of the equipment reviewed.

D3. Dependent on the treatment process and microbial inactivation mechanisms utilized, protocols evaluating medical waste treatment systems shall specifically delineate or incorporate, as applicable:

(a) Waste compositions that typify actual waste to be processed

(b) Waste types that provide a challenge to the treatment process

(c) Comparable conditions to actual use (i.e., process time, temperature, chemical concentration, pH, humidity, load density, load volume)

(d) Assurances that biological indicators (i.e., ampules, strips) are not artificially affected by the treatment process

(e) Assurances of inoculum traceability, purity, viability and concentration

(f) Dilution and neutralization methods that do not affect microorganism viability

(g) microorganism recovery methodologies that are statistically correct (i.e., sample collection, number of samples/test, number of colony forming units/plate)

(h) Appropriate microbial culturing methods (i.e., avoidance of microbial competition, the selection of proper growth media and incubation times)

E. *Technology Approval Process*

E1. To initiate the technology review process, the manufacturer (vendor) shall complete and submit the "Application for Evaluation and Approval of Medical Waste Treatment Technologies" information request form (See Attachment B) to the Agency. The manufacturer (vendor) shall:

(a) Provide a detailed description of the medical waste treatment equipment to be tested including manufacturer's instructions and equipment specifications, operating procedures and conditions including, as applicable, treatment times, pressure, temperatures, chemical concentrations, irradiation doses, feed rates, and waste load composition.

(b) Provide documentation demonstrating the treatment method meets microbial inactivation criteria and required testing protocols including a detailed description of the test procedures and calculations used in fulfilling required performance standards verifying microbial inactivation, of user verification methodology, and of microbial culturing protocols which ensure traceability, purity and concentration.

(c) Provide information on available parametric controls/monitoring devices, verifying microbial inactivation and ensuring operator non-interference.

(d) Provide documentation of applicable emission controls for suspected emissions.

(e) Provide information relating to waste residues including their potential hazards/toxicities and their specific mode of disposal or recycling.

(f) Provide documentation providing occupational safety and health assurance; and

(g) Provide information on energy efficiency and other potential benefits the treatment technology has to offer to the environment.

E2. The manufacturer (vendor) shall demonstrate that all required pathogen surrogates and resistant bacterial endospores are inactivated to criteria specified in Section A and Section C under all Agency specified challenge waste load compositions.

E3. The manufacturer (vendor) shall develop and demonstrate that site approval and user verification testing protocols are workable and valid.

E4. The manufacturer (vendor) shall demonstrate where technically practical, the relationship between biological indicator data and data procured from real-time parametric treatment monitoring equipment.

E5. The manufacturer (vendor) shall develop contingency response plans and protocols for use in the event of an emergency, accident, or equipment malfunction. The manufacturer (vendor) shall demonstrate that developed protocols are effective in providing operator safety from physical, chemical, or biological exposures during and after the event including decontamination procedures.

E6. The manufacturer (vendor) shall demonstrate evidence of USEPA pesticide registration for those treatment processes that employ a chemical agent to inactivate microorganisms.

E7. Upon demonstration to the Agency's satisfaction, technology approval is granted only under the conditions specified in the manufacturer's instructions and equipment specifications, operating procedures and conditions including, as applicable, treatment times, temperatures, pressure, chemical concentrations, irradiation doses, feed rates, and waste load composition. Revisions to these equipment and operating conditions, as warranted relevant to the Agency, will require re-application for approval to the Agency.

F. *Site Approval Process*

F1. To fulfill microbial inactivation requirements and information requirements for site approval, the equipment user shall:

      (a) Demonstrate that the equipment sited is the same equipment and process approved by the Agency as specified in Section E.

      (b) Demonstrate that required resistant bacterial endospores are inactivated as specified in Section A2 criteria under typical waste load and Agency specified challenge compositions.

      (c) Verify that user verification protocols adequately demonstrate microbial inactivation.

      (d) Verify the relationship between biological indicator data and data procured from real-time parametric treatment monitoring equipment.

F2. The site facility shall provide a written operations plan that includes:

      (a) The names or positions of the equipment operators

      (b) The waste types or categories to be treated

      (c) Waste segregation procedures required

      (d) Waste types prohibited for treatment

      (e) Equipment operation parameters

      (f) Microbial inactivation monitoring procedures

      (g) Shut-down, clean-out and maintenance procedures

      (h) Personal protective equipment requirements

      (i) Operator training requirements

F3. The site facility shall provide a written emergency and contingency response plan that includes:

      (a) A description of proper responses, including identification of system upsets (i.e., power failure, jamming, inadequate treatment conditions) and emergency conditions (i.e., fire, explosion, release of chemical or biohazardous materials)

      (b) A description of personal protective equipment requirements for routine, abnormal, and emergency operations

      (c) A description of all potential occupational safety and health risks posed by the equipment and its use

F4. The site facility shall submit to the Agency for their review:

      (a) Equipment model number and serial number

      (b) Equipment specification and operations manual

      (c) Certification that equipment is identical to the state authorized system

      (d) A copy of the facility's operations plan

      (e) A copy of the facility's emergency and contingency response plan

      (f) Certification documentation of operator training

F5. As a condition of site approval, the Agency shall have a right to inspect the facility and the right to revoke site approval if health and safety violations are discovered, if permit conditions are not being fulfilled, or if the facility is not adhering to its written plans.

F6. Any modifications to the medical waste treatment unit may require re-approval by the Agency and may involve further efficacy testing.

G. *User Verification*

G1. To verify that the medical waste treatment unit is functioning properly and that performance standards are achieved, the equipment user shall:

      (a) Demonstrate that required resistant bacterial endospores are inactivated to criteria as specified in Section A2 under standard operating procedures using protocols that have previously been approved by the Agency as specified under Section E and F.

   (b) Demonstrate adherence to the frequency of biological monitoring specified by the Agency.

   (c) Document and record all biological indicator and parametric monitoring data.

G2. To document microbial inactivation for steam sterilizers and autoclaves, the equipment operator shall:

   (a) Adopt standard written operating procedures which denote:
      (1) Sterilization cycle time, temperature, pressure
      (2) Types of waste acceptable
      (3) Types of containers and closures acceptable
      (4) Loading patterns or quantity limitations

   (b) Document times/temperatures for each complete sterilization cycle.

   (c) Use time–temperature sensitive indicators to visually denote the waste has been decontaminated.

   (d) Use biological indicators placed in the waste load (or simulated load) periodically to verify that conditions meet microbial inactivation requirements as specified in Section A2.

   (e) Maintain all records of procedure documentation, time-temperature profiles, and biological indicator results.

G3. Medical waste incinerators are to be operated, maintained, and monitored as specified in applicable site and operating permits.

H. *Small Medical Waste Treatment Devices*

   H1. All small medical waste treatment devices shall fulfill the requirements necessary for technology approval and shall meet the microbial inactivation requirements as defined in Section A.

   H2. Technology and siting approval are the responsibility of the manufacturer or equipment vendor. The manufacturer (vendor) shall provide to the Agency:

   (a) All information required for technology approval as defined in Section E

   (b) All information required of site approval for a typical site for which the equipment is designed as defined in Section F

   (c) All materials and documents required of the user to ensure proper use, safety, and effective treatment. These materials and documents would include:
      (1) An operations and maintenance manual
      (2) Information on proper use and potential misuse
      (3) Microbial inactivation testing instructions
      (4) Training/education manual
      (5) Available service agreements/programs

   H3. The manufacturer (vendor) shall furnish the user of the treatment device:

   (a) An operations and maintenance manual

   (b) Information on proper use and potential misuse

    (c) Microbial inactivation testing instructions

    (d) Training/education manual

    (e) Available service agreements/programs

H4. Upon the installation of the treatment device, the manufacturer shall compile a record of the buyer, the location, and the results of on-site challenge testing at time of purchase. This information shall be submitted annually to the Agency by the manufacturer (vendor) as the notification record of site registrations of equipment installed that previous year.

I. *Previously Approved Technologies*

  I1. Medical waste treatment equipment which is subject to these registration and technology approval requirements that has been installed and operated before January 1, 1994, shall comply with current efficacy standards by (date). By (date), pre-existing medical waste treatment equipment shall have been modified to meet current standards, taken out of service, or replaced by approved equipment.

  I2. Steam sterilizers, autoclaves, and incinerators are not included within the category of "emerging treatment technologies" and are not subject to these registration and technology approval requirements. Site and operation permits are still necessitated, as required, under applicable state regulations.

J. *Waste Residue Disposal*

  J1. Information on the characteristic(s) of all waste residues (liquids and solids), and the mechanism(s) and model(s) of their disposal shall be provided by the manufacturer on the "Application for Evaluation and Approval of Medical Waste Treatment Technologies." This information shall include:

    (a) Description of residues (i.e., liquid, solid, shredded, hazardous constituents)

    (b) Waste designation (i.e., hazardous, special, general)

    (c) Disposal mechanism (i.e., landfilling, incineration, recycling)

    (d) Recycling efforts, if anticipated, (i.e., waste types, amounts, percentages, name and location of recycling effort)

  J2. Information on waste residue disposal shall be provided by the user facility as required under site approval (Section F). This information shall include:

    (a) All information requested in Section J1

    (b) The disposal site (name and address)

    (c) The mechanism of disposal (i.e., landfilling or incineration)

    (d) The amounts of residue(s) anticipated to be disposed of (e.g., volume and weight per week)

  J3. If residue(s) are to be recycled, the following information shall be provided by the user facility as required under site approval (Section F). This information shall include:

    (a) The types of waste residue to be recycled

    (b) The amounts of waste residue to be recycled

    (c) The percentage of the total waste and waste residue to be recycled

    (d) The recycling mechanism used

    (e) The name and location of the recycler

J4. Previously untreated medical wastes used in the development and testing of prototypical equipment shall be considered potentially infectious and will be required to be disposed of as untreated medical waste.

J5. Prototypical equipment testing using non-infectious or previously treated medical waste (i.e., treated by an approved process such as steam sterilization) that has been inoculated with recommended pathogen surrogates can be disposed of as general solid waste after verification of microbial inactivation.

J6. All liquid and solid waste residues will be disposed of in accordance with applicable state and local regulations.

K. *Operator Training*

K1. To assure proper operation of the treatment process, the manufacturer (vendor) shall provide to the user as part of the treatment equipment purchase an operator training program which shall include:

  (a) A description of all mechanical equipment instrumentation, and power controls

  (b) A description of system operations including waste types acceptable, loading parameters, process monitors, treatment conditions, and residue disposal procedures

  (c) A description of all parametric controls and monitoring devices, their appropriate settings as correlated with biological indicators, and calibration requirements

  (d) A description of proper responses, including identification of system upsets (i.e., power failure, jamming, inadequate treatment conditions) and procedures to be followed during emergency conditions (i.e., fire, explosion, release of chemical or biohazardous materials)

  (e) A description of the procedures for equipment shut-down and clean-out for maintenance or other purposes

  (f) A description of personal protective equipment requirements for routine, abnormal, and emergency operations

  (g) A description of all potential occupational safety and health risks posed by the equipment and its use

K2. The facility shall develop a written equipment operations plan which shall include:

  (a) Delegation of responsibility for safe and effective equipment operation to operating personnel

  (b) A description of operating parameters that must be monitored to ensure microbial inactivation

  (c) A description of all process monitoring instrumentation and established ranges for all operating parameters

  (d) A description of the methods required to ensure process monitoring instrumentation is operating properly

  (e) A description of methods and schedules for periodic calibration of process monitoring instrumentation

  (f) A description of the procedures for equipment shut-down and clean-out for maintenance or other purposes

K3. The facility shall develop a written contingency and emergency response plan to include:

(a) A description of all potential occupational safety and health risks posed by the equipment and its use

(b) A description of proper responses for system upsets and emergency conditions

(c) A description of personal protective equipment requirements for routine, abnormal, and emergency operations

(d) A description of proper medical response if required

(e) A predesignated disposal site for untreated or inadequately treated medical waste if a mechanical failure precludes use of the treatment equipment

K4. The facility shall document and keep on record copies of all training for at least 3 years.

L. *Research and Development*

L1. The Agency may issue an Experimental Permit for medical waste treatment processes or techniques that are undergoing research and development if the applicant can provide evidence that:

(a) Environmental impact is minimal

(b) Occupational exposures are minimal.

L2. The Agency's "Application for Evaluation and Approval of Medical Waste Treatment Technologies" information request form shall be submitted and shall contain environmental and occupational safety data in addition to equipment specifications, residue management and disposal, and any available preliminary microbial inactivation data and protocols.

L3. All equipment testing shall preferably use non-infectious or previously treated medical waste (i.e., treated by an approved process such as steam sterilization) that has been inoculated with recommended pathogen surrogates listed in Section B. Waste residues generated can be disposed of as general solid wastes upon verification of microbial inactivation. Untreated medical wastes used in the development and testing of prototypical equipment shall be considered potentially infectious and will be required to be disposed of as untreated medical waste.

L4. All Experimental Permits have a duration not to exceed two years with a one-time renewal.

L5. Granting of an Experimental Permit does not assure future site approval on state approval of the process.

L6. Facilities with Experimental Permits cannot accept waste for monetary gain.

Attachment B: Application for Evaluation and Approval of Medical Waste Treatment Technologies

The "Application for Evaluation and Approval of Medical Waste Treatment Technologies" is provided as a guidance document to assist state agencies in reviewing new medical waste treatment technologies. The document is intended to serve only as a model for state development of initial application forms by providing a general format of pertinent technology review questions. Definitions and terms used in this document may require revision to conform with specific state legislative and regulatory requirements.

Application for Evaluation and Approval of Medical Waste Treatment Technologies

Complete the following questionnaire and return it along with the application. Please include any additional support data which may be applicable. Use additional paper if necessary. Reference with the related section number(s).

A. *General*

A1. Is the treatment technology best suited for on-site use at the point of generation, or is it adaptable for use as a commercial or regional treatment process receiving waste from several generators?

On-site_____   Commercial/regional_____   Both_____

A2. Is this treatment technology specified for use at small generator facilities such as physician, dental, or veterinary offices or clinics?

Yes_____   No_____

A3. Has this treatment technology been approved/disapproved in any other state? Is so, please indicate which states have issued a decision and submit copies of approvals/disapprovals.

A4. Has the use of this equipment ever resulted in any environmental or occupational safety violation (federal, state, or local)?

A5. Has the use of this equipment ever resulted in injuries, of any kind, or transmission of any disease to any person? Describe all such instances.

A6. Have you reviewed all applicable state solid and medical waste regulations for medical waste acceptance, treatment, and disposal?

A7. Have you inquired as to whether any other permits are required? Please enclose agency response and requirements with your application. List all required permits and enclose copies of any permit approvals.

NOTE: Local governments or other agencies may require permits.

B. *Level of Treatment*

B1. Does the level of microbial inactivation achieved by the treatment process meet the following definition?

"Inactivation of vegetative bacteria, fungi, lipophilic/hydrophilic viruses, parasites, and mycobacteria at a 6 $\log_{10}$ reduction or greater; and inactivation of *Bacillus stearothermophilus* spores or *Bacillus subtilis* spores at a 4 $\log_{10}$ reduction or greater."

Yes_____   No_____

If no, specify where the definition is unfulfilled.

C. *Characterization of Proposed Treatment Process*

    C1. Please check the appropriate categories that best describe the methods of this proposed technology. Proposed treatment technologies may incorporate several of the categories listed below.

Chemical	————	Heat	————
Mechanical	————	Shredder	————
Microwave	————	Grinder	————
Hammermill	————	Irradiation	————
Plasma arc	————	Radiowave	————
Encapsulation	————		
Other (specify)	————————————————————		

D. *Waste Compatibility with Proposed Treatment Process*   Please identify if the proposed system is compatible or noncompatible with the following types of waste.

Type of waste	Compatible	Noncompatible
D1. Cultures and stocks of infectious agents and associated biologicals	————	————
D2. Liquid human and animal waste including blood and blood products and body fluids	————	————
D3. Pathological waste	————	————
D4. Contaminated waste from animals	————	————
D5. Sharps	————	————
D6. Other ————————————	————	————

Please refer to the state medical waste regulations for further definition of the medical waste categories and prescribed medical waste management requirements.

    D7. What waste characteristics present the most challenge to the proposed treatment process:

Organic materials	————	
Liquids	————	
Density/compaction	————	
Other characteristics	————	Specify: ————————————

    D8. Describe by composition (i.e., material and percentage) those medical wastes that would pose the most challenge to the proposed technology. Why?

D9. Describe the physical or chemical components of medical wastes that would interfere, cause mechanical breakdown, or compromise the treatment process or microbial inactivation efficacy.

E. *Microbiological Test Procedures* Any proposed treatment method shall be capable of inactivating vegetative bacteria, fungi or yeasts, parasites, lipophilic/hydrophilic viruses, and mycobacteria at a 6 $\log_{10}$ reduction or greater. Bacterial spores shall be inactivated at a 4 $\log_{10}$ reduction or greater. A representative from each of the following microbial groups is required for testing.

E1. Listed below are several test organisms which have been used as microbiological indicators to determine the effectiveness of a given treatment method. If there are any data either to support or refute the inactivation of any of the biological indicators using the proposed treatment process under normal operating conditions, please check the appropriate space next to the indicator.

NOTE: If protocols utilized by the applicant to generate microbial inactivation data are deemed unacceptable by the Department, the Department reserves the right to request that the applicant resubmit data generated from Department-approved protocols. If data has not yet been procured to support the inactivation of the listed biological indicators below, please contact the Department before initiating efficacy testing to ensure research protocols are in accordance with the Department's requirements.

Vegetative bacteria

* *Staphylococcus aureus* (ATCC 6538)
* *Pseudomonas aeruginosa* (ATCC 15442)

Fungi

* *Candida albicans* (ATCC 18804)
* *Penicillium chrysogenum* (ATCC 24791)
* *Aspergillus niger*

Viruses

* Polio 2 or polio 3
* MS-2 bacteriophage (ATCC 15597-B1)

Parasites

* *Cryptosporidium* spp. oocycsts
* *Giardia* spp. cysts

Mycobacteria

- *Mycobacterium terrae*
- *Mycobacterium phlei*
- *Mycobacterium bovis* (BCG) ATCC 35743)

Bacterial spores

- *Bacillus stearothermophilus* (ATCC 7953)
- *Bacillus subtilis* (ATCC 19659)

E2. Were the results certified by an independent public health or certified testing laboratory?

Yes_____   No_____

If yes, indicate the name, address, and telephone number of the certifying laboratory and attach the test protocol, results and an explanation of any available data not supporting the reduction factors referenced above.

F. *By-Products and Discharges of the Treatment Process*
   F1. Please indicate all by-products and discharges (to air, water, or land) which may be generated as a result of this alternate treatment technology.

Stack emissions_____   Heat_____   Slag_____

Vapors or fumes_____   Ash_____   Liquid_____   Smoke_____

Aerosols_____   Leachate_____   Dust_____   Odor_____

Steam_____   Chemical residues_____

Other (specify)_____

   F2. If any of the above by-products or discharges are indicated, how will they be controlled?
   F3. If there are no by-products or discharges indicated, how was this determined?
   F4. Are any of these by-products or discharges USEPA-listed hazardous wastes (40 CFR Part 261), biohazardous, etc.?

No_____   Yes_____

If yes, explain necessary controls, personal protective equipment, storage, disposal, etc.

G. *Environmental Effects of the Treatment Process*
   G1. Are any negative effects on the environment anticipated from the use of the treatment process and/or disposal of the treated waste from the treatment process?
   G2. What environmental, occupational, and/or public health hazards would be associated with a malfunction of the treatment process? Specify.

G3. If the treatment process includes the use of water, steam, or other liquids, how will this waste discharge be handled (i.e., sewer, recycled, etc.)? Specify.

G4. What are the physical characteristics of the waste residues generated from the treatment process (i.e.., wet, dry, shredded, powdered, etc.)? Specify.

G5. How will the treated medical waste from this process be disposed of (i.e., landfill, incineration, recycled, etc.)? Specify.

G6. Are any by-products classified as hazardous waste (40 CFR Part 261)?

Yes_____ No_____ —Complete Item A6.

H. Occupational Hazards

H1. What are the potential hazards associated with the treatment process?

H2. What hazard abatement/reduction strategies will be used in during the operation of this treatment process (include engineering controls, personal protection equipment, etc.)?

H3. What training will the operator(s) of the treatment process receive?

I. Critical Factors of the Treatment Process

I1. What are the critical factors that influence the specific treatment technology? Specify.

I2. What are the consequences if these factors are not met? Specify.

I3. Explain the ease and/or difficulty of operation of the medical waste treatment system. Specify.

I4. What type of ongoing maintenance is required in the operation of the treatment system? Specify.

Maintenance manual attached? Yes_____ No_____

I5. What emergency measures would be required in the event of a malfunction? Specify.

I6. How are these measures addressed in an emergency plan or in the operations protocol?

I7. What is the maximum amount of waste to be treated by this process per cycle?

I8. How long is a cycle?

J. Chemical Inactivation Treatment Processes

J1. If the treatment process involves the use of chemical inactivation:
   (a) What is the name of the active ingredient?
   (b) What concentrations must be used and maintained?
   (c) At what pH is the chemical agent active?
   (d) What is the necessary contact time?
   (e) If there is any incompatibility with specific materials and surfaces, specify.
   (f) What is the pH of any end products (i.e., liquid effluents)?
   (g) List any additional factors or circumstances that may interfere with the chemical's inactivation potential.

J2. What is the active life of the chemical agent after it has been exposed to air or contaminated medical waste?

J3. Have studies been conducted relative to the long-term effectiveness of the chemical agent while in use? If yes, please attach a copy of the study and test results.

J4. What health and safety hazards may be associated with the chemical (present and long-term)? Specify.

    MSDS attached?   Yes_____   No_____

J5. Is the chemical agent registered for this specific use with the Environmental Protection Agency (USEPA) Pesticide Registration Division?
Yes_____   No_____

    If yes, provide the USEPA registration number _____ and a copy of the EPA-approved label instructions for use.

J6. Is the spent chemical agent classified as a hazardous waste by USEPA (40 CFR Part 261) or by other state criteria?   Yes_____   No_____

    If yes, specify whether by USEPA or by which state(s):

J7. Is an environmental impact study for the chemical agent available?
Yes_____   No_____
If yes, attach a copy of this information.

K. *Quality Assurance and Verification of Microbial Inactivation*

K1. How is the quality assurance of the treatment process addressed? Specify.

K2. What is the recommended frequency that a microbiological indicator should be used to confirm effectiveness of the system? Specify.

K3. Other than the biological indicators listed in Section E, what other indicators, integrators, or monitoring devices would be used to show that the treatment unit or process was functioning properly? (Please describe and explain.)

K4. How is it determined that the processed waste has received proper treatment? (Check the appropriate item.)
Temperature indicator:

    Visual only_____   Continuous_____   Both_____
    Pressure indicator:

    Visual only_____   Continuous_____   Both_____
    Time indicator:

    Visual only_____   Continuous_____   Both_____
    Chemical concentration indicator:

    Visual only_____   Continuous_____   Both_____

    Other: specify_____

K5. How have the treatment process monitors been correlated with biological indicators to ensure effective and accurate monitoring of the treatment process? Specify.

K6. What is the established process monitor calibration schedule, and what is its frequency of calibration?

K7. How are the process monitors interfaced to the system's operations to effect proper treatment conditions? Explain.

K8. How are the process monitor controls secured to prevent operator override of the process before treatment is adequately affected? Explain.

K9. What failure mode and effect analyses have been performed on the treatment system? Specify and provide.

L. *Post-treatment Residue Disposal, Reclamation, or Recycling*

L1. How will the treated medical wastes from this process be disposed of:

Burial in an approved landfill_____

Incineration_____

Recycled_____

L2. If the wastes are to be recycled, provide additional evidence regarding this strategy.

L3. If the wastes are to be recycled, what percentage of the treated waste will recycled? How will the remainder of the treated waste be disposed of?

M. *Potential Environmental Benefits*

M1. Has an energy analysis been conducted on the proposed technology?

Yes_____   No_____
If yes? Specify and provide results of that analysis.

M2. Has an economic analysis been performed on the proposed technology?

Yes_____   No_____
If yes? Specify and provide results of that analysis.

M3. How does this treatment technology improve on existing medical waste treatment and disposal methods? Specify.

M4. What is the potential of this proposed technology for waste volume reduction? Specify.

N. *Other Relevant Information and Comments*   All approvals or denials received from other states, counties or agencies concerning any aspect of equipment operation and efficacy; as well as all safety, competency or training requirements for the users/operators, etc. must also be included.

0. *Certification Statement*

**Application for Evaluation and Approval of
Medical Waste Treatment Technologies**
Certification Statement

I certify that the information requested and contained in this document is accurate and complete and that all existing documentation requested in this application for this system or similar systems is provided. The Vendor, identified below, agrees to provide [state agency] all results of all studies conducted by or for any state, company, agency or country, or any other person as defined at [state regulation], which the vendor conducts, or is in any way aware of, to determine the operational performance of any aspect of the equipment for which authorization to operate in this state is requested on the filing of this application. I am aware that regulated medical waste management systems to be operated in this state for regulated medical waste treatment and/or destruction must be identical to the system described in this application for authorization to operate in this state and for which operational data is presented in the application for [state agency] review. Any and all changes in the system and related equipment after this application submittal and [state agency] review and authorization to operate must be submitted in writing to [state agency] prior to use. The [state agency's] permitting conditions or other agency's authorizations granted to operate this system to treat and/or destroy regulated medical waste will be reviewed by [state agency] periodically to ensure specifically authorized regulated medical waste technology systems meet currently accepted standards for regulated medical waste management. [State Agency] may modify system operational or performance requirements for systems that received prior authorizations to operate, if warranted to protect human health and the environment.

I am further aware that on reviewing the completed application and the required attachments, [state agency] may have additional questions and require submissions of data and other information deemed necessary regarding this or related medical waste disposal systems. Failure to provide all existing requested information will result in delays in processing the request for authorization to operate. Failure to provide all required information as outlined in the application, or willfully withholding information, may be cause for [state agency] to deny or rescind authorization to operate if [state agency] determines that the information not submitted would have been in any way relevant to its review of this technology.

Name of system equipment:
        Model number:

Name of certifying person (must be a corporate officer):
        Title:

Signature of certifying person (must be a corporate officer):
        Date:

Name of person completing application:
        Title:

Name of vendor (company):
        Telephone:
        Fax:

Name of division:
        Address:
        City, state, & zip code:

Attachment C: Example: Microbial Inactivation Testing Protocol for a Grinder/Chemical Medical Waste Inactivation Process

Preface. The following protocol is provided as an example of the steps and procedures required to determine the level of microbial inactivation of a system that cannot ensure or provide integrity of the biological indicator carrier (i.e., test strip, ampule) through the treatment process to recovery. This protocol is not intended to be all inclusive or meet all the variables or constraints associated with the multiplicity of medical waste treatment technologies. However, the protocol includes the components and the processes that require consideration to ensure the data recovered and numeric calculations made accurately represent the true microbial inactivation level of the treatment process.

This example provides a protocol for a chemical inactivation/grinding medical waste treatment process that does not allow the retrieval of the biological indicator carrier. For each step in the protocol, an explanation or note is offered (in brackets) to provide rationale or background for the step or process described. For the protocol provided, adherence to good microbial and laboratory practices is essential for researcher and equipment operator safety and for the generation of accurate data.

Example: Microbial Inactivation Testing Protocol for a Grinder/Chemical Medical Waste Treatment Process

I. Materials

A. *Bacillus stearothermophilus* spores as a suspension of 2×10^{10} initial inoculum. *Note: B. stearothermophilus* spores were chosen as the spore of choice due to the thermophilic nature of *B. stearothermophilus* and its ability to optimally grow at elevated temperatures. Culturing collected waste samples at 60°C using *B. stearothermophilus* spores as a biological indicator reduces the number of potential cross contaminants that might arise on a culture plate. A spore suspension of 2×10^{10} initial inoculum was chosen to provide an adequate number of recoverable spores for determining a $4 \log_{10}$ reduction. Determination of this concentration may require trial runs to ascertain the recovery concentrations.

B. Surrogate waste load constructed to contain by weight: 5% organic material and 95% plastics, cellulose, and glass. Total weight of sample to be between 15 and 20 pounds. *Note:* The surrogate waste load used in this example was constructed to represent the typical medical waste composition that would be treated by this system at the user site location. Surrogate waste loads may also be constructed to replicate medical waste loads which challenge the efficacy of the system. The sample weight of the load was selected as being representative of the feed rate and typical loading conditions of the unit. Weight loads should be constructed to mimic conditions of actual use.

II. Protocols

A. Control run

1. Add 2×10^{10} *B. stearothermophilus* spore suspension to surrogate waste load. The spore suspension should be added as to not expose the researcher or equipment operator to the biological indicator. To minimize potential exposures and to adequately disperse the spore suspension throughout the load, the spore suspension could be transferred into four or more separate plastic screw-capped tubes. These tubes could subsequently be equally dispersed throughout the surrogate waste load.

2. Load inoculated surrogate waste into the previously cleaned (decon-

taminated) treatment unit and run unit without chemical inactivation agent. [The unit should be previously decontaminated to minimize cross contamination from spores originating from previous efficacy testing.]

3. Collect ten one (1) gram samples during the duration of the run (i.e., collect samples at the beginning of waste discharge through final discharge). *Note:* The amount, number and collection frequency of the sample collection will be determined previously by trial runs. The important consideration for this determination is to ensure that during the span of the run, the test data collected provide an accurate reflection of the level of microbial inactivation for the entire load.

4. Place the 1-gram samples immediately upon collection into pre-weighed (combination weight of both liquid and tube) plastic screw cap tubes containing an appropriate neutralizing solution and vortex vigorously for 5 minutes. *Note:* This step is required to neutralize chemical agent activity at the time the waste exits the unit and is necessary to determine actual microbial inactivation by the treatment process and minimize the inclusion of residual chemical activity that might be present. The amount, concentration, and exposure time of the selected neutralizing agent must be pre-determined so as to neutralize the specific chemical agent without inhibiting growth of the biological indicator. Collection tubes are pre-weighed, including neutralizing agent to determine the weight of the actual waste sample collected.

5. Construct an approximate 10-gram composite sample from the 10 representative samples collected in step 3. [This step provides for the evaluation of the level of microbial inactivation of the entire load without assaying each individual sample taken above.]

6. Decant, sieve, and filter as required to separate solid waste material from the neutralizing liquid. Save liquid effluent. [This step is required to wash bacterial spores from the collected waste sample. Protocols involved in this rinsing step will be determined by trial runs to ascertain the best mechanisms to adequately rinse and separate the solid waste components from the liquid rinse.]

7. Wash and vortex solid materials, a second time with neutralizing buffer. Decant, sieve, and filter as required to separate solid waste material from liquid. Combine liquid effluent with that obtained in step 6. [This step provides an extra wash to collect from the waste as many of the spores as possible.]

8. Filter liquid through Millipore filtration unit or equivalent to concentrate retrieved spores on membrane filter. Wash filter with 10 mls of citrate or other appropriate buffer. [This step concentrates retrieved spores to equal the number of spores from 10 grams waste/10 mls buffer or by factoring, the number of spores from 1 gram waste per 1 ml buffer. For example, plating one ml of the liquid would result in the number of cfu on the plate to be equal to the number spores per one gram of waste.]

(a) Triplicate plate 0.1 ml from the 10 ml concentrate in step 8 above; this dilution represents plate A. [This step equates to a total dilution of 1:10.]

(b) Add 1.0 ml of the 10 ml concentrate in step 8 above to 9.0 ml of buffer solution (this represents a 1:10 serial dilution and is represented as dilution tube B). Triplicate plate 0.1 ml of dilution tube B; this dilution represents plate B. [This step equates to a total dilution of 1:100.]

(c) Add 1.0 ml of dilution tube B above to 9.0 ml of buffer solution (this represents an additional 1:10 serial dilution and is represented as dilution tube C). Triplicate plate 0.1 ml of dilution tube C; this dilution represents plate C. [This step equates to a total dilution of 1:1000.]

(d) Add 1.0 ml of dilution tube C above to 9.0 ml of buffer solution. This represents an additional 1:10 serial dilution and is represented as dilution tube D). Triplicate plate 0.1 ml of dilution tube D; this dilution represents plate D. [This step equates to a total dilution of 1:10,000.]

B. Test run
 1. Follow protocols in Section II.A, except run the treatment unit with specified chemical inactivation agent concentrations.
 2. Upon washing the membrane filter in step II.8 with 10 ml of buffer:
    (a) Triplicate plate 1 ml of buffer in step 2 above via the pour plate method (i.e., 1 ml of spore concentrate into 10 to 12 ml of liquid agar). Vortex and pour into plate; this represents plate A'. [This step equates to no dilution factor, i.e., this number represents the number of spores per gram of waste.]
    (b) Triplicate plate 0.1 ml of buffer in step 2 above via the pour plate method (i.e., 0.1 ml of spore concentrate into 10 to 12 ml of liquid agar). Vortex and pour into plate; this represents plate B'. [This step equates to a 1:10 dilution factor.]
    (c) Add 1.0 ml of the buffer in step 2 above to 9.0 ml of buffer solution [this represents a 1:10 serial dilution and is represented as dilution tube C']. Triplicate plate 0.1 ml of dilution tube C'; this dilution represents Plate C'. [This step equates to a total dilution of 1:100.]

III. *Calculations*   Using the equations found in Section C3 of "State Guideline for Approval of Alternate Medical Waste Treatment Technologies" (see Attachment A), the following calculations are performed:

A. Calculate initial inoculum in spores per gram waste.
 1. 2×10^{10} spores/15 lb waste =
    2×10^{10} spores/6.8×10^{3} grams waste =
    3×10^{6} spores/gram waste = inoculum = IC)

$$IC = 3 \times 10^{6}$$

B. Calculate number of spores recovered.
  1. Step 1: "Control" data:

	a	b	c
Plate A	TMTC*	TMTC	TMTC
Plate B	TMTC	TMTC	TMTC
Plate C	TMTC	TMTC	TMTC
Plate D	200 cfu**	210 cfu	190 cfu

  *Too many to count.
  **Colony forming units.

Accounting for the dilution factor of 10,000 for plate D, the average recovery of viable "control" spores per gram equals 200 × 10,000 or 2,000,000 spores/gram or 2×10^6 spores/gram.

$$RC = 2 \times 10^6$$

  2. Step 2: "Test" results:

	a	b	c
Plate A	50 cfu	48 cfu	52 cfu
Plate B	5 cfu	4 cfu	6 cfu
Plate C	1 cfu	0 cfu	0 cfu

The average recovery of viable "test" per gram equals 50 spores per gram (no dilution factor).

$$RT = 5 \times 10^1$$

C. Calculate $\log_{10}$ reduction.
  1. Step 1: "Control" results:

$\log_{10}RC = \log_{10}IC - \log_{10}NR$; where

$\log_{10}RC = \log_{10}(2 \times 10^6 \text{ spores/gram}) = 6.301$
$\log_{10}IC = \log_{10}(3 \times 10^6 \text{ spores/gram}) = 6.477$
$\log_{10}NR = \log_{10}IC - \log_{10}RC$
$\log_{10}NR = 6.477 - 6.301 = 0.176$

$\log_{10}NR = 0.176$

2. Step 2: "Test" results and $\log_{10}$kill calculation:
   (a) $\log_{10}$kill $= \log_{10}$IT $- \log_{10}$NR $- \log_{10}$RT, where:

   $\log_{10}$IT $= \log_{10}$IC $= 6.477$
   $\log_{10}$NR $= 0.176$
   $\log_{10}$RT $= \log_{10}(5 \times 10^1) = 1.699$
   (b) $\log_{10}$reduction ($\log_{10}$kill), where:

   $\log_{10}$kill $= 6.477 - 0.176 - 1.699 = 4.602$
   $\log_{10}$kill $= 4.602$

Attachment D: List of Participants in Roundtable Discussions in New Orleans, Atlanta, and Washington, D.C.

Federal Agencies

Centers for Disease Control and Prevention

Richard Knudsen, Ph.D., Chief Biosafety
Centers for Disease Control and Prevention
1600 Clifton Road, NE
Mail Stop F05
Atlanta, GA 30333
Telephone: (404) 639-3238

Food and Drug Administration

Timothy Ulatowski
Associate Director for General Devices
Food and Drug Administration
1390 Pickard Dr.
Rockville, MD 20850
Telephone: (301) 427-1307/Fax: (301) 427-1977

National Institutes of Health

Edward A. Pfister, R.S., M.S.P.H.
Environmental Health Specialist
National Institutes of Health
Bldg No. 13, Room 3W64
9000 Rockville Pike
Bethesda, MD 20892
Telephone: (301) 498-7990

Ronald Trower
Occupational Health and Safety Specialist
National Institutes of Health
Bldg No. 13, Room 3KO4
9000 Rockville Pike
Bethesda, MD 20892
Telephone: (301) 496-2346

U.S. Department of Transportation

George E. Cushmac, Ph.D.
U.S. Department of Transportation
RSPA
Mail Stop DHM22
400 7th Street, SW
Washington, DC 20590-0001
Telephone: (202) 366-4545/Fax: (202) 366-3753

Eileen Martin
U.S. Department of Transportation
RSPA, Room 8100
Mail Stop DHM12
400 7th Street, SW
Washington, DC 20590

Phillip T. Olson, P.E. C.I.H.
U.S. Department of Transportation
RSPA, Room 8100
Mail Stop DHM22
400 7th Street, SW
Washington, DC 20590
Telephone: (202) 366-4545

Jennifer Posten
U.S. Department of Transportation
RSPA
Office of Hazardous Materials Standards
Washington, DC 20590
Telephone: (202) 366-4488

U.S. Environmental Protection Agency

Robin Biscaia
U.S. EPA Region I
RCRA Support Section
HRW CAN 3
One Congress Street
Boston, MA 02203
Telephone: (617) 573-5754/Fax: (617) 573-9662

Srinivas Gowda
U.S. EPA, Registration Division
(H7505C)
2401 M Street, SW
Washington, DC 20460
Telephone: (703) 305-6845/Fax: (703) 305-5786

Sid Harper
U.S. EPA Region IV, Office of Solid Waste
345 Courtland Street, NE
Atlanta, GA 30365
Telephone (404) 347-2091/Fax: (404) 347-5205

Kristina L. Meson
U.S. EPA, Office of Solid Waste
(OS-332)
401 M Street, SW
Washington, DC 20460
Telephone: (202) 260-5736/Fax: (202) 260-0225

Zig Vaituzis
U.S. EPA, Antimicrobial Branch
(H7505C)
Office of Pesticide Programs
401 M Street, SW
Washington, DC 20460
Telephone: (703) 305-7167/Fax: (703) 305-5786

Michaelle Wilson
Chief, Special Wastes Section
U.S. Environmental Protection Agency
Office of Solid Waste
401 M Street, SW
Washington, DC 20160

State Agencies

California

John P. Winn, R.E.H.S.
Environmental Health Specialist V
Supervisor, Medical Waste Management
California Department of Health Services
Environmental Management Branch
Environmental Health Services Section
601 North 7th Street
P.O. Box 942732
Sacramento, CA 94234-7320
Telephone: (916) 324-2206/Fax: (916) 323-9869

Delaware

Indra Batra
Delaware Department of Natural Resources
& Environmental Control
Division of Air & Waste Management
P.O. Box 1401
89 Kings Highway
Dover, DE 19903
Telephone: (302) 739-3822

Illinois

Douglas W. Clay, P.E.
Illinois EPA
2200 Churchill Road
P.O. Box 19276
Springfield, IL 62794-9276
Telephone: (217) 524-3300/Fax: (217) 524-3291

Louisiana

Charles H. Anderson
Sanitarian Program Manager
Office of Public Health
Louisiana Department of Health and Hospitals
P.O. Box 60630
325 Loyola Avenue
New Orleans, LA 70160
Telephone: (504) 568-8343/Fax: (504) 568-5119

Mary Lou Austin
Department of Environmental Quality
Solid Waste Division
7290 Bluebonnet
Baton Rouge, LA 70810
Telephone: (504) 765-0249

Henry B. Bradford, Jr., Ph.D.
Health Laboratory Director
Louisiana Department of Health and Hospitals
Office of Public Health
Division of Laboratory Services
325 Loyola Avenue, Room 709
New Orleans, LA 70112
Telephone: (504) 568-5375

Carolyn Dinger
Environmental Program Manager
Louisiana Department of Environmental Quality
Office of Solid and Hazardous Waste
Solid Waste Division
P.O. Box 82178
Baton Rouge, LA 70884-2178
Telephone: (504) 765-0249

Bobby G. Savoie
Office of the Secretary
Department of Health and Hospitals
1201 Capitol Access Rd. - 3rd Floor
Baton Rouge, LA 70802

Louis Trachtman, M.D.
Louisiana Department of Health and Hospitals
325 Loyola Avenue
New Orleans, LA 70112
Telephone: (504) 568-5050

Maine

Scott Austin
Maine Department of Environmental Protection
Bureau of Hazardous Material and Solid Waste Control
State House Station #17
Augusta, ME 04333-0017
Telephone: (207) 287-2651/Fax: (207) 287-7826

Maryland

Beverly A. Collins, M.D.
Department of Health and Mental Hygiene
Maryland Health Department
Office of Licensing and Certification Programs
4201 Patterson Avenue
Baltimore, MD 21215

Bill Dorrill, Deputy Director
Department of Health and Mental Hygiene
Maryland Health Department
Office of Licensing and Certification Programs
4201 Patterson Avenue
Baltimore, MD 21215

Patricia Meinhardt, M.D., M.P.H.
Maryland Department of Health
210 West Preston Street
Baltimore, MD 21201
Telephone: (301) 225-6677

Steven T. Wiersma, M.D., M.P.H.
Maryland Department of Environment
2500 Broening Highway
Baltimore, MD 21224
Telephone: (410) 631-3851/Fax: (410) 631-3198

Massachusetts

Howard Wensley, M.S., C.H.O.
Department of Public Health
Division of Community Sanitation
150 Freemont Street
Boston, MA 02111
Telephone: (617) 727-2660

Michigan

Lawrence Chadzynski, M.P.H.
Environmental Quality Specialist
Michigan Department of Public Health
Medical Waste Regulation
Division of Environmental Health
Bureau of Environmental and Occupational Health
3423 North Logan/Martin L. King Jr. Blvd.
P.O. Box 30195
Lansing, MI 48909
Telephone: (517) 335-8637

Samuel Davis, B.S., R.M. (AAM)
Michigan Department of Public Health
Bureau of Laboratory and Epidemiological Services
Quality Control Unit
Laboratory Services Section
Division of Administration
3500 North Logan Street
P.O. Box 30035
Lansing, MI 48909
Telephone: (517) 335-8074

New Jersey

Robert M. Confer, M.B.A.
Bureau Chief
Bureau of Medical Waste
Residuals Management and Statewide Planning
New Jersey Department of Environmental Protection & Energy
Division of Solid Waste Management
840 Bear Tavern Road
CN 414
Trenton, NJ 08625
Telephone: (609) 530-8599/Fax: (609) 530-8899

Rana A. Kazmi, Ph.D.
New Jersey Department of Health
Regulated Medical Waste Project
CN 369
3635 Quakerbridge Road
Trenton, NJ 08625
Telephone: (609) 588-3124/Fax: (609) 588-7431

Ronald Ulinsky
Department of Health
Public Health Sanitation and Safety
3635 Quaker Bridge Road
CN 369
Trenton, NJ 08625
Telephone: (609) 588-3124

New York

Ira F. Salkin, Ph.D.
Director, Regulated Waste Management
Wadsworth Center for Laboratories & Research
New York Department of Health
P.O. Box 509, Empire State Plaza
Albany, NY 12201-0509
Telephone: (518) 474-7413

North Carolina

Ernest Lawrence, Ph.D.
DEHNR—Solid Waste Section
401 Oberlin Road
Suite 150
Raleigh, NC 27605
Telephone: (919) 733-0692

Ohio

Alison Shockley
Ohio Environmental Protection Agency
Division of Solid and Infectious Waste Management
1800 Watermark Drive
Columbus, OH 43215
Telephone: (614) 644-2813/Fax: (614) 644-2329

Oklahoma

Harriett Muzljakovich
Oklahoma State Department of Health
Solid Waste Division
1000 NE 10th Street
Oklahoma City, OK 73117-1299
Telephone: (405)271-7155/Fax: (405)271-7079

Puerto Rico

Florilda Forestier
Puerto Rico Environmental Quality Board
Land Pollution Control Area
P.O. Box 11488
Santurce, Puerto Rico 00910
Telephone: (809) 274-962/Fax: (809) 767-8118

Yira Suarez
Puerto Rico Environmental Quality Board
431 Ponce de Leon Avenue
Hato Ray, Puerto Rico 00917
Telephone: (809) 274-8962

Rhode Island

Roger Greene
Assistant to the Director
Rhode Island Department of Environmental Management
9 Hayes Street
Providence, RI 02908-5003
Telephone: (401) 277-2771/Fax: (401) 277-6802

Diann L. Miele, M.S.
Environmental Scientist
Rhode Island Department of Health
206 Cannon Building
3 Capitol Hill
Providence, RI 02908-5097
Telephone: (401) 277-3424/Fax: (401) 277-6953

A. Joseph Sherry
Rhode Island Department of Health Laboratory
50 Orms Street
Providence, RI 02904
Telephone: (401) 274-1011

South Carolina

Jacob Baker
Department of Health and Environmental Control
Bureau of Solid and Hazardous Waste
2600 Bull Street
Columbia, SC 29201
Telephone: (803) 734-5213

Phillip R. Morris, Manager
South Carolina Department of Health
and Environmental Control
Infectious Waste Management Section
Bureau of Solid & Hazardous Waste Management
2600 Bull Street
Columbia, SC 29201
Telephone: (803) 734-5448/Fax: (803) 734-5199

Joann Bliek
South Carolina Department of Health and Environmental Control
Infectious Waste Section
2600 Bull Street
Columbia, SC 29201
Telephone: (803) 734-4834

Texas

Patricia Riley, D.V.M.
Texas Water Commission
Industrial and Hazardous Wastes/Waste Evaluation
P.O. Box 13087
Austin, TX 78711-3087
Telephone: (512) 908-6832/Fax: (512) 908-6410

Lynne M. Sehulster, Ph.D.
Texas Department of Health
Infectious Disease Epidemiology
1100 West 49th Street
Austin, TX 78756
Telephone: (512) 458-7328/Fax: (512) 458-7601

Virginia

Robert G. Wickline, P.E
Virginia Department of Environmental Quality
Monroe Building, 11th Floor
101 North 14th Street
Richmond, VA 23219
Telephone: (804) 225-2321/Fax: (804) 786-0320

Washington

Ned C Therien, M.P.H., M.S., RS
Food Program Specialist
Washington Department of Health
Environmental Health Programs
Office of Community Environmental Health Programs
Building 3 Airdustrial Center
P.O. Box 47826
Olympia, WA 47826
Telephone: (206) 438-7219 Fax: (206) 586-5529

Wayne L. Turnberg, R.S., M.S.P.H.
Washington Department of Ecology
3190 160th Avenue SE
Bellevue, WA 98008-5452
Telephone: (206) 649-7030/Fax: (206) 649-7098

West Virginia

Joe Wyatt, R.S.
West Virginia Bureau of Public Health
Office of Environmental Health Services
815 Quarrier Street, Suite 418
Charleston, WV 25301-2616
Telephone: (304) 558-2981/Fax: (304) 558-0691

Canada

Michael Brodsky
Chief, Environmental Bacteriology
Ministry of Health
81 Resources Road
Etobicoke, Ontario M9P 3T1

Gordon Donnelly, M.B.A., P.E.
Ministry of the Environment
2 St. Clair Avenue West, 14th Floor
Toronto, Ontario M4V IL5
Telephone: (416) 323-5130

Other Participants

Kimberly Browning
Environmental Analyst
SAIC
7600-A Leesburg Pike
Falls Church, VA 22043
Telephone: (703) 734-2587

Nelson 'Sig' Slavik, Ph.D.
President
Environmental Health Management Systems Inc.
P.O. Drawer 6309
South Bend, IN 46660
Telephone: (219) 272-8748

REFERENCES

1. Miele DJ and Greene R. Creating the model guidelines for states. *Med Waste Anal*, 2(9):7–10. June 1994.

12
HAZARDOUS DRUG WASTE MANAGEMENT[1,2]

In 1990 the American Society of Hospital Pharmacists (ASHP) published its most recent version of its technical assistance bulletin on handling cytotoxic and hazardous drugs (1). The information provided in this report was to later serve as the basis for revised recommentations by the Occupational Safety and Health Administration in its guidance document, "Controlling Occupational Exposure to Hazardous Drugs," published on April 14, 1995 (2). Prior to the ASHP publication, early occupational healthcare worker concerns focused on hazards associated with anticancer pharmaceutical agents. Anticancer drugs are also referred to as cytotoxic drugs (e.g., cell killing), antineoplastic drugs (e.g., agents used to treat neoplasms), or as chemotherapy drugs (e.g., chemicals used in anticancer therapy). In its 1990 technical assistance bulletin, the ASHP described the need to revise its earlier guideline published in 1985 for managing anticancer drugs in the workplace (3). This was based on the recognition that not all anticancer drugs exhibit cytotoxic characteristics (e.g., genotoxic, oncogenic, mutagenic, teratogenic, or hazardous), and that other pharmaceuticals not used in cancer therapy often do exhibit such hazardous characteristics. In addition, ASHP recognizes the term "cytotoxic" to be a less appropriate term to describe these drugs as new hazardous agents (e.g., genotoxic biologicals and some biotechnological agents) continue to be developed (1). This chapter addresses the management and disposal of this new broader

---

[1]Adapted from Turnberg WL. Antineoplastic drug waste management. Reprinted from the *Regulatory Analyst: Medical Waste,* 15(5):1–8, copyright 1993 by Auerbach Publishers, New York, New York. Reprinted by permission.
[2]Adapted from Turnberg WL. Antineoplastic drug waste management—An EPA perspective. *Medical Waste Analyst,* 2(6):6–8, copyright 1994 by Technomic Publishing Company, Lancaster, Pennsylvania. Reprinted by permission.

categorization of those therapeutic agents which are referred to by ASHP and OSHA as "hazardous drugs." Sections specific to antineoplastic drug waste management have also been included in the discussion.

Proper management of the waste generated in the course of hazardous drug preparation and administration must be considered by healthcare providers to protect those that may inadvertently come into contact with the waste downstream. Comprehensive hazardous drug waste management programs should recognize and address the legal aspects of hazardous drug waste disposal and other safety precautions that may extend beyond regulatory requirements. Although disposal considerations are based on their chemical nature, hazardous drug waste is often associated with biohazardous waste (e.g., blood and body fluids). This discussion, which is adapted from work previously published by the author (4,5), addresses current national standards, guidelines, and commonly accepted practices related to hazardous drug waste management and disposal.

HAZARDOUS DRUG TOXICITY

Antineoplastic drugs comprise a class of chemically unrelated drugs that are designed to inhibit tumor growth by killing actively growing cells (6). This action generally occurs without differentiation between normal and cancerous cells and may also involve interaction with cellular DNA, RNA and protein synthesis (7). Many of the drugs that have been approved by the Food and Drug Administration for clinical use or trial have been categorized based on their chemical mechanism of action (8). Many of these agents, when administered in therapeutic doses, have been linked to carcinogenic, mutagenic and teratogenic side effects (8).

More recently, the effects of chronic exposure to these drugs by medical personnel during routine preparation and administration has been examined. Primary routes of exposure include direct skin contact, inhalation of aerosols and ingestion (9–12). Chronic exposure has been implicated in the presence of mutagenic agents in urine of pharmacists and nurses (13–20), increased sister chromatid exchanges in circulating lymphocytes in nurses (21,22), increased association of fetal loss (23) and malformations in the offspring of nurses (24), and liver damage in nurses handling these agents (25).

Acute exposure resulting from direct skin contact or aerosol inhalation can be severe (26). Such exposures have been reported to produce local skin necrosis, cellulitis and irritation of the skin and mucous membranes, especially the eyes (8). Pharmacists have reported symptoms of lightheadedness, dizziness or facial flushing while preparing admixtures of these agents (27). Other reported symptoms among medical personnel have included shortness of breath, chest tightness and congestion resulting from drug preparation, and abdominal pain, diarrhea, and vomiting from direct skin contact (28). One report identified the eventual loss of a patient's hand function resulting from a needle-stick injury to the patient's finger with the antineoplastic agent mitomycin-C (29).

CONSEQUENCES OF IMPROPER DISPOSAL

The populations at risk from improperly disposed hazardous drug waste include those that may be exposed to the agents inadvertently once the waste drugs have entered the municipal waste stream. Exposure by the public to waste hazardous drugs would be expected to be rare based on current municipal waste management systems of collection and disposal. Possible exposure scenarios may include children or scavengers entering or rummaging through dumpsters or other waste containers, or human exposure to contaminated liquids that have leaked from waste containers or waste transport vehicles.

Waste industry workers, particularly municipal waste collectors, and healthcare custodial workers would be expected to experience the greatest, although low, potential for exposure to hazardous drugs improperly disposed of in waste. Antineoplastic drug residues may be associated with many waste items found in the general waste stream resulting from healthcare activities in hospitals or from homes, such as contaminated wipes, paper towels, disposable diapers, or improperly disposed hypodermic needles. To date, such exposures have not been documented, nor have potential side effects been examined. Nevertheless, the opportunity for acute exposure remains a concern among waste industry workers whenever hazardous drugs are disposed of unsafely or improperly.

Occupational exposure by waste industry workers to infectious waste in the municipal waste stream has been reported in the literature. A recent survey of 940 waste industry workers in Washington State found 10% of responding waste collectors reporting needle-stick injuries during the year preceding the survey (30). Although exposure to waste antineoplastic drugs was not addressed in this study, the survey emphasizes that human exposures can occur whenever high-risk wastes are disposed of improperly in the municipal waste stream.

HAZARDOUS DRUG WASTE CATEGORIES

Hazardous drug wastes may be recognized as comprising three categories:

1. *Bulk Contaminated Materials:* defined by the National Institutes of Health (NIH) to include (a) antineoplastic drug vials and syringes filled to a weight greater than 3% of the capacity of the container, and (b) intravenous solutions containing antineoplastic drugs.

2. *Trace Contaminated Materials:* defined by the NIH as wastes containing minimal or trace amounts of antineoplastic drugs. Examples include used syringes and needles, empty drug vials and ampules, empty intravenous bottles and tubing, disposable gloves, gowns, goggles and masks, cleaning cloths, alcohol wipes, and other objects that have come into contact with hazardous agents during preparation, administration, cleanup, or disposal.

3. *Contaminated Human Excreta and Body Fluids:* Urine, feces, vomitus, and blood of patients being treated with antineoplastic drugs have been found to contain high concentrations of either the antineoplastic agents or hazardous metabolites for up to 48 hours or more following treatment.

HOSPITAL ANTINEOPLASTIC DRUG PROFILE

In 1990 the U.S. Environmental Protection Agency (EPA) published "Guides to Pollution Prevention, Selected Hospital Waste Streams," in which various hospital hazardous waste streams were identified (31). The hospital waste profile developed by this study was based on audits of three hospitals. Procurement of antineoplastic drugs to produce chemotherapy solutions was generally reported to occur through a central clinic or pharmacy. On-hand quantities of the drugs were reported as sufficient to last for less than two weeks. Chemicals were mixed or compounded in a hood, which recirculates air through a filter. Chemotherapy wastes were reported to account for the largest volumes of hazardous waste generated by the audited facilities. (Other hazardous wastes included solvents, photographic chemicals, formaldehyde wastes, radioactive wastes, mercury, and other toxics and corrosives.)

The report indicated that only a small percentage of the waste contained concentrated amounts of antineoplastic agents. The waste was characterized as being made up largely of lightly contaminated items such as personal protective clothing, gauze pads, and hypodermic needles. The contaminated waste was placed into plastic bags or containers with an average generation rate of 2 to 8 cubic feet of waste per week for the hospitals audited. The waste was then either transported off-site to a class 1 (hazardous waste) landfill or incinerated as infectious waste.

HAZARDOUS WASTE REGULATIONS

Subtitle C of the Resource Conservation and Recovery Act (RCRA), the nation's hazardous waste management law, was passed in 1976 by the U.S. Congress to control hazardous wastes from "cradle to grave." Under its federal mandate, the EPA published criteria for defining hazardous waste on May 19, 1980, which were applicable to all generators of hazardous waste as defined by EPA, including hospitals, laboratories and other medical institutions that generate such waste (32).

Under Title 40 of the U.S. *Code of Federal Regulations* Part 261 (referred to as 40 CFR Part 261), the EPA established criteria for identifying and listing a waste as hazardous. Part 261, Subpart C categorizes hazardous wastes based on characteristics that include ignitability, corrosivity, reactivity, and toxicity as defined by the regulation. Part 261, Subpart D identifies specific lists of hazardous wastes. These lists include hazardous wastes from nonspecific sources (the F list), those from specific sources (the K list), acute hazardous wastes (the P list), and toxic wastes (the U list).

Under the RCRA hazardous waste regulations, generators are required to determine whether a given waste is hazardous based on the criteria established by the EPA regulations. If a generator determines that hazardous waste is being produced in sufficient quantities, the generator must notify the EPA and be assigned an EPA hazardous waste generator identification number. Once in the RCRA system, the generator would be subject to the full hazardous waste management requirements established by RCRA for all hazardous waste generated by the facility. Included would be requirements for record keeping and reporting, hazardous waste storage, personnel training, preparedness and prevention, contingency planning, emergency procedures, use and management of containers, and transportation.

RCRA allows for conditional exemptions from the hazardous waste requirements based on waste volume generated [40 CFR Part 261.5]. Small-quantity generators of hazardous waste may be "conditionally" exempt from most RCRA regulations if the total amount of hazardous waste generated by the facility is less than 220 pounds per month (100 kilograms), or less than 2.2 pounds per month (1 kilogram) for acute hazardous waste identified on RCRA's P list. However, small-quantity generators must still meet certain RCRA hazardous waste requirements. These include identifying whether a hazardous waste has been generated at the facility, and if so, ensuring that the waste is either treated or disposed in an on-site facility or transported to an off-site treatment, storage, or disposal facility that is permitted or otherwise authorized to receive such waste.

Hazardous waste generated from households have been excluded categorically from all RCRA hazardous waste management requirements. Under the federal system, household hazardous waste is subject to the municipal solid waste disposal requirements under RCRA Subtitle D and may be disposed of at an approved municipal solid waste landfill. However, state or local requirements or policies often restrict or discourage this disposal option. Applicable state or local authorities must be consulted to make this determination.

Under the federal system, hazardous waste from conditionally exempt small-quantity generators and households may be disposed of at a permitted municipal solid waste landfill. However, the federal municipal solid waste landfill requirements, codified at 40 CFR Part 258, restrict containerized liquids from being disposed of at a landfill unless the container is small (similar in size to that normally found in household waste) and the container is designed to hold liquids for use other than for storage. If a liquid hazardous waste from a small-quantity generator or household fails to meet the landfill liquids restriction, it must be either treated or disposed of through other systems permitted or authorized to treat or dispose of hazardous waste.

The Resource Conservation and Recovery Act established a national baseline standard for hazardous waste management which was to be implemented by the EPA. However, the act allowed and encouraged the EPA to authorize individual states to administer their own hazardous waste programs provided that a state meets or exceeds all authorities and requirements established under RCRA. For a complete understanding of hazardous waste requirements within a state that has been authorized by EPA to implement the RCRA hazardous waste program, the applicable state agency administering the hazardous waste program must be contacted.

ANTINEOPLASTIC DRUG WASTE DISPOSAL UNDER RCRA

There are seven antineoplastic drugs that are referenced on the Part 261, Subpart D hazardous waste list (U-listed toxic waste). These represent only a small percentage of the overall number of antineoplastic drugs available today, and include:

Agent	U List No.
Chlorambucil	U035
Cyclophosphamide	U058
Daunomycin	U059
Melphalan	U150
Mitomycin C	U010
Streptozotocin	U206
Uracil mustard	U237

Based on the federal requirements, only these seven antineoplastic drugs are subject to the hazardous waste management requirements established under RCRA. For medical facilities generating 220 pounds or more of hazardous waste per month (and therefore subject to all RCRA hazardous waste requirements), the smaller volumes of U-listed bulk antineoplastic drug waste or spill waste generated by the facility would be tallied with the facility's overall hazardous waste volume and subject to the hazardous waste disposal requirements.

Medical facilities generating smaller volumes of hazardous waste (e.g., less than 220 pounds per month) would only be subject to those parts of the RCRA regulations that apply to conditionally exempt small-quantity generators. These medical facilities would still be required to meet the small-quantity generator requirements, which include identifying whether hazardous wastes are generated, and if so, disposing of such wastes in legally permitted treatment, storage, or disposal facilities as allowed under RCRA. Disposal of conditionally exempt liquid hazardous waste into municipal solid waste landfills would be allowed only if the liquids were contained in small (household-sized) containers that were not designed specifically for waste storage (e.g., the original antineoplastic drug container).

Antineoplastic drug waste generated in the course of home healthcare would be exempt from the federal hazardous waste management requirements. However, state or local regulations or policies may place restrictions on such disposal from homes. State or local health or waste management agencies should be consulted to make this determination.

Some medical facilities that are small-quantity generators may choose to store their hazardous wastes on-site indefinitely. However, if a generator accumulates more than 2200 pounds of hazardous waste on site (1000 kilograms), the generator would no longer be viewed as a conditionally exempt small-quantity generator and would become subject to federal hazardous waste management requirements.

Another potential option for conditionally exempt small-quantity generators involves disposal of liquid antineoplastic drug wastes into a sanitary sewage system. Before a facility chooses this option, the local sewage utility and water pollution control agency should be contacted to determine whether or under what conditions such a practice would be allowed under state or local law.

Although only seven of the many antineoplastic agents are identified on RCRA's U list of hazardous wastes, all waste antineoplastic drugs used by medical facilities should be handled and disposed of following the same toxic waste disposal policies because of their similar toxic characteristics. This is supported by the National Institutes of Health in the report by Vaccari et al. (33), in which the authors state:

> We have stated that currently only seven antineoplastic agents appear on RCRA's U list, yet we have chosen to manage this entire category of agents as toxic waste. This policy was instituted only after serious consideration of the possible financial burden and inconvenience it may cause. We realize the toxic potential of these chemicals and feel very strongly that we have an obligation to take every reasonable precaution so that others are not exposed to these agents and the environment is protected.

The EPA has specifically addressed two disposal scenarios in written policies: (1) disposal requirements for *de minimis* or trace-contaminated waste containing U-listed antineoplastic agents (e.g., gloves, gowns), and (2) disposal requirements for bulk antineoplastic drug formulations (e.g., unused intravenous solutions). Each is discussed below.

Trace Quantity (*De Minimis*) Disposal

Questions have been made regarding the RCRA interpretation for *de minimis* quantities of a U-listed EPA hazardous substance. Should all waste that contains EPA U-listed antineoplastic drug contaminants be handled and regulated as a hazardous waste by facilities that are subject to the full hazardous waste requirements (e.g., those generating more than 220 pounds of hazardous waste per month)? An EPA policy letter addresses this issue directly. The policy letter, OSWER Directive 9441.1987(45), June 16, 1987, states:

> This responds to [your] letter of April 13, 1987, regarding the regulatory status of chemotherapy drugs and related supplies. In particular, you questioned whether the weight of the "empty" vial should be included in determining the amount of drug residues to be disposed.

> As you pointed out, several chemotherapy drugs are listed in 40 CFR 261.33(f) (commonly known as the U list). As such, these wastes are regulated under the EPA hazardous waste regulations (unless subject to the small quantity generator exclusion). Included in the listing are the following discarded commercial chemical products, off-specification species, container residues, and spill residues:

(1) Chlorambucil   (U0355)
(2) Cyclophosphamide   (U0588)
(3) Daunomycin   (U0599)
(4) Melphalan   (U1500)
(5) Mitomycin C   (U0100)
(6) Streptozotocin   (U2206)
(7) Uracil mustard   (U2237)

Under EPA regulations governing the management of hazardous wastes, any container used to hold these chemicals (such as vials) are considered hazardous wastes unless these containers meet the criteria of an "empty container." Under the empty container provision such vials are excluded from regulation if the material has been removed by pouring, pumping, and aspirating, and no more than 1 inch of residue remains in the bottom of the vial or no more than 3 percent by weight of the total capacity of the container remains in the container. (See 40 CFR 261.7).

The Agency is aware, however, that prudent practice dictates that materials contaminated with these chemicals (such as syringes, vials, gloves, gowns, aprons, etc.) not be handled after use. Therefore, to minimize exposure to these toxic chemicals, the Agency recommends that the entire volume of waste be weighed and that there be no attempt to remove any residue from the vial before disposal.

Chemotherapy drugs that are not listed hazardous wastes are not regulated by EPA. However, you should contact your State or local government regarding the management of these chemicals. Also, the National Institutes of Health (NIH) provides guidance on handling and management of antineoplastics.

Policy Signature: Jacqueline W. Sales, Chief, Regulation Development Section

Ed Rau, Chief, Hazardous and Solid Waste Management Section, Environmental Protection Branch of the National Institutes of Health (NIH) was contacted to describe the trace antineoplastic drug waste disposal policies of the NIH (33). As described by Dr. Rau, the NIH policy on antineoplastic waste management has not changed significantly since first published in 1984 in an article entitled "Disposal of antineoplastic wastes at the National Institutes of Health" (34). By that policy, the NIH continues to manage all of its antineoplastic drug waste in the same manner whether or not the waste has specifically been listed as hazardous waste by the EPA. The NIH manages its bulk-contaminated antineoplastic drug waste as RCRA hazardous waste. However, trace-contaminated waste is not managed as hazardous waste at the NIH, whether or not the contaminant appears on the EPA U list. Rather, the waste is incinerated with the facility's infectious (solid) waste in an off-site incinerator.

The NIH recognizes the issue of *de minimis* contamination as a gray area of hazardous waste regulatory interpretation, but has determined that waste contaminated with trace amounts of U-listed antineoplastic agents is not subject to RCRA hazardous waste requirements. The NIH presents its policy as the most practical, in part because of the unwillingness for hazardous waste disposal facilities to accept infectious or nuclear waste mixtures. The NIH also presents its policy as safe because all, as opposed to just the seven EPA U-listed antineoplastic drug wastes, are handled as toxic waste.

In its policy directive addressing trace-contaminated antineoplastic drug waste, the EPA has presented what appears to be a conflicting regulatory interpretation for U-listed antineoplastic waste disposal. On one hand, the policy pulls waste contaminated with U-listed antineoplastic substances into the RCRA hazardous waste management system. On the other hand, the restriction is eased by referring the reader to the more practical policies followed by the NIH.

Antineoplastic Drug Formulations Disposal

Antineoplastic drug formulations fall within the category of bulk-contaminated materials as defined by the NIH. The EPA's position on the management of formulations (e.g., intravenous solutions) containing antineoplastic U-listed drugs was addressed in a policy response letter on April 25, 1988. This EPA policy interpretation states:

> This letter is in response to (your) March 24, 1988, request for clarification of the status of certain antineoplastic drug wastes. Your request was for an interpretation of 40 CFR 261.33, with respect to excess antineoplastic drug formulations which are not needed and thus are discarded.
>
> If an antineoplastic drug is mixed with diluents, such as water or saline solution, the excess diluted or undiluted amount to be discarded is unused commercial chemical product. If the discarded unused commercial chemical product is listed in 40 CFR 261.33, the material is a listed hazardous waste regardless of dilution with water or saline because the product still would be the sole active ingredient. However, it is not considered a "spent material." Section 261.1(c)(1) defines a spent material as any material that has been used and as a result of contamination can no longer serve the purpose for which it was produced without processing. The portion antineoplastic drug, if diluted, has not yet been used for its intended function, nor is it contaminated.
>
> If an antineoplastic drug is mixed with diluents and with other pharmaceuticals for use, the unused mixed excess portion that is discarded is a solid waste. If the antineoplastic drug is listed in 40 CFR 261.33 the unmixed excess portion is a listed hazardous waste, provided the antineoplastic drug is the sole active ingredient in the mixed formulation. If it is not the sole active ingredient, the mixture would not be the listed hazardous wastes; however, the formulation may still be hazardous if it exhibits any of the hazardous waste characteristics.
>
> In all of the situations you described, the material (if it met the listing in 40 CFR 261.33) would have to be sent to a permitted or interim status hazardous waste management facility, if the facility generates more than 1 kg per month of acutely hazardous waste, or more than 100 kg/month of non-acutely hazardous waste. See 40 CFR 261.5(f) and (g) for the hazardous waste management options for conditionally exempt small quantity generators of hazardous waste.
>
> In addition, any state in which you generate, transport, treat, store, or dispose of these formulations may have regulations that are more stringent than the Federal hazardous

waste rules. You therefore should check with the State agencies to determine what regulations, if any, apply to handling these materials.

Letter Signature: Devereaux Barnes, Director, Characterization and Assessment Division

In conclusion, the federal requirements for antineoplastic drug waste management have been clarified and supplemented by this discussion. Because state regulations are often more stringent than their federal counterparts, the appropriate state agencies should be consulted for final regulatory determinations on antineoplastic drug waste management.

OSHA WORK-PRACTICE DISPOSAL GUIDELINES

Chronic and acute effects of antineoplastic and other hazardous drugs on healthcare professionals have been recognized by the U.S. Occupational Safety and Health Administration (OSHA) as an occupational hazard. Proper use of safety equipment (e.g., vertical laminar flow hoods, gloves, goggles) during the preparation and administration of antineoplastic agents has been observed to reduce exposure risks significantly among healthcare workers (14,16,20). It was this recognition that provided the basis for OSHA to issue comprehensive guidelines covering all aspects of chemotherapy preparation and administration, including waste disposal.

Federal antineoplastic drug management guidelines were published by OSHA in 1986 as "Instruction PUB 8-1.1, Work Practice Guidelines for Personnel Dealing with Cytotoxic (Antineoplastic) Waste" (9). On April 14, 1995, OSHA published its revised guideline that included recommendations for managing all hazardous drugs (with the exception of anesthetic agents) as opposed to only antineoplastic drugs (2). The revised guideline, "OSHA Instruction CPL 2-2.20B CH-4, Controlling Occupational Exposure to Hazardous Drugs," applies to all settings where employees are exposed to these agents while on-the-job. Such settings include physicians' offices, hospitals, and home health care agencies. The purpose of the guideline is to assist OSHA inspectors during inspections. Being a compliance directive, the guideline is not directly enforceable. Rather, it provides both the practitioner and the OSHA inspector with a recognized standard of practice that is enforceable under other general OSHA work place safety and health standards. Copies of the guideline may be obtained by contacting the applicable Regional OSHA Office identified in Appendix A (Federal and State OSHA Contacts) of this book.

The OSHA hazardous drugs (HD) guideline is presented in 9 sections that are organized as follows: (A) Introduction; (B) Categorization of Drugs as Hazardous; (C) Background: Hazardous Drugs as Occupational Risks; (D) Work Areas; (E) Prevention of Employee Exposure; (F) Medical Surveillance; (G) Hazard Communication; (H) Training and Information Dissemination; and (I) Recordkeeping.

Key sections of the hazardous drugs guideline that relate to hazardous drugs management and disposal are presented below. The reader should be aware that the sections presented here must be recognized in context with the complete hazardous drugs guideline which should be obtained and understood by the reader.

Categorizing Hazardous Drugs

It has long been recognized that certain anticancer drugs pose occupational risks to employees in the healthcare setting. In 1990 the American Society of Hospital Pharmacists published its most recent technical assistance bulletin that also recognized occupational hazards associated with certain non-anticancer drugs as well, and in doing so defined this new broader categorization of those therapeutic agents as "hazardous drugs" (1). Criteria identified by ASHP and recommended by OSHA to identify drugs that may represent occupational hazards include those agents that exhibit one or more of the following characteristics:

- Genotoxicity
- Carcinogenicity
- Teratogenicity or fertility impairment, and
- Serious organ or other toxic manifestation at low doses in experimental animals or treated patients

A listing of several commonly used agents is presented in Table 12.1. In presenting this list, OSHA states:

> (The list) is not all inclusive, should not be construed as complete, and represents an assessment of some, but not all, marketed drugs at a fixed point in time. (The list) was developed through consultation with institutions which have assembled teams of pharmacists and other health care personnel to determine which drugs should be handled with caution. These teams reviewed product literature and drug information when considering each product. Sources for this (list) are the Physicians' Desk Reference, Section 10:00 in the American Hospital Formulary Service Drug Information (35), IARC publications (particularly Volume 50); (36), the Johns Hopkins Hospital, and the National Institutes of Health, Clinical Center Nursing Department. No attempt to include investigational drugs was made, but they should be prudently handled as hazardous drugs until adequate information becomes available to exclude them. Any determination of the hazard status of a drug should be periodically reviewed and updated as new information becomes available. Importantly, new drugs should routinely undergo a hazard assessment.

In presenting hazardous drug criteria and this list, OSHA recognizes that no standardized reference has been developed, nor has consensus been reached on those agents presented in Table 12.1. In its guideline, OSHA advises that decisions for designating a pharmaceutical agent as a hazardous drug should be based on professional judgement by those trained in pharmacology/toxicology. OSHA directs the reader to a report by McDiarmid et al. which describes how such a list was developed by one institution (37). Several major considerations presented by OSHA include (37):

**TABLE 12.1 Some Common Drugs
Considered Hazardous**

Chemical/Generic Name	Source[a]
Altretamine	C
Aminoglutethimide	A
Azathioprine	ACE
L-Asparaginase	ABC
Bleomycin	ABC
Busulfan	ABC
Carboplatin	ABC
Carmsutine	ABC
Chlorambucil	ABCE
Chloramphenicol	E
Chlorotrianisene	B
Chlorozotocin	E
Cyclosporin	E
Cisplatin	ABCE
Cyclophosphamide	ABCE
Cytarabine	ABC
Dacarbazine	ABC
Dactinomycin	ABC
Daunorubicin	ABC
Diethylstilbestrol	BE
Doxorubicin	ABCE
Estradiol	B
Estramustine	AB
Ethinyl Estradiol	B
Etoposide	ABC
Floxuridine	AC
Fluorouracil	ABC
Flutamide	BC
Ganciclovir	AD
Hydroxyurea	ABC
Idarubicin	AC
Ifosfamide	ABC
Interferon-A	BC
Isotretinoin	D
Leuprolide	BC
Levamisole	C
Lomustine	ABCE
Mechlorethamine	BC
Medroxyprogesterone	B
Megestrol	BC
Melphalan	ABCE
Mercaptopurine	ABC
Methotrexate	ABC
Mitomycin	ABC

(*continued*)

TABLE 12.1 (*Continued*)

Chemical/Generic Name	Source[a]
Mitotane	ABC
Mitoxantrone	ABC
Nafarelin	C
Pipobroman	C
Plicamycin	BC
Procarbazine	ABCE
Ribavirin	D
Streptozocin	AC
Tamoxifen	BC
Testolactone	BC
Thioguanine	ABC
Thiotepa	ABC
Uracil Mustard	ACE
Vidarabine	D
Vinblastine	ABC
Vincristine	ABC
Zidovudine	D

[a]Sources:
A The National Institutes of Health, Clinical Center Nursing Department
B Antineoplastic drugs in the Physician's Desk Reference
C American Hospital Formulary, Antineoplastics
D Johns Hopkins Hospital
E International Agency for Research on Cancer

Source: From OSHA Instruction CPL 2-2.20B CH-4, Appendix 21-1, April 14, 1995

- Is the drug designated as Therapeutic Category 10:00 (Antineoplastic Agent) in the American Hospital Formulary Service Drug Information? (35)
- Does the manufacturer suggest the use of special isolation techniques in its handling, administration, or disposal?
- Is the drug known to be a human mutagen, carcinogen, teratogen or reproductive toxicant?
- Is the drug known to be carcinogenic or teratogenic in animals (drugs known to be mutagenic in multiple bacterial systems or animals should also be considered hazardous)?
- And, is the drug known to be acutely toxic to an organ system?

Hazardous Drug Safety and Health Plan

In Section E of the guideline (Prevention of Employee Exposure), OSHA recommends that a written Hazardous Drug Safety and Health Plan be developed whenever hazardous drugs are used in the workplace to 1) protect employees from health hazards associated with HDs, and 2) to keep exposures as low as reasonbly achievable. The element of waste management should be included in the plan. When developed, the plan should be readily accessible to all employees, contractors and trainees. Other elements that should be included in a plan are based upon ASHP recommendations (1) which are presented in the guideline.

Hazardous Drugs Management and Disposal

Key guideline sections specific to the management and disposal of hazardous drugs are presented below.

Section D—Work Areas

3. Disposal of Drugs and Contaminated Material

Contaminated materials used in the preparation and administration of HDs, such as gloves, gowns, syringes and vials, present a hazard to support and housekeeping staff. The use of properly labeled, sealed and covered disposal containers, handled by trained and protected personnel, should be routine, and is required under the Bloodborne Pathogens Standard (38) is such items are contamined with blood or other potentially infectious materials. HDs and contaminated materials should be disposed of in accordance with federal, state, and local laws. Disposal of some of these drugs is regulated by the EPA. Those drugs which are unused commercial chemical products and are considered by the EPA to be toxic wastes must be disposed of in accordance with 40 CFR part 261.33 (39). Spills can also represent a hazard; the employer should ensure that all employees are familiar with appropriate spill procedures.

Section E. Prevention of Employee Exposure

Personal Protective Equipment (PPE)

5. Personal Protective Equipment Disposal and Decontamination

All gowns, gloves, and disposable materials used in preparation should be disposed of according to the hospital's hazardous drug waste procedures and as described under this review's section on Waste Disposal.

Work Equipment

The ASHP recommends that HD-labeled plastic bags be available for all contaminated materials (including gloves, gowns, and paperliners), so that contaminated material can be immediately placed in them and disposed of in accordance with ASHP recommendations (1).

Work Practices

1. Labeling

In addition to standard pharmacy labeling practices, all syringes and IV bags containing HDs should be labeled with a distinctive warning label such as:

> Special Handling/Disposal Precautions.

In addition, those HDs covered under Hazard Communication Standard (HCS) must have labels in accordance with section (f) of the standard to warn employees handling the drug(s) of the hazards.

2. Needles

The ASHP recommends that all syringes and needles used in the course of preparation be placed in "sharps" containers for disposal without being crushed, clipped or capped (1,40).

6. Packaging HDs for Transport

. . . HDs that are shipped and which are subject to EPA regulation as hazardous waste are also subject to Department of Transportation regulations as specified in 49 CFR part 172.101.

Work Practices

. . . Syringes, IV bottles and bags, and pumps should be wiped clean of any drug contamination with sterile gauze. Needles and syringes should not be crushed or clipped. They should be placed in a puncture-resistant container then into the DH disposal bag with all other HD contaminated materials.

. . . Administration sets should be disposed of intact. Disposal of the waste bags should follow HD disposal requirements. Unused drugs should be returned to the pharmacy.

. . . All protective equipment should be disposed of upon leaving the patient care area.

This section also addresses home healthcare as follows:

The increased use of HDs in the home environment necessitates special precautions. Employees involved in home care delivery should follow the above work practices and employers should make administration and spill kits available. Home health care workers should have emergency protocols with them as well as phone numbers and addresses in the event emergency care becomes necessary (1). Waste disposal for drugs delivered for home use and other home contaminated material should also be considered by the employer and should follow applicable regulations.

Caring for Patients Receiving HDs

1. Personnel Protective Equipment

Personnel dealing with excreta, primarily urine, from patients who have received HDs in the last 48 hours should be provided with and wear latex or other appropriate gloves and disposable gowns, to be discarded after each use or whenever contaminated, as detailed under Waste Disposal. Eye protection should be worn if splashing is possible. Such excreta contaminated with blood, or other potentially infectious materials as well, should be managed according to the Bloodborne Pathogen Standard. Hands should be washed after removal of gloves or after contact with the above substances.

Waste Disposal

1. Equipment

Thick, leakproof plastic bags, colored differently from other hospital trash bags, should be used for routine accumulation and collection of used containers, discarded gloves, gowns, and any other disposable material. Bags containing hazardous chemicals (as defined by Section C of HCS), shall be labeled in accordance with Sectino F of the Hazard Communication Standard where appropriate. Where the Hazard Communication Standard does not apply, labels should indicate that bags contain HD-related wastes.

Needles, syringes, and breakable items not contaminated with blood or other potentially infectious materials, should be placed in a "sharps" container before they are stored in the waste bag. Such items that are contaminated with blood or other potentially infectious material must be placed in a "sharps" container. Similarly, needles should not be clipped or capped nor syringes crushed, if contaminated by blood or other potentially infectious material, such needles/syringes must not be clipped, capped, or crushed (except as on a rare instance where a medical procedure requires recapping). The waste bag should be kept inside a covered waste container clearly labeled "HD Waste Only." At least one such receptacle should be located in every area where the drugs are prepared or administered. Waste should not be moved from one area to another. The bag should be sealed when filled and the covered waste container taped.

2. Handling

Prudent practice dictates that every precaution be taken to prevent contamination of the exterior of the container. Personnel disposing of the HD waste should wear gowns and protective gloves when handling waste containers with contaminated exteriors. Prudent practice further dictates that such a container with a contaminated exterior be placed in a second container in a manner which eliminates contamination of the second container. HD waste handlers should also receive hazard communication training as discussed below in Section H (Training and Information Dissemination) of this guideline.

3. Disposal

Hazardous drug-related wastes should be handled separately from other hospital trash and disposed of in accordance with applicable EPA, state and local regulations for hazardous waste (39,33). This disposal can occur at either an incinerator or a licensed sanitary landfill for toxic wastes, as appropriate. Commercial waste disposal is performed by a licensed company. While awaiting removal, the waste should be held in a secure area in covered, labeled drums with plastic liners.

Chemical inactivation traditionally has been a complicated process that requires specialized knowledge and training. The MSDS should be consulted regarding specific advance on cleanup. IARC (41) and Lunn et al. (42) have validated inactivation procedures for specific agents that are effective. However, these procedures vary from drug to drug and may be impractical for small amounts. Care must be taken because of unique problems presented by the cleanup of some agents, such as by-product formation (43). Serious consideration should be given to alternative disposal methods.

Spills

Emergency procedures to cover spills or inadvertent release of hazardous drugs should be included in the facility's overall health and safety program . . .

Note: The OSHA guideline presents specific recommendations for spills that address 1) personnel contamination; 2) cleanup of small spills; 3) cleanup of large spills; 4) spills in biological safety cabinets; and 5) spill kits. Language specific to waste disposal includes:

Any broken glass fragments should be picked up using a small scoop (never the hands) and placed in a "sharps" container. The container should then go into a HD disposal bag, along with used absorbent pads and any other contaminated waste.

Storage and Transport

2. Receiving Damaged HD Packages

The ASHP recommends that broken containers and contaminated packaging mats be palced in a "sharps" container and then into HD disposal bags (1). The bags should then be closed and placed in receptacles as described under Waste Disposal.

The appropriate protective equipment and waste disposal materials should be kept in the area where shipments are received, and employees should be trained in their use and the risks of exposure to HDs.

Discussion

OSHA recognizes the danger of inadvertent contact by medical personnel or housekeeping staff with hazardous drug waste and has established guidelines to protect workers through proper handling, containerization, labeling, and disposal of such waste products. OSHA recommends that disposal be conducted to conform to federal, state and local restrictions, and specifically references EPA hazardous waste requirements (40 CFR 261.33 (f) and practices conducted by the NIH as reported in 1984 (33).

In its 1984 report, the NIH identifies itself as an EPA registered hazardous waste generator because of the types and volumes of hazardous waste it generates. The NIH disposes of bulk antineoplastic wastes as hazardous waste, either at an EPA-approved land internment site or by incineration in an EPA-permitted hazardous waste incinerator. The NIH policy for trace-contaminated waste disposal is by incineration using an off-site solid waste incinerator, without specifying the need to have this waste stream handled specifically as RCRA hazardous waste. This practice has been supported by others in the literature with the recommendation that sufficient incineration temperatures be attained to achieve destruction of the drugs.

The OSHA guidelines and the NIH report both address concerns regarding human excreta or other body substances contaminated with antineoplastic agents. Recommendations by the NIH and OSHA caution the caregiver to use personal protective equipment to avoid contact with contaminated human waste. However, specific disposal policies have not been addressed.

Although not required under RCRA, the NIH recommends that its policies be followed by conditionally exempt small-quantity generators rather than circumventing them with "systems that may prove hazardous or more costly in the long run."

OSHA GUIDELINE COMPLIANCE

In a 1990 report, Valanis et al. conducted a survey to compare antineoplastic drug-handling policies of 24 hospitals for consistency with the OSHA recommendations (44). The researchers found most hospitals to have policies for antineoplastic drug management, but that most policies varied between facilities and were generally less complete than the OSHA guidelines. Of the 20 hospitals handling antineoplastic drugs, most reported disposing of unused or outdated drugs and equipment (empty vials, syringes, tubing) in hazardous waste containers. Two-thirds of hospitals reported incinerating these wastes. Some facilities reported returning drugs to the manufacturer, and others reported pouring waste drugs down the drain.

In a 1989 survey of 51 hospitals in Washington State, 35 (69 percent) of respondents reported using antineoplastic drugs at their facilities (45). Of those, 28 hospitals (80 percent) reported incinerating the waste, 3 (9 percent) reported disposal with a registered hazardous waste transporter, 1 (3 percent) reported pouring waste down the drain, and 1 (3 percent) reported having no special disposal requirements beyond segregating the waste, double bagging, and disposing at a municipal landfill. Although this survey was not designed to compare compliance with OSHA guidelines, final disposal trends are apparent.

ANTINEOPLASTIC DRUG WASTE MINIMIZATION

The EPA, in its 1990 report on selected hospital waste streams, identified options for minimizing the volume of antineoplastic drug waste through implementation of administrative controls (31). These were identified as waste segregation, minimizing cleanup waste volume, and employee training. Methods to minimize waste that were outlined by the EPA include:

- Segregating chemotherapy wastes from other wastes through training and providing separate waste containers
- Disposing of noncontaminated garments with the general waste (although glove contamination should be assumed)
- Purchasing of drug volumes according to need
- Returning outdated drugs to the manufacturer
- Employing proper spill containment and cleanup procedures (e.g., use small versus large absorbent devices, when practical)

CONCLUSION

Policies for hazardous drug waste management, whether generated from medical facilities or during home healthcare, should be based both on an understanding of federal, state, and local regulations and a recognition of hazards that may require waste management steps beyond those required by regulation. Policy development should also take into consideration the facility's overall perception by the public and the need to maintain a positive image in the community.

The following steps are recommended for developing a formal hazardous drug waste management and disposal policy:

1. Conduct a facility audit to identify the categories (e.g., bulk waste, trace-contaminated waste, or excreta) and volumes of each category of hazardous drug waste material generated, and how it moves through the facility.
2. Determine whether or not the facility qualifies as a RCRA hazardous waste generator.
3. Contact local, state, and federal health and waste management authorities to obtain copies of statutes, rules, guidelines, and policies to identify compliance requirements and recommendations.
4. Contact other medical facilities and determine how their hazardous drug waste is managed.
5. Identify and evaluate the feasibility of the various disposal options available.
6. Based on the above, establish and implement a formal facility policy for hazardous drug waste that identifies procedures for waste identification, segregation, containment, labeling, transport, treatment (if necessary), disposal, contingency planning, and staff training.

REFERENCES

1. American Society of Hospital Pharmacists. ASHP technical assistance bulletin on handling cytotoxic and hazardous drugs. *Am J Hosp Pharm,* 47:1033–1049. 1990.
2. U.S. Department of Labor, Occupational Safety and Health Administration. Controlling Occupational Exposure to Hazardous Drugs. OSHA Instruction CPL 2-2.20B CH-4. April 14, 1995.
3. American Society of Hospital Pharmacists. ASHP technical assistance bulletin on handling cytotoxic drugs in hospitals. *Am J Hosp Pharm,* 42:131–137. 1985.
4. Turnberg WL. Antineoplastic drug waste management. *Med Waste: Regul Anal,* 15(5):1–7. February 1993.
5. Turnberg WL. Antineoplastic drug waste management: An EPA perspective. *Med Waste Anal,* 2(6):6–8. March 1994.
6. Doull J, Klaassen CD, and Amdur MO. *Casarett & Doull's Toxicology: The Basic Science of Poisons,* 2nd ed. Macmillan Publishing Co., Inc., New York. 1980.
7. Zimmerman P, Larsen R, Barkley E, and Gallelli J. Recommendations for the safe

handling of injectable antineoplastic drug products. *Am J Hosp Pharm*, 38. November 1981.

8. Harrison BR. Safe handling of cytotoxic drugs: A review. In: Perry, MC, Ed. *The Chemotherapy Source Book*. Williams & Wilkins, Baltimore, Maryland. 1992.

9. U.S. Department of Labor, Occupational Safety and Health Administration. Work Practice Guidelines for Personnel Dealing with Cytotoxic (Antineoplastic) Drugs. OSHA Instruction PUB 8-1.1. Office of Occupational Medicine, U.S. Department of Labor. January 29, 1986.

10. Neal A, Wadden RA, and Chiou WL. Exposure of hospital workers to airborne antineoplastic agents. *Am J Hosp Pharm*, 40:597–601. 1983.

11. U.S. Department of Health and Human Services. Recommendations for the Safe Handling of Parenteral Antineoplastic Drugs. NIH Publication 83-2621. 1983.

12. Neely J. Cytotoxic drugs. In: The Identification and Control of Health Hazards for Hospital Workers, pp. 3–29 to 3–35. Alberta Occupational Health and Safety, Edmonton, Alberta. February 1986.

13. Falck K, Grohn P, Sorsa M, Vainio H, Heinonen E and Holsti LR. Mutagenicity in urine of nurses handling cytostatic drugs. *Lancet*, 1:1250–1251. 1979.

14. Nguyen TV, Theiss JC and Matney TS. Exposure of pharmacy personnel to mutagenic antineoplastic drugs. *Cancer Res*, 42:4792–4796. 1982.

15. Macek K. Hospital personnel who handle anticancer drugs may face risks. *JAMA*, 247(1):11–12. 1982.

16. Anderson RW, Puckett WH, Dana WJ, Nguyen TV, Theiss JC and Matney TS. Risk of handling injectable antineoplastic agents. *Am J Hosp Pharm*, 39:1881–1887. 1982.

17. Bos RP, Leenaars AO, Theuws JLG and Henderson PT. Mutagenicity of urine from nurses handling cytostatic drugs. *Int Arch Occup Environ Health*, 50:359–369. 1982.

18. Gibson JF, Gompertz D, and Hedworth-Whitty RB. Mutagenicity of urine from nurses handling cytotoxic drugs. *Lancet*, pp. 100–101. January 14, 1984.

19. Rogers B and Emmett EA. Handling antineoplastic agents: urine mutagenicity in nurses. *IMAGE: J Nurs Scholarship*, 19(3):108–113. 1987.

20. Kolmodin-Hedman B, Hartvig P, Sorsa M, and Falck K. Occupational handling of cytostatic drugs. *Arch Toxicol*, 54:25–33. 1983.

21. Norppa H, Sorsa M, Vainio H, Grohn P, Heinonen E, Holsti L,and Nordman E. Increased sister chromatid exchange frequencies in lymphocytes of nurses handling cytostatic drugs. *Scand J Work Environ Health*, 6:299–301. 1980.

22. Waksvik H, Klepp O, and Brogger A. Chromosome analysis of nurses handling cytostatic agents. *Cancer Treatment Rep*, 65(7–8):607–610. July/August 1981.

23. Selevan SG, Lindbohm ML, Hornung RW, and Hemminki K. A study of occupational exposure to antineoplastic drugs and fetal loss in nurses. *N Engl J Med*, 313:1173–1178. 1985.

24. Hemminki K, Kyyronen P, and Lindbohm ML. Spontaneous abortions and malformations in the offspring of nurses exposed to anaesthetic gases, cytostatic drugs, and other potential hazards in hospital, based on registered information of outcome. *J Epidemiol Community Med*, 39:141–147. 1985.

25. Sotaniemi EA, Sutinen S, Arranto AJ, Sutinen S, Sotanieme KA, Lehtola J and Pelkonen RO. Liver damage in nurses handling cytostatic agents. *Acta Med Scand*, 214:181–189. 1983.

26. U.S. Department of Health and Human Services. Antineoplastic drugs: Guidelines for protecting the safety and health of healthcare workers. NIOSH Publication 88–119, pp. 5–13 to 5–24. September 1988.

27. Ladik CF, Stoehr GP, and Maurer MA. Precautionary measures in the preparation of antineoplastics. *Am J Hosp Pharm*, 37:1185–1186. 1980.

28. McDiarmid M and Egan T. Acute occupational exposure to antineoplastic agents. *J Occup Med*, 30(12):984–987. December 1988.

29. Duvall E and Baumann B. An unusual accident during the administration of chemotherapy. *Cancer Nurs*, 3(4):305–306. 1980.

30. Turnberg WL and Frost F. Survey of occupational exposure of waste industry workers to infectious waste in Washington state. *Am J Public Health*, 80(10):1262–1264. October 1990.

31. U.S. Environmental Protection Agency. Guides to Pollution Prevention, Selected Hospital Waste Streams. EPA/625/7–90/009. June 1990.

32. U.S. Environmental Protection Agency. Hazardous Waste Management System: Identification and Listing of Hazardous Waste—Final Rule, Interim Final Rule and Request for Comments. *Federal Register,* 45:33084. May 19, 1980.

33. Vaccari P, Tonat K, DeChristoforo R, Gallelli J, and Zimmerman P. Disposal of antineoplastic wastes at the National Institutes of Health. *Am J Hosp Pharm*, 41:87–92. January 1984.

34. Personal Communication with Ed Rau, Chief, Hazardous and Solid Waste Management Section, Environmental Protection Branch, National Institutes of Health. October 1995.

35. McEvoy GK, ed. American Hospital Formulary Service Drug Information. American Society of Hospital Pharmacists, Bethesda, Maryland. 1993.

36. International Agency for Research on Cancer. *IARC Monographs on the Evaluations of the Carcinogenic Risk of Chemicals to Humanas: Pharmaceutical Drugs.* Vol. 50. Lyon, France: IARC. 1990.

37. McDiarmid MA, Kolodner K, Humphrey F, Putman D, and Jacobson-Kram D. Baseline and phosphoramide mustard-induced sister-chromatid exchanges in pharmacists handling anti-cancer drugs. *Mutation Research,* 279:199–204. 1992.

38. U.S. Department of Labor, Occupational Safety and Health Administration. Occupational Exposure to Bloodborne Pathogens Standard. 29 CFR 1910.1030. 1991.

39. U.S. Environmental Protection Agency. Discarded commercial chemical products, off specification species, container residues, and spill residues thereof. 40 CFR 261.33(f). 1994.

40. U.S. Department of Health and Human Services. Public Health Service. National Institutes of Health. Recommendations for the Safe Handling of Cytotoxic Drugs. NIH Publication No. 92-2621. 1992.

41. Castegnaro M, Adams J, and Armour MA, eds. Laboratory decontamination and destruction of carcinogens in laboratory wastes: Some antineoplastic agents. International Agency for Research on Cancer. Scientific Publications No. 73. Lyon, France: IARC. 1985.

42. Lunn G, Sansone EB, Andrews AW, and Hellwig LC. Degradation and disposal of some antineoplastic drugs. *J Pharm Sciences,* 78:652–659. 1989.

43. Lunn G and Sansone EB. Validated methods for handling spilled antineoplastic agents. *Am J Hosp Pharm,* 46:1131. 1989.

44. Valanis B, Driscoll K, and McNeil V. Comparison of antineoplastic drug handling policies of hospitals with OSHA guidelines: A pilot study. *Am J Public Health*, 80(4):480–481. April 1990.

45. Turnberg WL. Survey of infectious waste management practices conducted by medical facilities in Washington state. In: Washington State Infectious Waste Project: Report to the Legislature. Washington Department of Ecology, Olympia, Washington. December 1989.

APPENDIX A

FEDERAL AND STATE OSHA CONTACTS

OSHA-APPROVED STATES AND TERRITORIES

States and territories with OSHA-approved programs are as follows:

Alaska

> Commissioner
> Alaska Department of Labor
> 1111 West Eighth Street
> Room 306
> Juneau, AK 99801
> Tel: (907) 465-2700

Arizona

> Director
> Industrial Commission of Arizona
> 800 West Washington
> Phoenix, AZ 85007
> Tel: (602) 542-5795

California

> Director
> California Department of Industrial Relations
> 455 Golden Gate Avenue
> Fourth Floor
> San Francisco, CA 94102
> Tel: (415) 703-4590

Connecticut

> Commissioner
> 200 Folly Brook Boulevard
> Wethersfield, CT 06109
> Tel: (203) 566-5123

Hawaii

> Director
> Hawaii Department of Labor and Industrial Relations
> 830 Punchbowl Street
> Honolulu, HI 96813
> Tel: (317) 232-2378

Indiana

> Commissioner
> Indiana Department of Labor
> State Office Building
> 402 West Washington Street
> Room W195
> Indianapolis, IN 46204
> Tel: (317) 232-2378

Iowa

> Commissioner
> Iowa Division of Labor Services
> 1000 East Grand Avenue
> Des Moines, IA 50319
> Tel: (515) 281-3447

Kentucky

> Secretary
> Kentucky Labor Cabinet
> 1049 U.S. Highway, 127 South
> Frankfort, KY 40601
> Tel: (502) 564-3070

Maryland

    Commissioner
    Maryland Division of Labor and Industry
    Department of Licensing and Regulation
    501 St. Paul Place, Second Floor
    Baltimore 21202-2272
    Tel: (410) 333-4179

Michigan

    Director
    Michigan Department of Labor
    Victor Office Center
    201 North Washington Square
    P.O. Box 30015
    Lansing, MI 48933
    Tel: (517) 373-9600

Minnesota

    Commissioner
    Michigan Department of Public Health
    3423 North Logan Street
    Box 30195
    Lansing, MI 48909
    Tel: (517) 335-8022

Nevada

    Director
    Nevada Department of Industrial Relations
    Capitol Complex
    1370 South Curry Street
    Carson City, NV 89710
    Tel: (702) 687-3032

New Mexico

    Secretary
    New Mexico Environmental Department
    Occupational Health and Safety Bureau
    1190 St. Francis Drive
    P.O. Box 26110
    Santa Fe, NM 87502
    Tel: (505) 827-2850

New York

  Commissioner
  New York Department of Labor
  State Office Building, Campus 12
  Room 457
  Albany, NY 12240
  Tel: (518) 457-2741

North Carolina

  Commissioner
  North Carolina Department of Labor
  4 West Edenton Street
  Raleigh, NC 27601
  Tel: (919) 733-0360

Oregon

  Administrator
  Department of Consumer and Business Services
  Labor and Industries Building
  350 Winter Street, NE, Room 430
  Salem, OR 97310
  Tel: (503) 378-3272

Puerto Rico

  Secretary
  Puerto Rico Department of Labor and Human Resources
  Prudencio Rivera Martinez Building
  505 Munoz Rivera Avenue
  Hato Rey, PR 00918
  Tel: (809) 754-2119

South Carolina

  South Carolina Department of Labor
  3600 Forest Drive
  P.O. Box 11329
  Columbia, SC 29211-1329
  Tel: (803) 734-9594

Tennessee

> Tennessee Department of Labor
> 710 James Robertson Parkway
> Suite A, Second Floor
> Nashville, TN 37243-0659
> Tel: (615) 741-2582

Utah

> Commissioner
> Industrial Commission of Utah
> P.O. Box 146600
> Salt Lake City, UT 84110-6600
> Tel: (801) 530-6898

Vermont

> Commissioner
> Vermont Department of Labor and Industry
> 120 State Street
> Montpelier, VT 05620
> Tel: (802) 828-2288

Virgin Islands

> Commissioner
> Virgin Islands Department of Labor
> 2131 Hospital Street
> Christiansted
> St. Croix, VI 00840-4666
> Tel: (809) 773-1994

Virginia

> Commissioner
> Virginia Department of Labor and Industry
> Powers-Taylor Building
> 13 South Thirteenth Street
> Richmond, VA 23219
> Tel: (804) 786-2377

Washington

Director
Washington Department of Labor and Industries
General Administration Building
P.O. Box 44001
Olympia, WA 98504-4001
Tel: (206) 956-4213

Wyoming

Administrator
Occupational Safety and Health Administration
Herschler Building, Second Floor East
122 West 25th Street
Cheyenne, WY 82002
Tel: (307) 777-7672

OSHA REGIONAL OFFICES*

Region I (CT,* MA, ME, NH, RI, VT*)

133 Portland Street
First Floor
Boston, MA 02114
Tel: (617) 565-7164

Region II (NJ, NY,* PR,* VI*)

201 Varick Street
Room 670
New York, NY 10014
Tel: (212) 337-2378

Region III (DC, DE, MD,* PA, VA,* WV)

Gateway Building, Suite 2100
33535 Market Street
Philadelphia, PA 19104
Tel: (215) 596-1201

*Asterisks indicate states and territories that operate their own OSHA-approved job safety and health programs (Connecticut and New York plans cover public employees only). States with approved plans must have a standard that is identical to, or at least as effective as, the federal standard.

Region IV (AL, FL, GA, KY,* MS, NC, SC,* TN*)

  1375 Peachtree Street, NE
  Suite 587
  Atlanta, GA 30367

Region V (IL, IN,* MI,* MN,* OH, WI)

  230 South Dearborn Street
  Room 32 44
  Chicago, IL 60604
  Tel: (312) 353-2220

Region VI (AR, LA, MN,* OK, TX)

  525 Griffin Street
  Room 602
  Dallas, TX 75202
  Tel: (214) 767-4731

Region VII (IA,* KS, MO, NE)

  911 Walnut Street, Room 406
  Kansas City, MO 64106
  Tel: (816) 426-5861

Region VIII (CO, MT, ND, SD, UT,* WY*)

  Federal Building, Room 1576
  1961 Stout Street
  Denver, CO 80294
  Tel: (303) 844-3061

Region IX (American Samoa, AZ,* CA,* Guam, HI,* NV,* Trust Territories of
the Pacific)

  71 Stevenson Street
  Room 420
  San Francisco, CA 94105
  Tel: (415) 744-6670

Region X (AK,* ID, OR,* WA*)
  1111 Third Avenue
  Suite 715
  Seattle, WA 98101-3212
  Tel: (206) 553-5930

OSHA AREA OFFICES
Alabama
    Birmingham   (205) 731-1534
    Mobile   (205) 441-6131

Alaska
    Anchorage   (907) 271-5152

Arizona
    Phoenix   (602) 640-2007

Arkansas
    Little Rock   (501) 324-6291

California
    San Francisco   (415) 744-7120

Colorado
    Denver   (303) 844-5285
    Englewood   (303) 843-4500

Connecticut
    Bridgeport   (203) 579-5579
    Hartford   (203) 240-3152

Florida
    Fort Lauderdale   (305) 424-0242
    Jacksonville   (904) 232-2895
    Tampa   (813) 626-1177

Georgia
    Savannah   (912) 652-4393
    Smyrna   (404) 984-8700
    Tucker   (404) 493-6644

Hawaii
    Honolulu   (808) 541-2685

Idaho
    Boise   (208) 334-1867

Illinois
    Calumet City   (708) 891-3800
    Des Plaines   (708) 803-4800
    North Aurora   (708) 896-8700
    Peoria   (309) 671-7033

Indiana
  Indianapolis   (317) 226-7290

Iowa
  Des Moines   (515) 284-4794

Kansas
  Wichita   (316) 269-6644

Kentucky
  Frankfort   (502) 227-7024

Louisiana
  Baton Rouge   (504) 389-0474

Maine
  Augusta   (207) 622-8417

Maryland
  Baltimore   (410) 962-2840

Massachusetts
  Braintree   (617) 565-6924
  Methuen   (617) 565-8110
  Springfield   (413) 785-0123

Michigan
  Lansing   (517) 377-1892

Minnesota
  Minneapolis   (612) 348-1994

Missisippi
  Jackson   (601) 965-4606

Missouri
  Kansas City   (816) 426-2756
  Saint Louis   (314) 425-4249

Montana
  Billings   (406) 657-6649

Nebraska
  Omaha   (402) 221-3182

Nevada
  Carson City   (702) 885-6963

New Hampshire
   Concord   (603) 225-1629

New Jersey
   Avenel   (908) 750-3270
   Hasbrouck Heights   (201) 288-1700
   Marlton   (609) 757-5181
   Parsippany   (201) 263-1003

New Mexico
   Albuquerque   (505) 766-3411

New York
   Albany   (518) 464-6742
   Bayside   (718) 279-9060
   Bowmansville   (716) 684-3891
   New York   (212) 264-9840
   Syracuse   (315) 451-0808
   Tarrytown   (914) 682-6151
   Westbury   (516) 334-3344

North Carolina
   Raleigh   (919) 856-4770

North Dakota
   Bismarck   (701) 250-4521

Ohio
   Cincinnati   (513) 841-4132
   Cleveland   (216) 522-3818
   Columbus   (614) 469-5582
   Toledo   (419) 259-7542

Oklahoma
   Oklahoma City   (405) 231-5351

Oregon
   Portland   (503) 326-2251

Pennsylvania
   Allentown   (215) 776-0592
   Erie   (814) 833-5758
   Harrisburg   (717) 782-3902
   Philadelphia   (215) 597-4955
   Pittsburgh   (412) 644-2903
   Wilkes-Barre   (717) 826-6538

Puerto Rico
  Hato Rey   (809) 766-5457

Rhode Island
  Providence   (401) 528-4669

South Carolina
  Columbia   (803) 765-5904

Tennessee
  Nashville   (615) 781-5423

Texas
  Austin   (512) 482-5783
  Corpus Christi   (512) 888-3257
  Dallas   (214) 320-2400
  Fort Worth   (817) 885-7025
  Houston   (713) 286-0583
  Houston   (713) 591-2438
  Lubbock   (806) 743-7681

Utah
  Salt Lake City   (801) 524-5080

Virginia
  Norfolk   (804) 441-3820

Washington
  Bellevue   (206) 553-7520

West Virginia
  Charleston   (304) 347-5937

Wisconsin
  Appleton   (414) 734-4521
  Madison   (608) 264-5388
  Milwaukee   (414) 297-3315

APPENDIX B

STATE MEDICAL WASTE CONTACTS*

Alabama

John Narramore, Chief
Special Wastes Branch
Land Division
Department of Environmental Management
1751 Congressman W.L. Dickinson Drive
P.O. Box 301463
Montgomery, AL 36130-1463
Tel: (334) 271-7761

Alaska

Glenn Miller, Program Manager
Solid and Hazardous Waste Management Section
Division of Environmental Quality
Department of Environmental Conservation
410 Willoughby Avenue, Suite 105
Juneau, AK 99801-1795
Tel: (907) 465-5150

---

*As of December 5, 1994. From U.S. Environmental Protection Agency, Office of Solid Waste, Washington, D.C., March 1995.

American Samoa

Togiba Tausaga, Director
American Samoa Environmental Protection Agency
Government of American Samoa
Pago Pago, AS 96799
Tel: [011] (684) 633-2304

Arkansas

James Shumate, Director of Medical Waste
Slot 22
Department of Health
4815 West Markham
Little Rock, AR 72205-3867
Tel: (501) 661-2920

California

Southern California

Vern Richard
Medical Waste Management Program
107 South Broadway, Room 3028
Los Angeles, CA 90012-4403
Tel: (213) 897-7470

Northern California

Michael Schott or Jack McGurk
Toxic Substances Control Program
California Department of Health Services
601 North Seventh Street
P.O. Box 942732
Sacramento, CA 94234-7320
Tel: (916) 327-6901/(916) 323-1167

Colorado

Public Assistance Hotline
Hazardous Materials and waste Management Division
Department of Health
4300 Cherry Creek Drive South
Denver, CO 80222
(303) 692-3320

Connecticut

    Maria Valez, Processing Manager
    Waste Engineering and Enforcement Division
    Bureau of Waste Management
    Connecticut Department of Environmental Protection
    79 Elm Street
    Hartford, CT 06106-5127
    Tel: (203) 424-3023

Delaware

    Indra Batra
    Solid Waste Management Branch
    Division of Air and Waste Management
    Delaware Department of Natural Resources
      and Environmental Control
    89 Kings Highway
    P.O. Box 1401
    Dover, DE 19903
    Telephone: (302) 739-3820

District of Columbia

    Patricia Van Buren, Chief
    Long Term Care Branch
    Health Facilities Division
    Department of Consumer and Regulatory Affairs
    614 H Street, NW
    Washington, DC 20001
    (202) 727-72000

Florida

    Tom Moore, Environmental Specialist
    Bureau of Solid and Hazardous Waste
    Division of Waste Management
    Department of Environmental Protection
    2600 Blair Stone Road, Twin Towers
    Tallahassee, FL 32399-2400
    Tel: (904) 488-0300

Georgia

Harold Gillespie, Unit Coordinator
Land Protection Branch
Environmental Protection Division
Department of Natural Resources
Atlanta Trade Port, Suite 104
4244 International Parkway
Atlanta, GA 30359-3902
Tel: (404) 362-2692

Guam

Fred Castro, Administrator
Guam Environmental Protection Agency
130 Rojas Street
D107 Harmon Plaza
Harmon, GU 96911
Tel: [011] (671) 646-8863

Hawaii

Helen Yoshimi
Hospital and Medical Facilities Branch
Hawaii Department of Health
P.O. Box 3378
Honolulu, HI 96801
Tel: (808) 586-4742

Idaho

Roger Perotto
Bureau of Communicable Disease Prevention
Department of Health and Welfare
450 West State Street
Boise, ID 83720
Tel: (208) 334-5930

Illinois

Beverly Albarracin
Division of Land Pollution Control
Illinois Environmental Protection Agency
2200 Churchill Road
P.O. Box 19276
Springfield, IL 62794-9276
Tel: (217) 524-3289

Indiana

Carmen Quintana
Office of Indiana State Health Commissioner
State Department of Health
1330 West Michigan Street
Indianapolis, IN 46204
Tel: (317) 383-6731

Iowa

Paul Lundy
Solid Waste Section
Department of Natural Resources
Henry Wallace State Office Building
900 East Grand Avenue
Des Moines, IA 50319-0034
Tel: (515) 281-8912

Kansas

Joe Cronin
Solid Waste Section
Bureau of Waste Management
Department of Health and the Environment
Forbes Field, Building 740
Topeka, KS 66620-0001
Tel: (913) 296-1667

Kentucky

Vicky Pettus, Manager
Resource Conservation and Local Assistant Branch
Bureau of Waste Management
Department of Environmental Protection
14 Reilly Road
Frankfort Office Park
Frankfort, KY 40601
Tel: (502) 564-6716

Louisiana

Gerald Mathes, Assistant Administrator
Solid Waste Division
Office of Solid and Hazardous Waste
Department of Environmental Quality
7290 Bluebonnet Drive
P.O. Box 82178
Baton Rouge, LA 70884-2178
Tel: (504) 765-0249

Maine

Scott Austin
Bureau of Hazardous Material and Solid Waste Control
Maine Department of Environmental Protection
State House Station 17
Augusta, ME 04333-0017
Tel: (207) 289-2651

Mariana Islands

Ignacio V. Cabrera, Environmental Specialist
Division of Environmental Quality
P.O. Box 1304
Saipan, MP 96950
Tel: [011] (670) 234-6114

Maryland

Dale Fuller
Special Medical Waste
Hazardous Waste Program
Maryland Department of the Environment
2500 Broening Highway
Baltimore, MD 21224
Tel: (410) 631-3344

Massachusetts

Howard Wensley, Director
Division of Community Sanitation
Massachusetts Department of Public Health
305 South Street
Jamaica Plain, MA 02103
Tel: (617) 727-2660

Michigan

Lawrence Chadzynski
Bureau of Health Care Facilities
Michigan Department of Public Health
3423 North Logan
P.O. Box 30195
Lansing, MI 48909
Tel: (517) 335-8637

Minnesota

Jackie Deneen
Solid Waste Section
Division of Groundwater and Solid Waste
Minnesota Pollution Control Agency
520 North Lafayette Road
St. Paul, MN 55155
Tel: (612) 296-7227

Mississippi

Mark Williams
Special Waste Section
Solid Waste Branch
Department of Environmental Quality
2380 Highway 80 West
P.O. Box 10385
Jackson, MS 39204
Tel: (601) 961-5171

Missouri

Richard Mansur
Solid Waste Management Program
Solid Waste Division
Department of Natural Resources
205 Jefferson Street, Jefferson Building
P.O. Box 176
Jefferson City, MO 65102
Tel: (314) 451-5401

Montana

Rick Thompson, Environmental Specialist
Solid and Hazardous Waste Program
Solid and Hazardous Waste Bureau
Department of Health and Environmental Sciences
Cogswell Building
P.O. Box 200901
Helena, MT 59620-0901
Tel: (406) 444-1430

Nebraska

Kelly Danielson, Program Specialist
Integrated Waste Management Section
Air and Waste Management Division
Department of Environmental Quality
1200 North Street, The Atrium, Suite 400
P.O. Box 98922
Lincoln, NE 68509-8922
Tel: (402) 471-4217

Nevada

David Emme, Supervisor
Solid Waste Branch
Waste Management Bureau
Division of Environmental Protection
Department of Conservation and Natural Resources
333 West Nye Lane
Carson City, NV 89710
Tel: (702) 687-4670, ext. 3001

New Hampshire

Carl Woodbury
Waste Management Specialist
Waste Management Compliance Bureau
Waste Management Division
Department of Environmental Services
6 Hazen Drive
Concord, NH 0301-6509
Tel: (603) 271-2925

New Jersey

> Robert Confer, Chief
> Bureau of Medical Waste and Technical Assistance
> Statewide Planning
> Division of Solid Waste Management
> New Jersey Department of Environmental Protection and Energy
> 840 Bear Tavern Road, CN 414
> Trenton, NJ 08625-0414
> Tel: (609) 530-8599

New Mexico

> Richard Stafford
> Solid Waste Bureau
> New Mexico Environmental Department
> P.O. Box 26110
> Santa Fe, NM 87502
> Tel: (505) 827-2866

New York

> Ira F. Salkin, Ph.D.
> Director, Regulated Medical Waste
> New York Department of Health
> P.O. Box 509, Empire State Plaza
> Albany, NY 12201-0509
> Tel: (518) 485-5386

North Carolina

> Ernie Lawrence, Waste Management Specialist
> Solid Waste Section
> Division of Solid Waste Management
> Department of Environmental Health and Natural Resources
> P.O. Box 27687
> Raleigh, NC 27611-7687
> Tel: (919) 733-0692

North Dakota

> Robert Wetsch, Manager
> Solid Waste Program
> Division of Waste Management
> Department of Health
> P.O. Box 5520
> Bismark, ND 58502-5520
> Tel: (701) 328-5166

Ohio

> Alison Shockley, Supervisor
> Infectious Waste Unit
> Division of Solid and Infectious Waste Management
> Ohio Environmental Protection Agency
> P.O. Box 163669
> Columbus, OH 43216-3669
> Tel: (614) 728-5335

Oklahoma

> Patricia Hoyle
> Solid Waste Management Division
> Oklahoma Department of Environmental Quality
> 1000 NE Tenth Street
> Oklahoma City, OK 73117-1212
> Tel: (405) 745-7115

Oregon

> Terrance Hollins, Administrative Specialist
> Waste Management and Cleanup Division
> Department of Environmental Quality
> 811 Southwest Sixth Avenue
> Portland, OR 97204
> Tel: (503) 229-6922

Pennsylvania

> Rod Hassinger, Chief
> General Permits and Beneficial Use Section
> Bureau of Waste Management
> Division of Municipal and Residual Waste
> Department of Environmental Resources
> 400 Market Street
> P.O. Box 8472
> Harrisburg, PA 17105-8472
> Tel: (717) 787-7381

Puerto Rico

> Russe Martinez, Chairman
> Puerto Rico Environmental Quality Board
> Office of the Governor
> Banco Nationale Plaza Building, Suite 431
> Ponce de Leon Avenue
> Hatorey, PR 00910
> Tel: (809) 767-8056

Rhode Island

> Roger Greene, Assistant Director
> Rhode Island Department of Environmental Management
> 9 Hayes Street
> Providence, RI 02908-5003
> Tel: (401) 277-2771

South Carolina

> Phil Morris, Director
> Infectious Waste Management Section
> Bureau of Solid and Hazardous Waste Management
> South Carolina Department of Health and
>   Environmental Control
> 2600 Bull Street
> Columbia, SC 29201
> Tel: (803) 896-4173

Tennessee

Sam Faleh
Solid Waste Management Division
Department of Environmental Conservation
401 Church Street
L&C Tower, Fourth Floor
Nashville, TN 37243-1535
Tel: (615) 532-0796

Texas

Patricia Riley, Medical Waste Specialist
Industrial and Hazardous Waste Division
Texas Natural Resource Conservation Commission
P.O. Box 13087
Austin, TX 7874-3087
Tel: (512) 239-6832

Utah

Ralph Bohn, Manager
Solid Waste Section
Division of Solid and Hazardous Waste
Department of Environmental Quality
288 North, 1460 West
P.O. Box 144880
Salt Lake City, UT 84114-4880
Tel: (801) 538-6170

Vermont

Dwight Moody
Solid Waste Management Division
Department of Environmental Conservation
Agency of Natural Resources
103 South Main Street, Laundry Building
Waterbury, VT 05671-0407
Tel: (802) 241-3444

Virgin Islands

Benjamin Nazario, Director
Division of Environmental Protection
Department of Planning and Natural Resources
Government of the Virgin Islands
1118 Watergut Homes
Christiansted Project
St. Croix, VI 00820
Tel: (809) 773-0565

Virginia

Robert Wickline
Office of Regulatory Planning and Development
Waste Division
Virginia Department of Environmental Quality
P.O. Box 10009
Richmond, VA 23240-0009
Tel: (804) 762-4213

Washington

Wayne Turnberg, Medical Waste Coordinator
Washington Department of Health
1511 3rd Avenue, Suite 700
Seattle, WA 98101-1549
Tel: (206) 587-5501

West Virginia

Joe Wyatt
Environmental Health Services
Infectious Medical Waste Program
Public Health Sanitation
Bureau of Public Health
815 Quarrier Street, Suite 418
Charleston, WV 25301-2616
Tel: (304) 558-2981

Wisconsin

Barb Gear, Medical Waste Coordinator
Solid Waste Management Section
Bureau of Solid and Hazardous Waste Management
Department of Natural Resources
101 South Webster Street
Madison, WI 53702
Tel: (608) 267-3548

INDEX